T0321151

The alpha-2 Adrenergic Receptors

The Receptors

Series Editor: *David B. Bylund*, University of Missouri,
Columbia, Missouri

Board of Editors

The alpha-2
Adrenergic Receptors

Edited by

Lee E. Limbird

Vanderbilt University
Nashville, Tennessee

HUMANA PRESS · CLIFTON, NEW JERSEY

Library of Congress Cataloging-in-Publication Data

The Alpha-2 adrenergic receptors.

(The Receptors)
Includes bibliographies and index.
1. Adrenergic receptors.   I. Limbird, Lee E.   II. Title: Alpha-two adrenergic
receptors.   III. Series.   [DNLM: 1. Receptors, Adrenergic, Alpha.   WL 102.8   A 4562]
QP364.7.A473   1988   612'.89   88-6833
ISBN 0-89603-135-7

Contents

v

Section 3: Biochemical Mechanisms of Receptor Action

Chapter 3
*Mechanisms for Inhibition of Adenylate Cyclase by
alpha-2 Adrenergic Receptors*
Peter Gierschik and **Karl H. Jakobs**

Section 4: Correlation of Receptor Binding and Function

Chapter 4

Structure–Activity Relationships for alpha-2 Adrenergic Receptor Agonists and Antagonists

Robert R. Ruffolo, Jr., Robert DeMarinis, Margaret Wise, and **J. Paul Hieble**

Chapter 5
Functions Mediated by alpha-2 Adrenergic Receptors
**Robert R. Ruffolo, Jr., Andrew J. Nichols,
and J. Paul Hieble**

Section 5: Receptor Regulation

Chapter 6
Regulation of alpha-2 Adrenergic Receptors
Paul A. Insel and **Harvey J. Motulsky**

Section 6: Future Vistas

Chapter 7
*What Happens Next? A Hypothesis Linking the
Biochemical and Electrophysiological Sequelae of
alpha-2 Adrenergic Receptor Occupancy with the
Diverse Receptor-Mediated Physiological Effects*
Lori L. Isom and **Lee E. Limbird**

Contributors

DAVID B. BYLUND • *Department of Pharmacology, School of Medicine, University of Missouri, Columbia, Missouri*

ROBERT M. DEMARINIS • *Departments of Pharmacology and Medicinal Chemistry, Smith Kline & French Laboratories, Swedeland, Pennsylvania*

PETER GIERSCHIK • *Pharmakologisches Institut, Universitat Heidelberg, FRG*

J. PAUL HIEBLE • *Departments of Pharmacology and Medicinal Chemistry, Smith Kline & French Laboratories, Swedeland, Pennsylvania*

PAUL A. INSEL • *Departments of Pharmacology and Medicine, University of California, San Diego, La Jolla, California*

LORI L. ISOM • *Department of Pharmacology, Vanderbilt University, Nashville, Tennessee*

KARL H. JAKOBS • *Pharmakologisches Institut, Universitat Heidelberg, FRG*

LEE E. LIMBIRD • *Department of Pharmacology, Vanderbilt University, Nashville, Tennessee*

HARVEY J. MOTULSKY • *Departments of Pharmacology and Medicine, University of California, San Diego, La Jolla, California*

ANDREW J. NICHOLS • *Departments of Pharmacology and Medicinal Chemistry, Smith Kline & French Laboratories, Swedeland, Pennsylvania*

JOHN W. REGAN • *Departments of Medicine and Biochemistry, Howard Hughes Medical Institute, Duke University Medical Center, Durham, North Carolina*

ROBERT R. RUFFOLO, JR. • *Departments of Pharmacology and Medicinal Chemistry, Smith Kline & French Laboratories, Swedeland, Pennsylvania*

MARGARET WISE • *Departments of Pharmacology and Medicinal Chemistry, Smith Kline & French Laboratories, Swedeland, Pennsylvania*

SECTION 1
HISTORICAL PERSPECTIVE

Chapter 1

alpha-2 Adrenergic Receptors

A Historical Perspective

David B. Bylund

1. Perspective

The actions of epinephrine, an adrenal hormone and central neurotransmitter, and norepinephrine, a peripheral sympathetic and central neurotransmitter, are mediated through alpha-1, alpha-2, and beta-adrenergic receptors. In addition, a variety of drugs also produce their effects by interacting with these receptors. alpha-2 Adrenergic agonists, such as clonidine and guanabenz, are widely used to treat hypertension. Clonidine is also used to ameliorate the symptoms of withdrawal from opiate drugs. Local administration of clonidine has been used successfully in the treatment of glaucoma. Mianserin, an effective antidepressant drug, is an alpha-2 adrenergic antagonist. The utility of other alpha-2 adrenergic antagonists as antidepressant medications is currently a subject of intense study. Recently alpha-2 adrenergic agonists have been demonstrated to ameliorate the cognitive defects exhibited by aged nonhuman primates, and it has been suggested that these drugs may be helpful for patients with Alzheimer's disease (Arnsten and Goldman-Rakic, 1985). alpha-2 Adrenergic drugs, in addition to their therapeutic usefulness, have played a major role in the development of current concepts of neurotransmission and adrenergic physiology and pharmacology.

The entity that we now call the alpha-2 adrenergic receptor has been the subject of much study and speculation during the past 20 years. The initial subclassification of adrenergic receptors

1

into the alpha and beta subtypes was done by Ahlquist (1948) on the basis of their pharmacology and not on the basis of their function (excitatory or inhibitory) as various investigators before him had suggested. Ahlquist studied the effects of five catecholamines on eight different physiological functions and clearly showed that the order of potency of the catecholamines for five of the physiological functions was markedly different from the order of potency for the other three functions. He postulated that these two different orders of potency represented the involvement of two independent populations of receptors, which he termed alpha and beta. Subsequently, beta-adrenergic receptors were divided into beta-1 and beta-2 subtypes (Lands et al., 1967). This was also a pharmacologic definition based on a comparison of the relative potencies of 12 agonists in several isolated organ systems. This subclassification of beta-adrenergic receptors (illustrated in Fig. 1A) has been substantiated both by the development of subtype selective antagonists and by binding studies (Minneman et al., 1979).

In contrast to the relatively simple subclassification of beta-adrenergic receptors by pharmacologic criteria, the subdivision of alpha-adrenergic receptors has followed a more circuitous route. On the basis of pharmacologic characteristics, the suggestion was made in 1965 that there are different types of alpha-adrenergic receptors. As noted by Rossum (1965), "the results obtained with both agonists and antagonists provide evidence that the structural requirements for drugs to react with and to activate alpha-

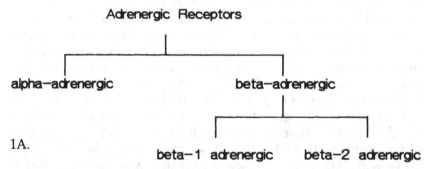

1A.

Fig. 1. These four panels (A–D) illustrate the historical development of the classification of adrenergic receptors from 1967 to 1980. By 1967 (1A) the beta-adrenergic receptors had been divided further into beta-1 and beta-2 subtypes. The 1974 scheme (1B) suggested that alpha adrenergic receptors could be differentiated by their location. The 1977 classification (1C) was based only on pharmacologic characteristics. The 1980 scheme (1D) emphasized possible biochemical correlates to the pharmacologic classification scheme.

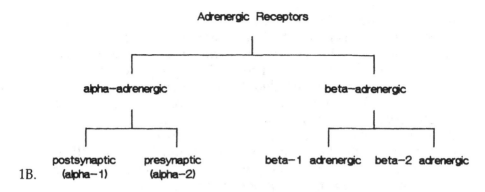

1B.

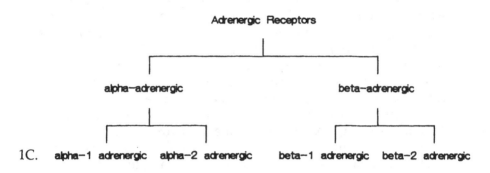

1C.

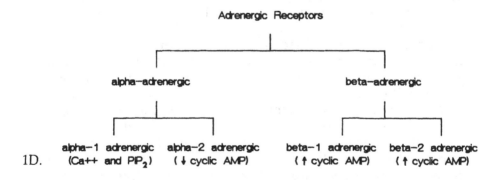

1D.

receptors in the vas deferens and the rabbit intestine are different." Some 8 years later the terms alpha-1 and alpha-2 were first suggested for alpha-adrenergic receptor subtypes, by analogy with the beta-1, beta-2 subdivision (Delbarre and Schmitt, 1973).

The historical development of the concept of an alpha-2 adrenergic receptor is (perhaps unfortunately) intricately interwoven into the development of the presynaptic receptor concept. The concept of presynaptic receptors has its roots in experiments by Brown and Gillespie (1956, 1957). Using the perfused cat spleen they observed that the overflow (output) of nor-epinephrine caused by nerve stimulation in the perfusate was increased in the presence of an irreversible alpha-adrenergic antagonist. They correctly concluded that this effect was a direct result of the blockade of the alpha-adrenergic receptors. "The increase in the output/stimulus of noradrenaline after dibenamine or dibenzyline (phenoxybenzamine) we have attributed to the action of this drug in blocking receptor sites for noradrenaline" (Brown and Gillespie, 1957). However, they had an incorrect understanding of the mechanism and thought that the receptor sites "utilized" the neurotransmitter, apparently by taking it up into the tissue. "It would appear that the most important factor in the disappearance of the sympathetic transmitter after its liberation is its utilization at the receptors upon which it acts" (Brown and Gillespie, 1957). These experiments stimulated a great deal of work and discussion over the next few years. Several additional explanations were advanced to explain the increased overflow of norepinephrine by alpha-adrenergic antagonists, including blockade of neuronal uptake, blockade of extraneuronal uptake, blockade of metabolism, and alterations in blood flow. Although none of these explanations proved to be adequate, as late as 1970 it was still "generally accepted that the increased outflow of nor-adrenaline obtained in the presence of phenoxybenzamine is due to the prevention of reuptake of released noradrenaline" (Langer, 1970).

Up to this time, it had been assumed that the amount of neurotransmitter released per impulse was constant. However in 1970 several laboratories began to question that assumption. For example, the observation in the CNS that "clonidine reduced the disappearance of noradrenaline" led to the suggestion of "a negative feedback mechanism, evoked by the noradrenaline receptor stimulation" (Anden et al., 1970). Similarly in the sympathetic nervous system, in attempting to explain the "increased flow of transmitter in the presence of phenoxybenzamine" the possibility was not excluded that "phenoxybenzamine may increase the release of noradrenaline per stimulus" (Langer, 1970; also see Haggendal, 1970). Strong experimental evidence to support this hypothesis was provided the following year when two laboratories published data showing that the release of dopamine beta-hydroxylase was increased by phenoxybenzamine (De Potter et al., 1971; Johnson et al., 1971). "The enhanced release of dopamine beta-hydroxylase during stimulation in the presence of phenoxybenzamine suggests that the increased outflow of nor-epinephrine may be due to an increase in transmitter release, which is coupled with the extrusion of increased amounts of soluble dopamine beta-hydroxylase" (Johnson et al., 1971).

Thus, although it was established that the increased overflow of norepinephrine following phenoxybenzamine was caused by an increased release of norepinephrine, the mechanism still remained to be defined. Following the initial suggestion that there was a local feedback mechanism of some type operating in the synaptic cleft, the concept of presynaptic receptor regulation of norepinephrine release was suggested independently by four groups in 1971 (Farnebo and Hamberger, 1971; Kirpekar and Puig, 1971; Langer et al., 1971; Starke, 1971). For example, Kirpekar and Puig (1971) suggested that "an alpha site on the presynaptic nerve terminal has an inhibitory role in the release of noradrenaline. It is suggested that noradrenaline released by nerve stimulation acts on these alpha sites of the presynaptic membrane to inhibit its own release."

With the idea of presynaptic and postsynaptic alpha receptors now in place, it was logical to consider whether or not these receptors were identical. Work in Langer's laboratory using phenoxybenzamine (Dubocovich and Langer, 1974) and in Starke's laboratory using clonidine (Starke et al., 1974) showed that the presynaptic and the postsynaptic alpha receptors were not identical. This led Langer (1974) to propose that "perhaps the postsynaptic alpha receptor that mediates the response of the effector organ should be referred to as alpha-1 while the presynaptic alpha receptor that regulates transmitter release should be called alpha-2" (see Fig. 1B).

The next major advance was made by Berthelson and Pettinger (1977), who generalized the alpha-2 adrenergic receptor concept to include other receptors, such as those in neuroendocrine systems, which were similar to the presynaptic receptor in terms of the rank order of potencies for agonists and antagonists. They suggested that there are "at least two pharmacologically distinct types of alpha-adrenergic receptors" (see Fig. 1C). They also noted a correlation between function and pharmacology and suggested "the existence of two functionally distinct types of alpha-adrenergic receptors based on the receptor responses to various adrenergic agonists" (alpha-1, excitatory; alpha-2, inhibitory). "However as more information is gained concerning the pharmacologic characteristics of all alpha-adrenergic receptors, it may become apparent that some excitatory receptors should be included in the alpha-2 classification." Thus, they concluded that although there may be functional correlates, the classification of receptors should be on a pharmacologic basis.

Shortly thereafter other authors (Wikberg, 1979; Fain and Garcia-Sainz, 1980) noted that there were biochemical correlates to the pharmacologic classification (*see* Fig. 1D). They suggested "the hypothesis that alpha-1 receptors mediate effects secondary to an elevation of intracellular calcium and involve increased turnover of phosphatidylinositol. In contrast, the alpha-2 receptors mediate effects which are due to inhibition of adenylate cyclase" (Fain and Garcia-Sainz, 1980). It is important to note that this was not an attempt to classify receptors based on their biochemical response, but an attempt to define the biochemical mechanism of receptor subtypes that had been classified previously on the basis of their pharmacologic characteristics. This proposal was based on earlier work by Sabol and Nirenberg (1979), who suggested that the alpha-receptor in the NG108-15 neuroblastoma × glioma hybrid cell line that is negatively coupled to adenylate cyclase had the properties of an alpha-2 adrenergic receptor, and the suggestion of Jones and Michell (1978) that alpha-1 adrenergic receptor activation was coupled to elevation in cytosolic calcium and phosphatidyl-inositol turnover.

The idea that alpha-adrenergic receptor stimulation might be related to an inhibitory effect on the adenylate cyclase system was first noted in the early 1960s. "Possibly the alpha-receptor is unrelated to the adenyl cyclase system, but the possibility of a depressant effect on this system should be explored" (Sutherland, 1965). Several years later this idea was stated as a hypothesis (Robison et al., 1967) and received experimental support with the report that in both isolated rat pancreatic islets and isolated rat fat cells, the combination of epinephrine plus propranolol (to activate only alpha receptors) reduced cyclic AMP levels to below control values. "These data indicate that the effects of epinephrine on cyclic 3'5'AMP accumulation are mediated via alpha- and beta-adrenergic receptor sites. Stimulation of the alpha-adrenergic receptor inhibits and stimulation of the beta adrenergic receptor increases cyclic 3'5'AMP synthesis" (Turtle and Kipnis, 1967; *see also* Burns and Langley, 1970). However the general applicability of this hypothesis was unclear since it was well known that in "many tissues in which the (alpha-adrenergic) effects of the catecholamines had been studied, an inhibitory effect of adenyl cyclase or a decrease in the intracellular level of cyclic AMP had never been observed" (Robison et al., 1970).

Although a correlation between biochemical mechanism and the alpha-adrenergic receptor subtypes may eventually prove to be useful, at the present time the attempt to do this is limited by our relative ignorance of the actual mechanisms for alpha-adrenergic receptors in most tissues. As noted previously (Bylund and U'Prichard, 1983), it appears that there may be a variety of biochemical mechanisms for transducing the effects of alpha-1 and alpha-2 adrenergic receptor stimulation, and thus it may not be useful to limit the biochemical mechanisms involved in alpha-adrenergic receptors to those shown in Fig. 1D. Recent evidence has emphasized that for the alpha-2 adrenergic receptor, several biochemical mechanisms may be involved. In some tissues alpha-2 adrenergic effects are seen in the absence of a detectable decrease in cyclic AMP levels. In other situations there may be a decrease in cyclic AMP levels, although this decrease may not be responsible for a particular physiologic effect of alpha-2 receptor stimulation. "Although lowering intracellular cyclic AMP concentrations may be sufficient to evoke a physiological effect in some target tissues, in other tissues such a signal may be neither necessary nor sufficient to elicit the physiological function typically attributed to alpha-2 adrenergic agonists" (Limbird and Sweatt, 1985). For example, secondary aggregation in platelets appears to be mediated by the Na^+/H^+ antiporter (Sweatt et al., 1985). Another example is the alpha-2 adrenergic sensitization of forskolin-stimulated cyclic AMP levels in HT29 cells. In this cell line, pretreatment with an alpha-2 agonist causes a 10- to 20-fold increase in subsequent stimulation of cyclic AMP synthesis by forskolin. This effect has been shown to be mediated by alpha-2 adrenergic receptors, but is not a result of an inhibition cyclic AMP synthesis. However, it may be mediated through the guanine nucleotide regulatory protein (N_i) (Jones et al., 1987).

Methods for labeling alpha-2 adrenergic receptors using the radioligand binding technique were developed in the mid 1970s following the development of radioligand binding assays for other neurotransmitter receptors such as the opiate receptor and the beta-adrenergic receptor. [3H]Dihydroergocryptine was found to specifically label alpha-adrenergic receptors in the rabbit uterus (Williams and Lefkowitz, 1976). Independently Snyder and coworkers used [3H]clonidine and [3H]WB4101 to label alpha-adrenergic receptors in the brain (Greenberg et al., 1976). It was initially thought that the presynaptic alpha receptors would be labeled by [3H]clonidine, whereas the postsynaptic receptors would be labeled by [3H]WB4101. However it was found that

[^{3}H]clonidine binding was slightly increased rather than decreased in brains from animals treated with the neurotoxin 6-hydroxydopamine that was presumed to have destroyed the presynaptic nerve terminals (U'Prichard et al., 1977). These data were interpreted as an indication of denervation sensitivity and that clonidine was actually labeling postsynaptic receptors. Since there were differences in the affinities of agonist and antagonists in inhibiting the binding of these two ligands, it was first proposed that there may be two distinct but interconvertible conformational states of the postsynaptic alpha-adrenergic receptor (Greenberg et al., 1976; U'Prichard et al., 1977; Greenberg and Snyder, 1978). However, as it became clearer that alpha-2 adrenergic receptors could also be located postsynaptically, subsequent investigations in several laboratories supported the alternate interpretation that [^{3}H]WB4101 selectively labeled central alpha-1 adrenergic receptors and that [^{3}H]clonidine selectively labeled alpha-2 adrenergic receptors. It was then shown that [^{3}H]dihydroergoeryptine labeled both alpha-1 and alpha-2 adrenergic receptors with equal affinity in the brain (Peroutka et al., 1978; Miach et al., 1978).

   Although methods were developed to label selectively alpha-2 adrenergic receptors (rather than both alpha-1 and alpha-2) using [^{3}H]dihydroergocryptine (Hoffman et al., 1979), most investigators have used the subtype-selective ligands. With the introduction of [^{3}H]yohimbine in 1980 (Motulsky et al., 1980; Yamada et al., 1980), many laboratories developed an interest in alpha-2 adrenergic receptor binding. The human platelet quickly became the "favored tissue" for these binding studies and has become the standard against which other tissues were compared. In addition to the partial agonists, [^{3}H]clonidine and [^{3}H]p-aminoclonidine, other alpha-2 adrenergic radioligands in common use include the full agonist [^{3}H]UK 14,304 and the antagonists [^{3}H]yohimbine, [^{3}H]rauwolscine, and [^{3}H]idazoxan (RX781094). By 1980 alpha adrenergic radioligand binding studies were generally accepted as providing additional and direct evidence supporting the alpha-1 and alpha-2 subdivision. "Ligand binding studies with labeled catecholamine antagonists have supported the hypothesis of separate alpha-2 and alpha-2 receptors" (Fain and Garcia-Sainz, 1980).

   Receptor binding studies have been instrumental in the development of models of alpha-2 receptor action. Using the platelet as a model system, and by analogy to the beta-adrenergic receptor, Hoffman and Lefkowitz (1980) developed a two-state model for the alpha-2 adrenergic receptor. According to this model, agonists bind with high affinity to the alpha-2(H) state of the receptor, but with low affinity to the alpha-2(L) state. On the other hand, antagonists interact at both states with equal affinity. The general applicability of this model still remains to be determined (Bylund and U'Prichard, 1983).

Recent evidence from radioligand binding studies has suggested that there may be subtypes of alpha-2 adrenergic receptors (Bylund, 1985, 1987; Nahorski, 1985). This subclassification of alpha-2 receptor is based on pharmacologic evidence and appears to be comparable to the subdivision of beta-adrenergic receptors into beta-1 and beta-2 subtypes. The validity of this subclassification will await studies using classical pharmacologic approaches and/or the purification of the receptor subtypes.

A working hypothesis for the classification for adrenergic receptors is presented in Fig. 2. It is proposed that beta-adrenergic, alpha-2 adrenergic, and alpha-1 adrenergic receptors are the three major adrenergic receptor subtypes. This is a pharmacologic classification, but importantly, each of the three subtypes appears to be coupled to a different guanine nucleotide binding protein: beta receptors to N_s; alpha-2 receptors to N_i; and alpha-1 receptors to N_x. The subdivision of beta-adrenergic receptors into

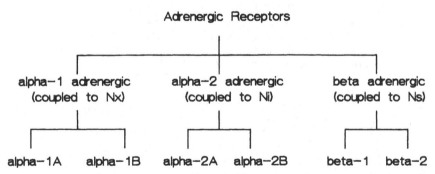

Fig. 2. This classification scheme for adrenergic receptors proposes that there are three major pharmacologic subtypes of adrenergic receptors, with each apparently coupled to a different guanine nucleotide binding regulation protein (N_s, stimulatory N protein; N_i, inhibitory N protein; N_x, unidentified N protein). Each of these three major subtypes may have additional subtypes. About 80% of beta-adrenergic receptors of the heart are beta-1, whereas 75% of beta receptors of the lung are beta-2. The alpha-2 adrenergic receptors of the human platelet, the HT29 human colonic cell line, and the porcine lung are classified as alpha-2A, whereas the OK opossum kidney cell line and the rat lung are alpha-2B. Alpha-1A and -1B subtypes have been proposed by Morrow and Creese (1986).

beta-1 and beta-2 is well accepted. The subdivision of alpha-2 receptor into alpha-2A and alpha-2B has been proposed (see above), as has the subdivision of alpha-1 receptors into alpha-1A and alpha-1B (Morrow and Creese, 1986).

During the past 15 years, considerable progress has been made in understanding alpha-2 adrenergic receptors. One shortcoming is that generally the study of receptor binding and biochemical mechanism of action have proceeded independently and more rapidly than has the study of the physiological function. Although it is clear that the density of alpha-2 adrenergic receptors is regulated by various factors, a good understanding of the physiologic significance of this regulation is lacking. The determination of the primary sequence of the alpha-2 adrenergic receptor from cDNA may serve to widen this gap. It is hoped that more laboratories will pursue the concurrent study of both the physiologic and biochemical aspects of this fascinating receptor.

REFERENCES

Ahlquist, R. P. (1948) A study of adrenotropic receptors. *Am. J. Physiol.* **153**, 586–600.

Anden, N.-E., Corrodi, H., Fuxe, K., Hokfelt, B., Hokfelt, T., Rydin, C., and Svensson, T. (1970) Evidence for a central noradrenaline receptor stimulation by clonidine. *Life Sci.* **9**, 513–523.

Arnsten, A. F. T. and Goldman-Rakic, P. S. (1985) α_2-Adrenergic mechanisms in prefrontal cortex associated with cognitive decline in aged nonhuman primates. *Science* **230**, 1273–1276.

Berthelson, S. and Pettinger, W. A. (1977) A functional basis for classification of α-adrenergic receptors. *Life Sci.* **21**, 595–606.

Brown, G. L. and Gillespie, J. S. (1956) Output of sympathin from the spleen. *Nature* **178**, 980.

Brown, G. L. and Gillespie, J. S. (1957) The output of sympathetic transmitter from the spleen of the cat. *J. Physiol.* **138**, 81–102.

Burns, T. W. and Langley, P. E. (1970) Lipolysis by human adipose tissue: The role of cyclic 3',5'-adenosine monophosphate and adrenergic receptor sites. *J. Lab. Clin. Med.* **75**, 983–997.

Bylund, D. B. (1985) Heterogeneity of alpha-2 adrenergic receptors. *Pharmacol. Biochem. Behav.* **22**, 835–843.

Bylund, D. B. (1987) Subtypes of alpha-2 Adrenergic Receptors in Human and Rat Brain, in *Epinephrine in the Central Nervous System* (J. Stolk, D. U'Prichard, and K. Fuxe, eds.) Oxford University Press, in press.

Bylund, D. B. and U'Prichard, D. C. (1983) Characterization of α_1 and α_2-adrenergic receptors. *Int. Rev. Neurobiol.* **24**, 343–431.

Delbarre, B. and Schmitt, H. (1973) A further attempt to characterize sedative receptors activated by clonidine in chickens and mice. *Eur. J. Pharmacol.* **22**, 355–359.

De Potter, W. P., Chubb, I. W., Put, A., and De Schaepdryver, A. F. (1971)

Facilitation of the release of noradrenaline and dopamine-β-hydroxylase at low stimulation frequencies by α-blocking agents. *Arch. Int. Pharmacodyn.* **193**, 191–197.

Dubocovich, M. L. and Langer, S. Z. (1974) Negative feed-back regulation of noradrenaline release by nerve stimulation in the perfused cat's spleen: Differences in potency of phenoxybenzamine in blocking the pre- and post-synaptic adrenergic receptors. *J. Physiol.* **237**, 505–519.

Fain, J. N. and Garcia-Sainz, J. A. (1980) Role of phosphatidylinositol turnover in alpha$_1$ and of adenylate cyclase inhibition in alpha$_2$ effects of catecholamines. *Life Sci.* **26**, 1183–1194.

Farnebo, L.-O and Hamberger, B. (1971) Drug-induced changes in the release of [^{3}H]-noradrenaline from field stimulated rat iris. *Br. J. Pharmacol.* **43**, 97–106.

Greenberg, D. A. and Snyder, S. H. (1978) Pharmacological properties of [^{3}H]dihydroergokryptine binding sites associated with alpha noradrenergic receptors in rat brain membranes. *Mol. Pharmacol.* **14**, 38–49.

Greenberg, D. A., U'Prichard, D. C., and Snyder, S. H. (1976) Alpha-noradrenergic receptor binding in mammalian brain: Differential labeling of agonist and antagonist states. *Life Sci.* **19**, 69–76.

Haggendal, J. (1970) Some Further Aspects on the Release of the Adrenergic Transmitter, in *New Aspects of the Storage and Release Mechanisms of Catecholamines* (Schumann, H. J. and Kroneberg, G., eds. Springer-Verlag, Berlin.

Hoffman, B. B. and Lefkowitz, R. J. (1980) Radioligand binding studies of adrenergic receptors: New insights into molecular and physiological regulation. *Ann. Rev. Pharmacol. Toxicol.* **20**, 581–608.

Hoffman, B. B., De Lean, A., Wood, C. L., Schocken, D. D., and Lefkowitz, R. J. (1979) Alpha-adrenergic receptor subtypes: Quantitative assessment by ligand binding. *Life Sci.* **24**, 1739–1746.

Johnson, D. G., Thoa, N. B., Weinshilboum, R., Axelrod, J., and Kopin, I. J. (1971) Enhanced release of dopamine-β-hydroxylase from sympathetic nerves by calcium and phenoxybenzamine and its reversal by prostaglandins. *Proc. Natl. Acad. Sci. USA* **68**, 2227–2230.

Jones, L. M. and Mitchell, R. H. (1978) Stimulus-response coupling at α-adrenergic receptors. *Biochem. Soc. Trans.* **6**, 672–688.

Jones, S. B., Towes, M. L., Turner, J. T., and Bylund, D. B. (1987) Alpha-2 adrenergic receptor mediated sensitization of forskolin-stimulated cyclic AMP production. *Proc. Natl. Acad. Sci. USA* **84**, 1294–1298.

Kirpekar, S. M. and Puig, M. (1971) Effect of flow-stop on noradrenaline release from normal spleens and spleens treated with cocaine, phentolamine or phenoxybenzamine. *Br. J. Pharmacol.* **43**, 359–369.

Lands, A. M., Arnold, A., McAuliff, J. P., Luduena, F. P., and Brown, T. G., Jr. (1967) Differentiation of receptor systems activated by sympathomimetic amines. *Nature* (Lond.) **214**, 597–598.

Langer, S. Z. (1977) Presynaptic receptors and their role in the regulation of transmitter release. *Br. J. Pharmacol.* **60**, 481–497.

Langer, S. Z. (1974) Presynaptic regulation of catecholamine release. *Biochem. Pharmacol.* **23**, 1793–1800.

Langer, S. Z. (1970) The metabolism of [^{3}H]noradrenaline released by electrical stimulation from the isolated nictitating membrane of the cat and from the vas deferens of the rat. *J. Physiol.* **208**, 515–546.

Langer, S. Z., Adler, E., Energo, A., and Stefano, F. J. E. (1971) The role of the alpha receptors in regulating noradrenaline overflow by nerve stimulation. *Proc. Intl. Congress Physiol. Sci.* **25**, 335.

Limbird, L. E. and Sweatt, J. D. (1985) α_2-Adrenergic Receptors: Apparent Interaction with Multiple Effector Systems, in *The Receptors* Vol. II (P. M. Conn, ed.) Academic, Florida.

Miach, P. J., Dausse, J.-P., and Meyer, P. (1978) Direct biochemical demonstration of two types of α-adrenoreceptor in rat brain. *Nature* **274**, 492–494.

Minneman, K. P., Hedberg, A., and Molinoff, P. B. (1979) Comparison of beta adrenergic receptor subtypes in mammalian tissues. *J. Pharmacol Exp. Ther.* **211**, 502–508.

Morrow, A. L. and Creese, I. (1986) Characterization of α_1-adrenergic receptor subtypes in rat brain: A reevaluation of [^{3}H]WB4104 and [^{3}H]prazosin binding. *Mol. Pharmacol.* **29**, 321–330.

Motulsky, H. J., Shattil, S. J., and Insel, P. A. (1980) Characterization of α_2-adrenergic receptors on human platelets using [^{3}H]yohimbine. *Biochem. Biophys. Res. Commun.* **97**, 1562–1570.

Nahorski, S. R., Barnett, D. B. and Cheung, Y.-D. (1985) α-Adrenoceptor-effector coupling: Affinity states or heterogeneity of the α_2-adrenoceptor? *Clin. Sci.* **68**, 39s–42s.

Peroutka, S. J., Greenberg, D. A., U'Prichard, D. C., and Snyder, S. H. (1978) Regional variations in alpha adrenergic receptor interactions of [^{3}H]dihydroergokryptine in calf brain: Implications for a two-site model of alpha receptor function. *Mol. Pharmacol.* **14**, 403–412.

Robison, G. A., Butcher, R. W., and Sutherland, E. W. (1970) On the Relation of Hormone Receptors to Adenyl Cyclase, in *Fundamental Concepts in Drug–Receptor Interactions* (Danielli, J. F., Moran, J. F., and Triggle, D. J., eds.) Academic, Florida.

Robison, G. A., Butcher, R. W., and Sutherland, E. W. (1967) Adenyl cyclase as an adrenergic receptor. *Ann. NY Acad. Sci.* **139**, 703.

Rossum, J. M. (1965) Different types of sympathomimetic α-receptors. *J. Pharm. Pharmacol.* **17**, 202–216.

Sabol, S. L. and Nirenberg, M. (1979) Regulation of adenylate cyclase of neuroblastoma $\times$ glioma hybrid cells by α-adrenergic receptors. *J. Biol. Chem.* **254**, 1913–1920.

Starke, K. (1971) Influence of α-receptor stimulants on noradrenaline release. *Natur Wissenschaften.* **58**, 420.

Starke, K., Montel, H., Gayk, W., and Merker, R. (1974) Comparison of the effects of clonidine on pre- and postsynaptic adrenoceptors in the rabbit pulmonary artery. *Naunyn Schmildebergs Arch. Pharmacol* **285**, 133–150.

Sutherland, E. (1965) Adenyl Cyclase and Hormone Action, in *Pharmacology of Cholinergic and Adrenergic Transmission* (Koelle, G. B., Douglas, W. W., and Carlsson, A., eds.) MacMillan, New York.

Sweatt, J. D., Johnson, S. L., Cragoe, E. J., and Limbird, L. E. (1985) Inhibitors of Na^+/H^+ exchange block stimulus-provoked arachidonic acid release in human platelets. *J. Biol. Chem.* **260**, 12910–12919.

Turtle, J. R. and Kipnis, D. M. (1967) An adrenergic receptor mechanism for the control of cyclic 3'5' adenosine monophosphate synthesis in tissues. *Biochem. Biophys. Commun.* **28**, 797–802.

U'Prichard, D. C., Greenberg, D. A., and Snyder, S. H. (1977) Binding char-
    acteristics of a radiolabeled agonist and antagonist at central nervous
    system alpha noradrenergic receptors. *Mol. Pharmacol.* **13**, 454–473.
Wikberg, J. E. S. (1979) The pharmacological classification of adrenergic
    alpha-1 and alpha-2 receptors and their mechanisms of action. *Acta
    Physiol. Scand.* (suppl.) **468**, 1–110.
William, L. T. and Lefkowitz, R. J. (1976) Alpha-adrenergic receptor
    identification by [^{3}H]dihydroergocryptine binding. *Science* **192**, 791–793.
Yamada, S., Yamamura, H. I., and Roeske, W. R. (1980) Alterations in cen-
    tral and peripheral adrenergic receptors in deoxycorticosterone/sale hy-
    pertensive rat. *Life Sci.* **27**, 2405–2416.

SECTION 2
CHARACTERIZATION OF
THE RECEPTOR AND
ITS BINDING SITE

Chapter 2

Biochemistry of alpha-2 Adrenergic Receptors

John W. Regan

1. Radiolabeled Ligands for the Reversible Labeling of alpha-2 Adrenergic Receptors

Physiologic and pharmacologic studies provide the present basis for the classification of alpha-adrenergic receptors (Langer, 1974; Berthelsen and Pettinger, 1977; Wikberg, 1978; Starke, 1981). This classification calls for the existence of discrete alpha-1 and alpha-2 adrenergic receptors with the extant possibility of further subdivision among the alpha-2 adrenergic receptors (Cheung et al., 1982; McGrath and Reid, 1985; Bylund, 1985). Until recently, demonstrating the actual physical existence of the unique macromolecules representing these alpha-adrenergic receptor subtypes has proved elusive. The difficulty in studying their biochemistry is primarily because most membrane-bound hormone receptors, including alpha-adrenergic receptors, are present only at very low concentrations in the cells of target tissues (i.e., frequently less than 0.001% of the total cellular protein). Therefore, a sensitive and reliable means of quantifying receptor activity was required before the biochemical characterization and purification of alpha-adrenergic receptors could progress. The means for the direct measurement of alpha-adrenergic receptor activity was acquired in 1976 with the development of [^{3}H]dihydroergocryptine ([^{3}H]DHE, Williams and Lefkowitz, 1976). Dihydroergocryptine,

15

a potent alpha-adrenergic agonist, was radiolabeled with tritium to a relatively high specific activity (25 Ci/mmol). The high affinity and specificity of [³H]DHE (K_d ~10 nM) coupled with the exquisite sensitivity for the detection of radioactivity (liquid scintillation counting) allowed quantitation of the low levels of alpha-adrenergic receptors present in membranes prepared from rabbit uterus (~150 fmol/mg protein).

Although [³H]DHE was the first radioligand used to label alpha-2 adrenergic receptors, [³H]DHE itself binds with equal affinity at both alpha-1 and alpha-2 adrenergic receptors, and thus, lacked the specificity for the selective determination of alpha-2 adrenergic receptor activity; however, its use in conjunction with selective competitors corroborated functional studies and provided compelling evidence for the existence of unique alpha-1 and alpha-2 adrenergic receptors (Miach et al., 1978; Hoffman et al., 1979; Haga and Haga, 1980). Since then several radioactive adrenergic ligands have been used for the reversible labeling of the alpha-2 adrenergic receptor. These radiolabeled agonists and antagonists are listed in Tables 1 and 2, respectively. At present the antagonists [³H]yohimbine and [³H]rauwolscine enjoy the most widespread use for the identification of alpha-2 adrenergic receptors. Since the earlier literature on these radioligands has been reviewed (Bylund and U'Prichard, 1983), present consideration will be limited primarily to recent work done with these compounds. Likewise most of the literature concerning the binding of [³H]epinephrine, [³H]clonidine, [³H]p-aminoclonidine ([³H]PAC), and [³H]DHE has also been thoughtfully reviewed by Bylund and U'Prichard (1983) and will not be examined in this review. Aside from the initial reports, some of the remaining compounds (i.e., [³H]guanfacine, [³H]dihydroergonine, and [³H]phentolamine) have not been used for further studies and will not be discussed here. Drawbacks to the use of the latter compounds involve either singularly, or in combination, the following: lack of alpha-2 adrenergic selectivity, high nonspecific binding, and commercial unavailability.

The remaining compounds, [³H]UK 14,304, [³H]idazoxan, and ¹²⁵I-amino-phenethyl-rauwolscine, are the most recently introduced alpha-2 adrenergic radioligands. In significant ways each of these compounds departs from the other alpha-2 adrenergic radioligands. For example, the clonidine analog [³H]UK 14,304 is reportedly the first radiolabeled full agonist that is not a catecholamine. [³H]Idazoxan represents a new chemical class of alpha-2 adrenergic antagonists, and ¹²⁵I-aminophenethyl-rauwolscine is the first alpha-2 adrenergic ligand to be radioiodinated. These

Table 1
Agonist Radioligands for Alpha-2 Adrenergic Receptors

References	Compound	Specific activity[a]	alpha-Adrenergic selectivity	Commercial availability[b]
U'Prichard and Snyder, 1977 Garcia-Sevilla and Fuster, 1986	[³H]Epinephrine	60–90	Nonselective	A, D
U'Prichard et al., 1977 Tanaka and Starke, 1979	[³H]Clonidine	20–60	alpha-2	A, D
Rouot and Snyder, 1979	[³H]p-Aminoclonidine	40–60	alpha-2	D
Timmermans et al., 1982	[³H]Guanfacine	24	alpha-2	—
Jarrott et al., 1982 Loftus et al., 1984 Turner et al., 1985 Neubig et al., 1985	[³H]UK 14,304	60–90	alpha-2	D

[a]Ci/mmol.
[b]A, Amersham; D, New England Nuclear-Dupont (from 1986 product catalogs).

Table 2
Antagonist Radioligands for Alpha-2 Adrenergic Receptors

References	Compound	Specific activity[a]	alpha-Adrenergic selectivity	Commercial availability[b]
Williams and Lefkowitz, 1976 Miach et all, 1978 Wood et al., 1979	[³H]Dihydroergocrytine	20–50	Nonselective	A, D
Jakobs and Rauschek, 1978	[³H]Dihydroergonine	27	Nonselective	—
Steer et al., 1979 Lynch and Steer, 1981	[³H]Phentolamine	23	Nonselective	—
Motulsky et al., 1980	[³H]Yohimbine	70–90	alpha-2	A, D
Perry and U'Pritchard, 1981	[³H]Rauwolscine	70–90	alpha-2	A, D
Lanier et al., 1986b	[¹²⁵I]-Rauwolscine-pAPC	2200	alpha-2	—
Howlett et al., 1982 Lane et al., 1983 McLaughlin and Collins, 1986	[³H]Idazoxan	30–50	alpha-2	A

[a] Ci/mmol.
[b] A, Amersham; D, New England Nuclear-Dupont (from 1986 product catalogs).

novel features hold promises of improving upon existing alpha-2 adrenergic radioligands. In the following sections these promises will be critically examined with respect to the possible advantages and disadvantages they may confer over [³H]PAC, [³H]yohimbine, and/or [³H]rauwolscine.

1.1. [³H]UK 14,304

UK 14,304 (5-bromo-6-[2-imidazoline-2-ylamino]-quinoxaline) is an alpha-2 adrenergic agonist with high affinity and selectivity for alpha-2 adrenergic receptors (Grant and Scrutton, 1980; Cambridge, 1981). It is structurally related to PAC and clonidine, which are also imidazolines; however, in contrast to the latter compounds, UK 14,304 is a full agonist at platelet alpha-2 adrenergic receptors (Grant and Scrutton, 1980). UK 14,304 has been radiolabeled with tritium to a specific activity of ~80 Ci/mmol and is available commercially.

The binding of [³H]UK 14,304 to membranes prepared from rat cerebral cortex (Loftus et al., 1984), human platelets (Neubig et al., 1985), and a human cell line (Turner et al., 1985) has been characterized. In general the binding of [³H]UK 14,304 has properties similar to those of [³H]PAC. In rat brain, [³H]UK 14,304 bound with high affinity (K_d = 1.4 nM) and with a relatively low level of nonspecific binding (Loftus et al., 1984). The number of binding sites labeled by [³H]UK 14,304 was in the range of what one might expect if binding had been measured with [³H]yohimbine (~200 fmol/mg protein), but it was claimed that [³H]UK 14,304 bound selectively to the high affinity state of the alpha-2 adrenergic receptor. An effect by GTP on the binding of [³H]UK 14,304 to rat brain membranes was not investigated, although it was shown that Mn^{2+} had a biphasic effect on binding activity, first increasing binding with low concentrations of Mn^{2+} (10–100 μM) and then decreasing binding with high concentrations (300–2000 μM). The inhibition of [³H]UK 14,304 binding by competitors was similar to the results obtained with other imidazolines (i.e., [³H]clonidine and [³H]PAC) in that agonists were more potent than antagonists. Among antagonists phentolamine (IC_{50} = 13 nM) was significantly more potent than yohimbine (IC_{50} = 190 nM), and, surprisingly, yohimbine was more potent than rauwolscine (IC_{50} = 330 nM). The affinity of prazosin (IC_{50} = 2700 nM) in competition for [³H]UK 14,304 binding was low and was consistent with known alpha-2 adrenergic pharmacology.

The binding of [³H]UK 14,304 to membranes prepared from human colonic adrenocarcinoma cells (HT29) grown in tissue culture was similar to the results obtained with rat brain membranes (Turner et al., 1985). Under the control conditions, the binding of

[³H]UK 14,304 could be resolved into two components: one site had an apparent K_d of 0.14 nM and comprised 33% of the total sites, and the other site was of lower affinity, having a K_d of 6.1 nM. At concentrations corresponding to the K_d values of the high- and low-affinity binding sites the level of nonspecific binding was 20 and 60%, respectively (Bylund, personal communication). The total number of sites labeled by [³H]UK 14,304 (360 fmol/mg protein) was similar to the total number of sites labeled by [³H]yohimbine (330 fmol/mg protein) and was significantly greater than the number of sites labeled by [³H]PAC (160 fmol/mg protein). The addition of MgCl$_2$, NaCl, or GTP had marked affects on the binding of [³H]UK 14,304. Thus, in the presence of 1 mM MgCl$_2$, there was a ~30% decrease in the total number of sites labeled by [³H]UK 14,304, and the binding became monophasic, modeling to one class of high-affinity binding sites (K_d = 0.26 nM). On the other hand, in the presence of 30 mM NaCl, the high-affinity component of [³H]UK 14,304 binding was virtually eliminated and most of the [³H]UK, 14,304 bound to a site with a K_d of ~3 nM. Finally, in the presence of 100 μM GTP, the specific binding of [³H]UK 14,304 could not even be determined. In contrast to the majority of results obtained with [³H]PAC, and with the binding of [³H]UK 14,304 in rat brain membranes, the potency of agonists was not significantly greater than the potency of antagonists. Competition binding sites with [³H]UK 14,304 in HT29 membranes showed that the affinity of yohimbine (IC$_{50}$ = 4 nM) was greater than the affinity of phentolamine (IC$_{50}$ = 49 nM), which is the reverse of results that have been generally obtained with [³H]agonists. Reasons for this discrepancy are not apparent.

In another study, the binding of [³H]UK 14,304 has been compared directly with the binding of [³H]yohimbine and of [³H]PAC to a receptor-enriched membrane preparation from human platelets (Neubig et al., 1985). Using a buffer containing 10 mM MgCl$_2$, [³H]UK 14,304 bound a single class of sites of high affinity (K_d = 0.9 nM). In addition, although it could not be quantified, it appeared that [³H]UK 14,304 was specifically binding to another site of lower affinity. The maximum number of sites labeled by [³H]UK 14,304 was ~65% of the total number of sites labeled by [³H]yohimbine. [³H]PAC labeled approximately the same number of sites as labeled by [³H]UK 14,304, but the binding of [³H]PAC could be resolved into two components; one with a K_d of 0.6 nM, and another with a K_d of 7.9 nM. In competition studies with [³H]yohimbine, unlabeled PAC bound to a third site of much lower affinity (K_i ~400 nM). Similar results have also been reported by U'Prichard et al. (1983). In the latter study the different affinity states of PAC binding have been described as SH (superhigh), H (high), and L (low). Thus, the SH and H sites can be directly identified with [³H]PAC, whereas all three sites can be resolved in competition studies using [³H]yohimbine and unlabeled PAC. ([³H]Yohimbine, labels all three sites with equal affinity.) Unlabeled UK 14,304, however, decreased [³H]yohimbine binding in a manner that was best described by interactions with only two sites. These sites consisted of a high-affinity component, which was analogous to the site identified by the direct binding of [³H]UK 14,304, and a low-affinity component that corresponds to the L sites identified in competition studies with unlabeled PAC. It must be assumed therefore that [³H]UK 14,304 did not distinguish between the SH and H sites and bound to them with equal affinity. Alternatively, the significance of the SH and H sites identified by [³H]PAC might be questioned since the three-site fit for the inhibition of [³H]yohimbine binding by unlabeled PAC was only of marginal statistical significance and could be adequately explained with a two-site model. If the latter supposition is accepted, then the results obtained from competition studies with PAC and UK 14,304 are in essential agreement: both ligands interact with a single high-affinity state and a single low-affinity state of the alpha-2 adrenergic receptor. The direct binding of [³H]UK 14,304 in the presence of $MgCl_2$ supports this two-site model; the binding of [³H]PAC does not. In weighing the evidence the practical aspects of obtaining reliable parameter estimates for the direct binding of radioligands to low-affinity sites should be considered. In particular the use of glass fiber filtration assays and too few assays at high concentrations of radioligand could lead to serious overestimates of the actual affinity of a low-affinity binding site. It seems possible therefore that the SH and H sites identified by [³H]PAC might represent the high- and low-affinity forms of the alpha-2 adrenergic receptor previously identified in competition studies. In any event it would seem prudent that the interpretation of the existing data should place greater emphasis on competition binding data. In addition it would be helpful if future studies paid more attention to validation of methods used for the direct quantitation of low-affinity binding sites.

In summary it appears that [³H]UK 14,304 will prove to be a useful ligand for the characterization of the high-affinity state of alpha-2 adrenergic receptors. In most respects it shares the same binding characteristics as [³H]PAC. Thus, both radioligands label approximately the same proportion of the total number of alpha-2 adrenergic receptors (as defined by the binding of [³H]yohimbine), and they are both quite sensitive to the presence of metal ions and guanine nucleotides. [^{3}H]UK, 14,304 may have an advantage over [^{3}H]PAC in that it binds with slightly higher affinity and it appears to be a full agonist. In addition, under steady-state conditions and in the presence of $MgCl_2$, the kinetics of [^{3}H]UK 14,304 binding appear to be less complex than those of [^{3}H]PAC. However, both ligands have complex association and dissociation curves, and more studies will be needed to fully understand the binding of these agonist ligands.

1.2. [^{3}H]Yohimbine/[^{3}H]Rauwolscine

The reversible binding of [^{3}H]yohimbine to alpha-2 adrenergic receptors was first characterized using human platelet membranes and intact human platelets (Motulsky et al., 1980; Mukherjee, 1981; Daiguji et al., 1981; Garcia-Sevilla et al., 1981; Macfarlane et al., 1981; Brodde et al., 1982; Limbird et al., 1982). These studies showed that in the range of 20–37°C, [^{3}H]yohimbine bound, in a saturable manner, to a single class of high-affinity binding sites. The K_d values were in the range of 1–3 nM, and the density of binding sites varied with the individual protocol, but were approximately 200 sites per intact platelet or ~200 fmol/mg of membrane protein. [^{3}H]Yohimbine binding showed the expected pharmacology of an alpha-2 adrenergic receptor. Thus, in competition binding studies the following order of potency for antagonists was obtained: rauwolscine ≥ yohimbine > phentolamine>> corynanthine > prazosin. For agonists the order of potency was: PAC > clonidine > (−)-epinephrine > (−)-norepinephrine >> isoproterenol. Stereoselectivity was shown: the (−)-isomers of both epinephrine and norepinephrine being approximately 10-fold more potent than the (+)-isomers. Unlike some of the results obtained with [^{3}H]DHE (Motulsky and Insel, 1982), all the competitors decreased [^{3}H]yohimbine binding to the same extent. At concentrations of [^{3}H]yohimbine in the range of its K_d, the amount of nonspecific binding was ~15% of the total binding in membranes and ~25% of the total binding in intact platelets.

Several studies have reported that metal ions affect the specific binding of [^{3}H]yohimbine. For example magnesium (as $MgCl_2$) at a concentration of ~5 mM decreased the affinity of [^{3}H]yohimbine to human platelet alpha-2 adrenergic receptor approximately two-fold (Daiguji et al., 1981; Cheung et al., 1982). In contrast sodium ions increased both the affinity and the maximal binding of [^{3}H]yohimbine to human platelet membranes (Limbird et al., 1982). The increase in affinity caused by sodium ions was found to be caused primarily by a faster rate of association of [^{3}H]yohimbine with the receptor, with little apparent change in the rate of dissociation. In another study the combination of 200 mM NaCl and 10 μM Gpp(NH)p was shown to increase the binding of [^{3}H]yohimbine to rat brain membranes by 100% (Woodcock and Murley, 1982). The increase in binding was accounted for solely by an increase in the B_{max} of [^{3}H]yohimbine binding with no change in K_d. Interestingly this same effect could be demonstrated with membranes prepared from rat hypothalamus, but not with renal cortical membranes. The latter findings point to alpha-2 adrenergic receptor heterogeneity between tissues and might explain the different effects of sodium on the binding parameters of [^{3}H]yohimbine in human platelet membranes. However, as discussed below other explanations are available.

Recently the effects of sodium and guanine nucleotides on the binding of [^{3}H]yohimbine to membranes prepared from rat cerebral cortex have been examined (Cheung et al., 1984). It was found that the effects of NaCl and Gpp(NH)p on [^{3}H]yohimbine binding varied with the methods used to prepare the membranes. Furthermore, the magnitude of the effect was directly correlated with the amount of endogenous norepinephrine that was retained during the preparation of the membranes. Thus, membrane preparations that retained higher concentrations of norepinephrine were found to have lower initial values for the maximal binding of [^{3}H]yohimbine, and they were affected to a greater extent by NaCl and Gpp(NH)p. Regardless of the methods used to prepare the membranes, the maximal binding of [^{3}H]yohimbine was roughly the same in the presence of NaCl and Gpp(NH)p. It was postulated that endogenous norepinephrine formed a very high-affinity complex with the alpha-2 adrenergic receptor and an N protein. Thus occupancy of the receptor by the endogenous norepinephrine precluded receptor occupancy by [^{3}H]yohimbine unless Gpp(NH)p or NaCl were present. The Gpp(NH)p acted to increase [^{3}H]yohimbine binding by disrupting the receptor–N protein interaction causing the receptor to adopt a low-affinity binding state with regards to the binding of the endogenous norepinephrine. NaCl increased [^{3}H]yohimbine binding presumably by a direct action on the receptor itself to cause an acceleration of the dissociation of endogenous norepinephrine. These findings emphasize that consideration must be given to the methods of membrane preparation, especially if comparisons are to be made that could involve apparent changes in receptor density. For example, differences between tissues in their sensitivity to

NaCl and to guanine nucleotides have been used to support the concept of alpha-2 receptor heterogeneity; however, instead of heterogeneity these differences might only reflect differences in the amount of retained norepinephrine. Thus with renal membranes, where the effects of NaCl and Gpp(NH)p were not observed, the content of endogenous norepinephrine may be very low; on the other hand it may also represent an intrinsic difference in the nature of the high-affinity complex formed between norepinephrine the alpha-2 adrenergic receptor, and N_i.

The binding of [³H]rauwolscine, developed shortly after the introduction of [³H]yohimbine, was initially characterized with membranes prepared from bovine cerebral cortex (Perry and U'Prichard, 1981). Not surprisingly, given what was known about the pharmacology of the unlabeled isomers, the binding properties of [³H]rauwolscine were very similar to those of [³H]yohimbine. Thus, [³H]rauwolscine labeled a single class of high-affinity binding sites (K_d = 2.5 nM) with the pharmacologic characteristics of alpha-2 adrenergic receptors. The binding of [³H]rauwolscine and [³H]yohimbine has been compared directly in several tissues. With human platelet membranes, [³H]rauwolscine bound with slightly lower affinity and with higher nonspecific binding than [³H]yohimbine (Motulsky and Insel, 1982). With rat cerebral cortical membranes, however, [³H]rauwolscine bound with significantly higher affinity than [³H]yohimbine and with about the same degree of nonspecific binding (Cheung et al., 1982). The K_i values obtained with various competing drugs were virtually identical regardless of whether the binding was being determined with [³H]yohimbine or [³H]rauwolscine. A comparison of K_d values obtained with these two ligands indicates that [³H]rauwolscine also binds with higher affinity to renal membranes.

[³H]Yohimbine and [³H]rauwolscine have been used successfully for the direct identification of alpha-2 adrenergic receptors in a variety of tissues. Tables 3 and 4 list some of these tissues. In general for most tissues where biochemical and physiological studies have provided pharmacologic evidence for the presence of alpha-2 adrenergic receptors, the binding of either one or both of these radioligands has been demonstrated. There are at this time, however, some exceptions. One tissue includes the rat pancreas, in which nearly irrefutable evidence points to the presence of alpha-2 adrenergic receptors on cells of the islets of Langerhans (Nakaki et al., 1981; Yamazaki et al., 1982). Good evidence also indicates that alpha-2 adrenergic receptors are present on the melanophores (pigment cells) of frogs (Berthelsen and Pettinger,

Table 3

[³H]Yohimbine: Characterization of Binding to Membranes from Assorted Tissues

Reference	Species/tissue	K_d^a	B_{max}^b	
Mukherjee, 1981	Human platelet	1.7	191	fmol/mg protein
Daiguji et al., 1981	Human platelet	1.2	180	fmol/mg protein
Garcia-Sevilla et al., 1981	Human platelet	3.0	188	fmol/mg protein
Brodde et al., 1982	Human platelet	2.0	221	fmol/mg protein
Cheung et al., 1982	Human platelet	0.6	148	fmol/mg protein
Limbird et al., 1982	Human platelet	4.8	173	fmol/mg protein
Hoffman et al., 1982	Human platelet	1.5	138	fmol/mg protein
Motulsky et al., 1980	Human platelet (intact)	2.7	207	sites/platelet
Macfarlane et al., 1981	Human platelet (intact)	2.9	200	sites/platelet
Glusa and Markwardt, 1983	Human platelet (intact)	1.9	202	sites/platelet
Kerry et al., 1984	Human platelet	3	258	sites/platelet
Glusa and Markwardt, 1983	Rabbit platelet	6.1	66	sites/platelet
Kerry et al., 1984	Rabbit platelet (intact)	19	270	sites/platelet
Glusa and Markwardt, 1983	Dog platelet (intact)	2.7	120	sites/platelet
Kerry et al., 1984	Rat platelets (intact)	15	42	sites/platelet

Table 3 (*continued*)

Reference	Species/tissue	K_d^a	B_{max}^b	
Kerry et al., 1984	Guinea pig platelet (intact)	NA	<10	sites/platelet
Cheung et al., 1982	Rat brain	4.7	125	fmol/mg protein
Rouot et al., 1982	Rat brain	10.	254	fmol/mg protein
Dickinson et al., 1986	Human brain (cortex)	2.6	107	fmol/mg protein
Petrash and Bylund, 1986	Human brain (cortex)	0.46	137	fmol/mg protein
Brodde et al., 1983	Calf brain (cortex)	3.3	4.0	pmol/g wet wt
Dickinson et al., 1986	Rabbit brain (cortex)	5.1	220	fmol/mg protein
Schmitz et al., 1981	Rat kidney	20	170	fmol/mg protein
Snavely and Insel, 1982	Rat kidney (cortex)	20	239	fmol/mg protein
Woodcock and Johnston, 1982	Rat kidney	10	120	fmol/mg protein
Matsushima et al., 1986	Rat kidney (basolateral membranes)	8.8	1130	fmol/mg protein
Brodde et al., 1983	Guinea pig kidney	5.2	4.7	pmol/g wet wt
Dickinson et al., 1986	Human kidney	1.6	95	fmol/mg protein

Reference	Tissue	K_d[a]	Density[b]	Units
Dickinson et al., 1986	Rabbit kidney	6.6	149	fmol/mg protein
Dickinson et al., 1986	Human spleen	1.6	217	fmol/mg protein
Dickinson et al., 1986	Rabbit spleen	7.3	331	fmol/mg protein
Dickinson et al., 1986	Guinea pig spleen	6.3	135	fmol/mg protein
Dickinson et al., 1986	Cat spleen	7.0	143	fmol/mg protein
Latifpour et al., 1982	Rat lung (neonatal)	1.5	304	fmol/mg protein
Feller and Bylund, 1984	Pig lung (5 wk)	0.22	129	fmol/mg protein
Agrawal and Daniel, 1985	Rat mesenteric artery	34	427	fmol/mg protein
Tsukahara et al., 1983	Bovine pial artery	18	687	fmol/mg protein
Bobik, 1982	Dog aorta	1.1	52	fmol/mg protein
Barnes et al., 1983	Dog trachea	2.7	51	fmol/mg protein
Chang et al., 1983	Rabbit ileal enterocyte	28	104	fmol/mg protein
Nakaki et al., 1983	Rat jejunum and ileal epithelium	6.0	37	fmol/mg protein
Dickinson et al., 1986	Human colon	1.7	94	fmol/mg protein
Tharp et al., 1981	Human adipocyte	3.9	145	fmol/mg protein

[a]Apparent equilibrium dissociation constant, nM

[b]Receptor density.

Table 4

[³H]Rauwolscine: Characterization of Binding to Membranes
from Assorted Tissues

Reference	Species/tissue	K_d^a	B_{max}^b	
Cheung et al., 1982	Human platelet	0.6	144	fmol/mg protein
Cheung et al., 1982	Rat brain	1.8	108	fmol/mg protein
Diop et al., 1983	Rat brain	1.3(H)[c]	48(H)	fmol/mg protein
		12(L)	187(H)	fmol/mg protein
Tanaka et al., 1983	Rat brain	3.6(H)	121(H)	fmol/mg protein
		212(L)	1500(L)	fmol/mg protein
Asakura et al., 1985	Rat brain	1.3(H)	46(H)	fmol/mg protein
		20(L)	212(L)	fmol/mg protein
Broadhurst and Wyllie, 1986[d]	Rat brain	1.8	220	fmol/mg protein
Summers et al., 1983	Human brain	2.1	135	fmol/mg protein
Perry and U'Pritchard, 1981	Bovine brain (cortex)	2.5	170	fmol/mg protein
Tanaka et al., 1983	Rat kidney	2.8	226	fmol/mg protein
Neylon and Summers, 1985	Rat kidney	2.3	8.8	pmol/g wet wt
McPherson and Summers, 1983	Rat kidney (glomeruli)	1.6	216	fmol/mg protein
Neylon and Summers, 1985	Human kidney	1.0	0.7	pmol/g wet wt
Neylon and Summers, 1985	Rabbit kidney	2.6	5.4	pmol/g wet wt
Neylon and Summers, 1985	Dog kidney	3.0	4.5	pmol/g wet wt
Neylon and Summers, 1985	Mouse kidney	2.8	12.8	pmol/g wet wt
Bottari et al., 1983	Human uterus (myometrium)	5.3	205	fmol/mg protein

[a]Apparent equilibrium dissociation, nM.
[b]Receptor density.
[c](H), high-affinity site; (L), low-affinity site.
[d]A low-affinity site, possibly representing binding to serotonin receptors, was also characterized.

1977), lizards (Carter and Shuster, 1982), and fish (Karlsson et al., 1985), but thus far they have not been characterized by direct radioligand binding. Binding studies also have apparently not been done on toad bladder, which interestingly, is the first tissue in which alpha-adrenergic stimulation was shown to decrease the synthesis of cyclic AMP (Turtle and Kipnis, 1967). Although the biochemical responses in the toad bladder have not been characterized with subtype selective ligands, the fact that stimulation leads to an inhibition of the activity of adenylate cyclase suggests that alpha-2 adrenergic receptors are present.

One tissue in which alpha-2 adrenergic receptors are purported to exist, and for which numerous binding studies have been conducted, is the rat submandibular gland. It was reported initially, in a study using [^{3}H]DHE, that adult rat submandibular gland contained a moderate density (~300 fmol/mg protein) of postsynaptic alpha-2 adrenergic receptors (Arnett and Davis, 1979). This conclusion was derived from the results of competition binding studies, which showed yohimbine and clonidine (K_i values = 180 and 140 nM, respectively) to be more potent than prazosin (K_i = 1520 nM). Although consistent with an alpha-2 adrenergic classification, the data were not submitted to further analysis to ascertain whether or not alpha-1 adrenergic receptors contributed to the binding of [^{3}H]DHE. Binding studies conducted with [^{3}H]clonidine, however, suggested that alpha-2 adrenergic receptors were either not present or were present at very low concentrations (<20 fmol/mg protein) in the submandibular glands of adult rats (Bylund and Martinez, 1980; Pimoule et al., 1980). However, a brief treatment of the rats with reserpine resulted in a marked increase in the number of [^{3}H]clonidine binding sites (~150 fmol/mg protein, Bylund and Martinez, 1980). Since the pharmacologic characteristics of this enhanced [^{3}H]clonidine binding showed prazosin to have very low affinity (K_i = 5200 nM) and yohimbine to have much higher affinity (K_i = 63 nM), it was concluded that a specific increase in the density of alpha-2 adrenergic receptors had been achieved. Seemingly contradictory results were obtained when it was reported that specific [^{3}H]yohimbine binding could not be detected in these same rat submandibular glands in which [^{3}H]clonidine binding was elevated (Bylund and Martinez, 1981). Further reports also suggested that these putative alpha-2 adrenergic receptors did not seem to be coupled to adenylate cyclase since decreases in cyclic AMP were not obtained following agonist stimulation (Bylund et al., 1982). The situation was not clarified when it was claimed that specific [^{3}H]yohimbine and [^{3}H]PAC binding could be detected concomitantly if young (3-wk-old) rats were used as the source of submandibular glands (Feller and Bylund, 1984). Under these circumstances [^{3}H]yohimbine labeled a similar number of sites as did [^{3}H]PAC (205 versus 256 fmol/mg protein, respectively). These and other data showing that the binding of both [^{3}H]yohimbine and [^{3}H]PAC could be influenced by Mg^{2+} and GTP led to the conclusion that alpha-2 adrenergic receptors were being identified by these radioligands (Feller and Bylund, 1984). This conclusion, however, should be tempered since the pharmacologic data also seem to indicate that both [^{3}H]PAC and [^{3}H]clonidine bind to the same sites regardless of whether the submandibular glands are from young or adult rats. Therefore, unless there are peculiar aspects to the physiological development of these binding sites, one would expect to find [^{3}H]yohimbine binding in the submandibular gland of reserpinized adult rats. Since this is not the case, then either the [^{3}H]PAC and [^{3}H]clonidine sites are not the same, in which case the [^{3}H]clonidine sites probably do not represent alpha-2 adrenergic receptors, or the [^{3}H]PAC and [^{3}H]clonidine sites are the same, in which case the binding of [^{3}H]yohimbine is to a separate site. If the latter interpretation is correct, then the binding of [^{3}H]yohimbine could be to an alpha-2 adrenergic receptor that subsequently disappears during the development of the rat submandibular gland. However, further biochemical and/or functional evidence will be necessary to establish the true identity of these radioligand binding sites in rat submandibular gland.

1.3. ^{125}I-Rauwolscine-pAPC

Difficulties in the identification of receptors by radioligand binding may arise from a variety of causes. Common causes include the possibilities that the receptor density is very low or that the quantity of tissue is limited. These difficulties can often be overcome through the use of a radioligand that has a higher specific radioactivity. Until recently the only radionuclide incorporated into alpha-2 adrenergic ligands was tritium, and the highest available specific radioactivity was ~85 Ci/mmol. Now, however, a radioiodinated derivative of rauwolscine has been prepared with a specific radioactivity of ~2200 Ci/mmol (Lanier et al., 1986b). Following the preparation of aminophenethyl-rauwolscine, radioiodination of the phenyl ring was accomplished by reaction with Na^{125}I and chloramine T. Referring to this iodinated ligand as ^{125}I-rau-pAPA, it was determined that ^{125}I-rau-pAPC bound with high affinity (K_d, 1–3 nM), and in a saturable manner, to membranes prepared from rat kidney, rat liver, and human platelets. At concentrations of ^{125}I-rau-pAPC in the range of its K_d, specific binding was ~70–80% of total binding. The specific binding was also inhibited by adrenergic ligands with an alpha-2 adrenergic specificity.

The increase in sensitivity gained by the use of ^{125}I over ^{3}H comes not only from the increase in specific radioactivity, but also from the higher efficiency in detecting gamma radiation (~80%) as compared with beta radiation (~50%). Thus, in principle ^{125}I-rau-pAPC confers a 41-fold increase in sensitivity compared to [^{3}H]rauwolscine: 25.6-fold from the increased specific radioactivity and 1.6-fold from the higher counting efficiency. This means that instead of using ~100 µg of protein per assay, which would give ~1000 cpm of specific [^{3}H]rauwolscine binding, one should be able to use ~2.5 µg of protein and get ~1000 cpm of specific ^{125}I-rau-pAPC binding (based upon K_d values of 2 nM, final concentrations of 2 nM for the radioligands, and an alpha-2 adrenergic receptor density of ~200 fmol/mg protein). Whether or not this level of sensitivity can be obtained with ^{125}I-rau-pAPC is presently unknown since it was not specifically investigated (Lanier et al., 1986b).

One potential obstacle to achieving a high level of sensitivity is the K_d of ^{125}I-rau-pAPC for alpha-2 adrenergic receptors (~2 nM). Since the K_d of [^{3}H]rauwolscine is also ~2 nM, to get a similar degree of receptor occupancy it is necessary to add more radioactivity for ^{125}I-rau-pAPC than for [^{3}H]rauwolscine to obtain the same final concentration of radioligand in the assay. This increased level of radioactivity can lead to problems with the measurement of small amounts of specific binding. For example if the filters used to separate bound and free radioligand bind a significant fraction of the radioligand, then a constant level of nonspecific binding will be present regardless of the amount of tissue added to the assay. Furthermore this nonspecific binding will be increased 41-fold for a radioiodinated ligand as compared with a tritiated ligand (i.e., by the same increment of increased sensitivity obtained for the measurement of specific binding). In the case of ^{125}I-rau-pAPC it was reported that ~0.5% of the radioligand bound to the filter. It would be expected, therefore, that ~4000 cpm of nonspecific binding would be present excluding the contribution by the tissue itself (assuming 100-µL assay volumes and a final radioligand concentration of 2 nM). Thus pushing ^{125}I-rau-pAPC to the limit by measuring binding on 2.5 µg of tissue protein, which would yield ~1000 cpm of specific binding (see previous discussion and assumptions), would be hampered to some extent by a low ratio of specific-to-nonspecific binding. In addition, since nonspecific binding increases with radioligand concentration, whereas specific binding saturates, this problem would be exacerbated at saturating concentrations of ^{125}I-rau-pAPC.

Notwithstanding possible liabilities, ^{125}I-rau-pAPC does offer the potential of measuring alpha-2 adrenergic receptor activity in situations in which tissue availability is limited. Additionally further experimentation should lead to the development of techniques that will minimize problems associated with nonspecific binding. With respect to radioiodinated ligands in general, maximum advantage is realized if their K_d values are in the low picomolar range. In this regard a congener of ^{125}I-rau-pAPC has been described that has an affinity for alpha-2 adrenergic receptors in this range (Lanier et al., 1986a). These developments in radioiodinated ligands should provide investigators with means of detecting alpha-2 adrenergic receptor activity at levels that are well beyond the limits of the presently available tritiated ligands.

1.4. [^{3}H]Idazoxan

Idazoxan (2-(2-(1,4-bezodioxanyl)-2-imidazoline, also known as RX781094, RS49415, and compound "170 150", is an alpha-2 adrenergic antagonist that has recently become available in a tritiated form. Initial studies indicated that a radiolabeled form of idazoxan might prove to be a superior ligand for the measurement of alpha-2 adrenergic receptor activity. However, since the number of published studies characterizing [^{3}H]idazoxan binding is limited, it will be necessary to examine some of the results obtained in functional studies using the unlabeled isomer in order to judge the potential of this radioligand.

Agonist stimulation of alpha-2 adrenergic receptors in the central nervous system is well known to cause hypotension and sedation. Less well known, however, is the evidence that stimulation of alpha-2 adrenergic receptors located postjunctionally in the peripheral vasculature can lead to a transient hypertension. In pithed rats, idazoxan antagonized this hypertensive response induced by the alpha-2 adrenergic agonists B-HT 933, B-HT 920, and clonidine (Mouille et al., 1981). The order of potency for antagonism of this alpha-2 adrenergic receptor-mediated pressor response was yohimbine > phentolamine > idazoxan (Alabaster and Brett, 1983).

The potency of idazoxan was also compared with some other alpha-adrenergic antagonists at alpha-2 adrenergic receptors present in rat vas deferens. In this tissue the neuronal release of norepinephrine, induced by electric field stimulation, will cause contractions of the smooth muscle. Agonist occupation of presynaptic receptors, however, will decrease the release of norepinephrine and will inhibit these contractions. Thus, the ability of clonidine to inhibit electrically evoked contractions was reversed by alpha-adrenergic antagonists with the following potency order: idazoxan > phentolamine > yohimbine >> prazosin. In a comparison with the effects of these same agents on the antagonism of alpha-1 adrenergic responses, it was concluded that idazoxan had an approximate sixfold greater selectivity for alpha-2 adrenergic receptors than did yohimbine (Chapleo et al., 1981; Doxey et al., 1983b). Different results were obtained, however, when similar experiments were done using rabbit vas deferens (Lattimer and Rhodes, 1985). Therefore although idazoxan was approximately five times more potent than yohimbine in the rat vas deferens, it was 10–50 times less potent than yohimbine in the rabbit vas deferens. These findings suggest species differences with respect to idazoxan's functional interaction at alpha-2 adrenergic receptors, and they are supported by the results obtained from binding studies. Thus in competition studies using [^{3}H]rauwolscine, idazoxan was threefold more potent than yohimbine using membranes prepared from rat brain, but it was about four times less potent than yohimbine with membranes from rabbit spleen (Alabaster and Brett, 1983). The latter finding probably does not represent a simple tissue difference since idazoxan was ~10-fold less potent than yohimbine in its ability to compete with [^{3}H]yohimbine binding in membranes prepared from rabbit cerebral cortex (Dickinson et al., 1986).

Other experimental models for functional studies of peripheral presynaptic alpha-2 adrenergic receptors include the electrically stimulated rat heart and guinea pig ileum. Stimulation of the sympathetic cardioaccelerator nerves to the heart results in a tachycardia that can be blocked by agonist occupation of presynaptic alpha-2 adrenergic receptors. The mechanism of this blockade is analogous to that in the rat vas deferens; i.e., inhibition of the neuronal release of norepinephrine. In guinea pig ileum, electric field stimulation produces a twitch response that is caused by the neuronal release of acetylcholine and activation of postsynaptic muscarinic receptors on the ileal smooth muscle. Again alpha-2 adrenergic agonists can block this twitch response by activation of prejunctional (presynaptic) alpha-2 adrenergic receptors that inhibit the release of acetylcholine. In the rat heart preparation, idazoxan was found to be equipotent (Doxey et al., 1984) or slightly less potent (Timmermans et al., 1984) than yohimbine. Using the guinea pig ileal preparation, idazoxan was again found to be equipotent with yohimbine in reversing the ability of norepinephrine to inhibit the twitch response (Timmermans et al., 1984). The high potency of idazoxan for prejunctional alpha-2 adrenergic receptors in guinea pig ileum is complemented by radioligand binding studies to membranes prepared from guinea pig spleen. Thus idazoxan (K_i ~8 nM) was found to be slightly more potent than yohimbine (K_i ~ 13 nM) in its ability to compete for [^{3}H]yohimbine binding to what are probably postjunctional alpha-2 adrenergic receptors (Dickinson et al., 1986).

The activity of idazoxan at alpha-2 adrenergic receptors present in hamster adipocytes has also been studied (Pushpendran and Garcia-Sainz, 1984). In this tissue, stimulation of beta-adrenergic receptors by epinephrine causes an increase in cyclic AMP production; however, because of the inhibition of adenylate cyclase by the simultaneous stimulation of alpha-2 adrenergic receptors, the net increase in cyclic AMP concentration is small. Blockade of alpha-2 adrenergic receptors by either idazoxan or yohimbine releases the inhibition of adenylate cyclase, which leads to a marked increase of cyclic AMP production in the presence of epinephrine. The effects of idazoxan and yohimbine were dose dependent, with idazoxan being ~10-fold more potent. Together with the previous findings, these results indicate that idazoxan is a potent antagonist at peripheral pre- and postsynaptic alpha-2 adrenergic receptors in the rat, hamster, and guinea pig; however, in rabbits idazoxan is less potent, especially when compared with yohimbine.

Functional as well as radioligand binding studies show that idazoxan is a potent alpha-2 adrenergic antagonist in the central nervous system. In both rabbits and dogs, idazoxan blocked decreases in blood pressure caused by the central actions of clonidine (Dabire et al., 1981; Hannah et al., 1983). Similarly, in rats, idazoxan blocked the hypotensive effect of clonidine with a potency that was approximately four times greater than that of rauwolscine (Timmermans et al., 1984). In the latter study it was also shown that idazoxan blocked the sedative actions of clonidine that appear to be mediated by activation of central alpha-2 adrenergic receptors as well.

Norepinephrine turnover in the brain is believed to be another functional measure of central alpha-2 adrenergic receptor activity. In principle, activation of presynaptic alpha-2 adrenergic receptors decreases the release of norepinephrine that leads to a corresponding decrease in norepinephrine turnover in the brain. Conversely, blockade of presynaptic alpha-2 adrenergic receptors leads to increases in the central turnover of norepinephrine. As expected idazoxan increased the turnover of norepinephrine in the rat brain in a dose-dependent manner (Walter et al., 1984). In this regard idazoxan was more specific than yohimbine, which also increased the turnover of dopamine; however, the actual significance of this differential effect is presently unknown.

The ability of idazoxan to compete for [³H]clonidine, [³H]prazosin, [³H]rauwolscine, and [³H]yohimbine binding to membranes prepared from several animal species has been investigated. For example [³H]clonidine and [³H]prazosin were used to assess the affinity of idazoxan at alpha-2 and alpha-1 adrenergic receptors, respectively (Howlett et al., 1982; Hannah et al., 1983; Dabire et al., 1983; Timmermans et al., 1984). It was found that the K_i of idazoxan in competition for the binding of [³H]clonidine was 6 nM in membranes prepared from rat cerebral cortex, whereas it was 550 nM in competition for the binding of [³H]prazosin (Howlett et al., 1982). This 94-fold selectivity of idazoxan for central alpha-2 adrenergic receptors exceeded that obtained for yohimbine, which was 14-fold selective. It is noted, however, that the affinities of yohimbine and rauwolscine tend to be lower when assessed with [³H]clonidine as compared with either [³H]rauwolscine or [³H]DHE. Thus, when [³H]rauwolscine was used to measure alpha-2 adrenergic receptor activity in rat brain membranes, the affinities of idazoxan and yohimbine were more nearly the same and so were their relative selectivities for alpha-2 adrenergic receptors (Alabaster and Brett, 1983). Similar results were also obtained when binding was measured with [³H]yohimbine (Dickinson et al., 1986). Therefore, although idazoxan had high affinity for human brain alpha-2 adrenergic receptors (K_i = 11 nM), it had much lower affinity for rabbit brain alpha-2 adrenergic receptors (K_i = 74 nM).

The binding of [³H]idazoxan has been studied with membranes prepared from rat cerebral cortex (Howlett et al., 1982; Langer et al., 1983; Doxey et al., 1983a) These reports demonstrate that [³H]idazoxan binds with high affinity (K_d ~4 nM) and can be inhibited by ligands with an alpha-adrenergic order of potency (idazoxan ≥ phentolamine > yohimbine >> prazosin). It is interesting that in competition with [³H]idazoxan, phentolamine is more potent than yohimbine. This is the same result that is often obtained when [³H]PAC is used as a radioligand, but it is op-

posite to that obtained when either [3H]yohimbine, [3H]rauwol-
scine, or [3H]DHE are used as the radioligands. This difference in
the order of potency of antagonists may be related to the fact that
idazoxan, like PAC, is an imidazoline.

The effects of various salts on [3H]idazoxan binding have
been examined (Lane et al., 1983). Under the control conditions
(50 mM Tris HCl, pH 7.8), the binding of [3H]idazoxan to rat
cerebral cortical membranes was of lower affinity (K_d = 12 nM)
than reported previously. Additionally, there was evidence that
[3H]idazoxan bound to distinct high- and low-affinity classes of
binding sites. In the individual presence of a variety of salts
(NaCl, KCl, $MgCl_2$, $CaCl_2$) the affinity of [3H]idazoxan was increased ~threefold, and the data were consistent with an interaction with a single class of high-affinity sites. If either $MgCl_2$ (1.2
mM) or $CaCl_2$ (1.3 mM) alone were included in the buffer, there
was no effect on affinity but there was a remarkable decrease in
the maximal binding of [3H]idazoxan (~70%). However, when
salts were present together (in a physiological salt solution), the
maximal binding of [3H]idazoxan was similar to control, and the
K_d was ~5 nM. A satisfying hypothesis to account for these diverse effects of mono and divalent metal cations on [3H]idazoxan
binding is unavailable.

Another tissue in which [3H]idazoxan binding has been examined is intact human platelets (Elliott and Rutherford, 1983).
Over a range of concentrations from 5 to 60 nM, the binding of
[3H]idazoxan was to a single class of sites with a K_d of 39 nM (the
incubation buffer contained 0.1% EDTA and 150 mM NaCl, pH
7.5). Nonspecific binding was ~50% of total binding (at 14 nM)
and the maximum number of binding sites (64 fmol per 10^8 platelets) was similar to that obtained with [3H]yohimbine (54 fmol per
10^8 platelets). The rank order of antagonist potencies obtained
from competition studies was characteristic of an alpha-2 adrenergic receptor, but it was different from the results obtained with
membranes prepared from rat cerebral cortex. Thus yohimbine
showed the highest potency, followed by idazoxan, phentolamine, and prazosin.

For the moment, it must be concluded that [3H]idazoxan of-
fers no particular advantage over [3H]rauwolscine for general use
in the measurement of alpha-2 adrenergic receptor activity. In
particular it seems to bind with lower affinity than [3H]rauwol-
scine, its specific radioactivity is less than that of [3H]rauwolscine,
it has a higher level of nonspecific binding, and its binding appears to be more sensitive to buffer composition. Although the
number of studies using [3H]idazoxan are limited, data obtained

from functional studies, and from competition binding studies using nonradioactive idazoxan, indicate that the affinity of idazoxan for alpha-2 adrenergic receptors is lower than that of rauwolscine in tissues other than rat brain. One oft-repeated claim is that idazoxan shows greater selectivity than rauwolscine for alpha-2 adrenergic receptors as compared with alpha-1 adrenergic receptors. Although this may be true, and could be of critical importance in functional studies, it is of less significance for radioligand binding studies. Thus in terms of its K_d for alpha-1 adrenergic receptors, the affinity of [^{3}H]rauwolscine is such that binding to alpha-1 adrenergic receptors would not be measured by most conventional binding methodologies. Recently derivatives of idazoxan have been prepared that show greater selectivity and potency for alpha-2 adrenergic receptors than idazoxan itself (Gadie et al., 1984; Doxey et al., 1984). These compounds appear to offer possibilities for improving upon existing alpha-adrenergic radioligands.

2. Radiolabeled Ligands for the Irreversible Labeling of alpha-2 Adrenergic Receptors

Radioligands that are capable of forming stable covalent bonds with the ligand binding site are of great value in trying to understand the molecular structure of receptors. Generally these so-called affinity probes are compounds that bear electrophilic functional groups that can react with the nucleophilic amino acid residues of proteins. Included among these probes are the photoaffinity ligands, which have the unique property that the reactive species can be specifically generated with light. This latter feature gives the user a measure of control that is nearly indispensable when it comes to radiolabeling membrane-bound receptors.

The ability to label receptors in membranes has been one of the major goals in the development of affinity probes. If accomplished, affinity labeling can provide valuable data on the structural identity of the receptor while the receptor is still in its native environment. However, even if they cannot be used successfully with membranes, affinity probes can be of great value with purified receptor preparations. Thus, regardless of the extent of the purification, as long as specific labeling of a protein or relevant macromolecule can be demonstrated, affinity probes provide the most convenient means for identification of the receptor. In fact even after complete purification, affinity labeling provides a means for understanding the structural and functional organization of the receptor, especially if the receptor is a multimer of subunits. Under such circumstances an affinity probe can be used to locate the binding site subunit and possibly the amino acid residues that comprise the ligand binding site itself.

2.1. [^{3}H]Phenoxybenzamine

Radioactive affinity ligands that have been used for the covalent labeling of alpha-2 adrenergic receptors are listed in Table 5. The first of these, [^{3}H]phenoxybenzamine, is an alpha-adrenergic antagonist that contains a chemically reactive beta-haloalkylamine functionality. Poor control over the reactivity of this group to effect preferential labeling of the receptor protein, over other nonspecific proteins, limits the general utility of [^{3}H]phenoxybenzamine. Another significant limitation is that [^{3}H]phenoxybenzamine is relatively nonselective for alpha-adrenergic receptor subtypes, and it can interact with other neurotransmitter receptors as well. These factors dictate that careful attention must be given to the specificity and pharmacologic characteristics of the labeling obtained with [^{3}H]phenoxybenzamine.

[^{3}H]Phenoxybenzamine was first shown to specifically label alpha-1 adrenergic receptors from rat liver (Kunos et al., 1983). Using a membrane preparation enriched in alpha-1 adrenergic receptors, an M_r 80,000 protein was labeled by [^{3}H]phenoxybenzamine. The labeling could be blocked by prazosin, but not by yohimbine, indicating the alpha-1 adrenergic nature of this protein. Although alpha-2 adrenergic receptors are present in rat liver membranes, they did not appear to be labeled under these conditions.

More recently, [^{3}H]phenoxybenzamine was shown to label a similar 80,000 protein in membranes prepared from DDT$_1$ MF-2 cells (Cornett and Norris, 1986). The labeling was shown to have the pharmacologic characteristics as well as the stereoselectivity expected of the alpha-1 adrenergic receptor. Two other proteins, with M_r values of 33,000 and 21,000, were also labeled by [^{3}H]phenoxybenzamine, but this labeling could only be blocked with nonradioactive phenoxybenzamine. Confirmation of the M_r 80,000 protein as the alpha-1 adrenergic receptor has been established by complete purification from DTT$_1$ MF-2 cells (Lomasney et al., 1986).

Human platelets, which lack alpha-1 adrenergic receptors, have been used as a source of alpha-2 adrenergic receptors for la-

Table 5
Radiolabeled Affinity Ligands for alpha-2 Adrenergic Receptors

Reference	Compound	Type	Species/Tissue	M_r^a
Regan et al., 1984	[³H]Phenoxybenzamine	Affinity	Human platelet	61,000
Shreeve et al., 1985	[³H]Phenoxybenzamine	Affinity	Human platelet	85,000
Regan et al., 1986a	[³H]Phenoxybenzamine	Affinity	Human platelet	64,000
Jaiswal and Sharma, 1985	[³H]p-Azidoclonidine	Photoaffinity	Rat adrenocortical carcinoma	64,000
Kawahara et al., 1985	[³H]p-Azidoclonidine	Photoaffinity	Rat brain (cortex)	—
Regan et al., 1986b	³H SKF 102229	Photoaffinity	Human platelet	64,000
Lanier et al., 1986a	¹²⁵I-Rauwolscine-AzPC	Photoaffinity	Pig brain	62,000

^{a}Apparent molecular weight.

beling by [³H]phenoxybenzamine. Initial attempts with membrane preparations, however, failed because of extensive labeling of nonreceptor proteins. On the other hand, following solubilization and partial purification of the alpha-2 adrenergic receptor, [³H]phenoxybenzamine labeled a protein with a characteristic alpha-2 adrenergic specificity (Regan et al., 1984). Figure 1 shows more recent results using a similar alpha-2 adrenergic receptor preparation. In the presence of [³H]phenoxybenzamine alone, predominant bands with M_r values of 64,000, 41,000 and 33,000 were obtained. Covalent labeling of the 64,000 and 41,000 bands by [³H]phenoxybenzamine was prevented, however, when labeling was done in the presence of phentolamine. Pharmacologic studies further established that labeling of the M_r 64,000 protein could be blocked with the agonist, (−)-epinephrine, but not with (+)-epinephrine or with the alpha-1 selective antagonist prazosin (Regan et al., 1984). Positive identification of the M_r 64,000 protein as the alpha-2 adrenergic receptor was obtained following its complete purification and subsequent labeling with [³H]phenoxybenzamine (Regan et al., 1986a). Additionally, the M_r 41,000 protein was identified as a degradation product of the M_r 64,000 receptor protein (Regan et al., 1986a).

It is noted that in preparations derived from both DTT₁ MF-2 cells (Cornett and Norris, 1986) and human platelets (Regan et al., 1984), the major protein labeled by [³H]phenoxybenzamine is one with an M_r of 33,000. Although neither an alpha-1 nor an alpha-2 adrenergic receptor, alkylation of this M_r 33,000 protein by [³H]phenoxybenzamine is blocked by prior treatment of the receptor preparations with nonradioactive phenoxybenzamine. This protein may represent a common structural element of cells; however, it was not identified in hepatic membranes labeled with [³H]phenoxybenzamine (Kunos et al., 1983). DDT₁ MF-2 cells are derived from smooth muscle cells and presumably share with platelets similar contractile proteins. It is possible, therefore, that the M_r 33,000 protein labeled by [³H]phenoxybenzamine might be involved in smooth muscle contraction. In this light, it would be interesting to determine if some of the antihypertensive actions of phenoxybenzamine are mediated by interactions with this M_r 33,000 protein.

Results from other studies using [³H]phenoxybenzamine contend that human platelet alpha-2 adrenergic receptors have an M_r of 85,000 (Shreeve et al., 1985). These studies made use of a combination of isoelectric focusing and SDS-PAGE to overcome problems associated with the nonspecific labeling of membrane proteins by [³H]phenoxybenzamine. Thus, human platelet mem-

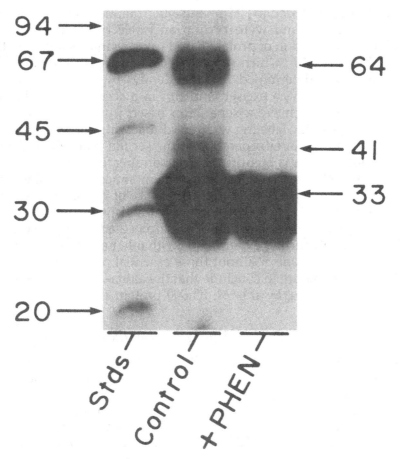

Fig. 1. ^{3}H-Phenoxybenzamine labeling of partially purified alpha-2 adrenergic receptors. Human platelet alpha-2 adrenergic receptors were purified ~1500-fold by sequential affinity and heparin-agarose chromatography (Regan et al., 1986a). Approximately 12 pmol of receptor binding activity (~300 pmol/mg protein) was desalted by Sephadex G-50 chromatography into a buffer consisting of 5 mM NaPO$_4$, 50 mM NaCl, 0.05% digitonin (pH 6.5). The desalted receptor was split into two 500-μL aliquots and 2.5 μL of [^{3}H]phenoxybenzamine (NEN-Dupont, specific activity 23 Ci/mmol) was added to give a final concentration of 200 nM. To determine nonspecific labeling, 2.5 μL of phentolamine (+ phen, above) was added to one of the samples to give a final concentration of 20 μM. The samples were incubated overnight at 4–6°C and then for 20 min at 30°C. The samples were then desalted as before, lyophilized, and redissolved in SDS-PAGE sample buffer (Laemmli, 1970). Following electrophoresis on a 10% polyacrylamide slab gel, the gel was prepared for fluorography using Enlightening (NEN-Dupont), dried, and autoradiographs were obtained. Exposures were made for 2 wk at −80°C using Kodak X-OMAT AR film.

branes were exposed to [^{3}H]phenoxybenzamine, were solubilized with digitonin, and were run on an isoelectric focusing gel to achieve a separation of protein based upon charge. Material in the vicinity of pH 4.6, which represented the isoelectric point of the receptor as determined independently by the binding of [^{3}H]yohimbine, was excised and run in a second dimension on SDS-PAGE to achieve a separation based upon size. Although very little specific labeling was found following separation in the first dimension, after separation in the second dimension, a peak of radioactivity was identified corresponding to a protein with an M_r of 85,000. When the labeling of membranes by [^{3}H]phenoxybenzamine was performed in the presence of 200 nM yohimbine, labeling of this M_r 85,000 protein was diminished, suggesting that this might represent labeling of the alpha-2 adrenergic receptor. Unfortunately further experiments with other adrenergic competitors were not done. Without these essential pharmacologic data it seems premature to conclude that the alpha-2 adrenergic receptor is represented by this M_r 85,000 protein.

2.2. [^{3}H]p-Azidoclonidine

The development of photoaffinity ligands for alpha-2 adrenergic receptors has proceeded at a slower pace relative to some of the other neurotransmitter and hormone receptors (Fedan et al., 1984). The first ligand introduced specifically for this purpose was [^{3}H]p-azidoclonidine (U'Prichard and Ernsberger, 1983). In spite of its commercial availability, however, relatively few studies have made effective use of this compound. The most thorough study employing [^{3}H]p-azidoclonidine is that by Kawahara et al. (1985). This study found that the reversible binding of [^{3}H]p-azidoclonidine was saturable, of high affinity, and of the appropriate pharmacology for an alpha-2 adrenergic ligand. It was also claimed that covalent labeling of alpha-2 adrenergic receptors was obtained. Although 100 μM (−)-norepinephrine could decrease the covalent labeling of proteins by [^{3}H]p-azidoclonidine, no other results were reported on the pharmacologic characteristics of this labeling. Using partially purified alpha-2 adrenergic receptor, it was observed that (−)-epinephrine could also decrease the covalent labeling of proteins by [^{3}H]p-azidoclonidine (Regan, unpublished results). Moreover, the latter studies found that the inactive isomer, (+)-epinephrine, gave the same result. Furthermore, as determined by the results of SDS-PAGE, phentolamine did not block the covalent labeling of any proteins by [^{3}H]p-azidoclonidine. Although it is apparent that catecholamines are capable of interfering with the covalent insertion of [^{3}H]p-azido-clonidine into some protein(s), to conclude that this labeling identifies an alpha-2 adrenergic receptor is inappropriate.

Two other groups also claim specific labeling of alpha-2 adrenergic receptors by [^{3}H]p-azidoclonidine (Jaiswal and Sharma, 1985; Shreeve et al., 1985). Using affinity-purified alpha-2 adrenergic receptors from a rat adrenocortical carcinoma cell line, Jaiswal and Sharma (1985) found that [^{3}H]p-azidoclonidine labeled a protein with an M_r of ~64,000 and that the labeling could be prevented by the presence of 100 μM p-aminoclonidine. No other agents were used to further characterize the pharmacologic specificity of this labeling. A need for further characterization of this labeling is indicated given that [^{3}H]p-azidoclonidine was used as the radiolabel, that p-amino-clonidine was used to define nonspecific labeling, and that the affinity column was made from and eluted with p-aminocloni-dine.

[^{3}H]p-Azidoclonidine was used by Shreeve et al. (1985) to corroborate results obtained using [^{3}H]phenoxybenzamine to label an M_r 85,000 protein (see previous discussion on [^{3}H]phen-oxybenzamine). These investigators presented data that purportedly show specific labeling of an M_r 85,000 protein by [^{3}H]p-azidoclonidine. Again, a proper pharmacologic characteri-zation of the labeling was not reported to substantiate the claim that this protein represents the alpha-2 adrenergic receptor. Addi-tionally an examination of the data shows that several peaks of radioactivity were decreased when labeling was done in the pres-ence of the competitor (phenoxybenzamine), and it appears that an arbitrary decision was made on which one to call the alpha-2 adrenergic receptor. Under these circumstances, i.e., extensive specific and nonspecific labeling, it is surprising that the tech-niques of isoelectric focusing and SDS-PAGE, used by these au-thors for [^{3}H]phenoxybenzamine labeling were not also applied to characterize the labeling of proteins by [^{3}H]p-azidoclonidine.

With respect to p-azidoclonidine, the existing studies that re-port photoaffinity labeling of alpha-2 adrenergic receptors lack the critical evidence needed to back up this assertion. This is un-expected since the reversible binding of [^{3}H]p-azidoclonidine in membranes is good, is characteristically alpha-2 adrenergic, and is of relatively high affinity. Specific factors contributing to the difficulties associated with the use of [^{3}H]p-azidoclonidine are difficult to identify. In membranes, the combination of a low re-ceptor density and the relatively low specific radioactivity of [^{3}H]p-azidoclonidine are certainly factors. In detergent-solubilized preparations of alpha-2 adrenergic receptors there is the potential problem of lack of an effective interaction of the receptor with the inhibitory guanine nucleotide binding protein. In the absence of this interaction one would expect the alpha-2 adrenergic receptor to be in a low-affinity state with respect to the binding of agonists. This would mean that high concentrations of $[^3H]p$-azidoclonidine would be needed for significant occupancy of the receptor. Even so, the kinetics of the binding (i.e., rapid dissociation) might preclude a high yield of photoinsertion into the receptor during exposure to UV light. Further possibilities are that the azido group is not in a position for productive photoaffinity labeling of the binding site or that the product of the labeling, although covalent, is not stable and is therefore reversible.

2.3. $[^3H]$SKF 102229

More recently other photoaffinity ligands have been developed for use with alpha-2 adrenergic receptors. One of these, SKF 102229, is a specific alpha-2 adrenergic antagonist that has been radiolabeled with tritium to a specific activity of ~80 Ci/mmol (Regan et al., 1986b). Studies with the unlabeled isomer in membranes showed it to have an apparent equilibrium dissociation constant (K_i) of ~15 nM (Regan et al., 1985). Photolysis of human platelet membranes in the presence of unlabeled SKF 102229 produced a dose-dependent blockade of the specific binding of $[^3H]$yohimbine. This irreversible blockade of $[^3H]$yohimbine binding could be prevented if photolysis was done in the presence of alpha-2 adrenergic ligands, but it could not be prevented in the presence of alpha-1 or beta-adrenergic ligands. Using the tritiated isomer $[^3H]$SKF 102229, reversible binding studies have yielded inconsistent results in part because of high nonspecific binding; however, $[^3H]$SKF 102229 has been used successfully to label irreversibly alpha-2 adrenergic receptors in membranes prepared from human platelets and rabbit kidney (Regan et al., 1986b). Figure 2 shows an autoradiograph obtained following SDS-PAGE of human platelet membranes photolyzed in the presence of various concentrations of $[^3H]$SKF 102229. Clearly at the higher concentrations of $[^3H]$SKF 102229 the nonspecific labeling of proteins is a problem. This is true in spite of the fact that 5 mM 2-mercaptoethanol was present to act as a scavenger of the nonspecific photolabeling. Nevertheless specific labeling of a protein with an M_r of 64,000 was obtained since labeling of this protein was prevented when photolysis was done in the presence of a high concentration of phentolamine. Furthermore the ratio of the specific-to-nonspecific labeling could be improved as the concentration of

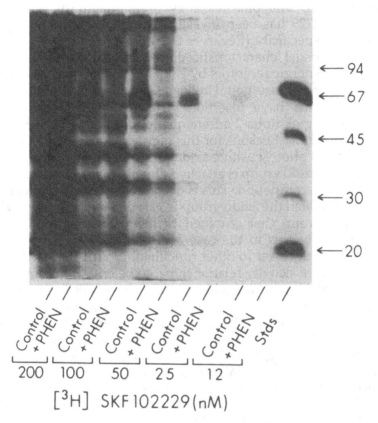

Fig. 2. Photoaffinity labeling of alpha-2 adrenergic receptors in human platelet membranes with increasing concentrations of [^{3}H]SKF 102229. Membranes were prepared from outdated human platelets, and photoaffinity labeling was done as previously described (Regan et al., 1986b). Membranes were incubated for 60 min on ice in a final volume of 500 μL and with the final concentrations of [^{3}H]SKF 102229 listed above. The buffer contained 50 mM Tris-HCl, 5 mM 2-mercaptoethanol, and 1 mM EDTA (pH 7.2). Following exposure to UV light (254 nm), the samples were centrifuged and the membranes were prepared for SDS-PAGE on a 10% gel according to Laemmli (1970). After electrophoresis the gel was prepared for fluorography and autoradiographs were obtained following an exposure for 2 wk at −80°C. Total labeling is shown above as control (C) and nonspecific labeling as phentolamine (phen). The nonspecific labeling was obtained in the presence of a final concentration of 10 μM phentolamine.

[^{3}H]SKF 102229 was decreased toward its K_d. Using partially purified as well as purified preparations of human platelet alpha-2 adrenergic receptors, excellent specific photolabeling of the M_r 64,000 protein has been obtained with essentially no nonspecific labeling of other proteins (Regan et al., 1986b; Regan, unpublished). The pharmacology of photoaffinity labeling by [^{3}H]SKF 102229 has been investigated and it shows an alpha-2 adrenergic specificity (Regan et al., 1986b).

An unusual characteristic of the photoaffinity labeling of alpha-2 adrenergic receptors by [^{3}H]SKF 102229 concerns its high yield of covalent insertion. Unlike most azido-based photoaffinity ligands, which have insertion yields of 5–15%, the yield of photolabeling of alpha-2 adrenergic receptors by [^{3}H]SKF 102229 approaches 80%. Reasons for this high yield may relate to the location of the photosensitive azido group in the molecule. In SKF 102229, the azido group is attached directly to the aromatic moiety of the parent compound. This is in contrast to many photoaffinity ligands in which the azido group is part of a larger extrinsic alkyl-aryl-azido group that is joined to the parent compound. The significance of this in the case of SKF 102229 is that the azido group is much closer to other functional groups that are known, from structure–activity relationships, to be critical for binding activity. This closer relationship may in turn result in a more intimate association of the azido group with elements of the receptor that are responsible for ligand binding. Additionally the possible involvement of nucleophilic amino acid residues with ligand binding could contribute to a higher yield of photoaffinity labeling by SKF 102229. With regard to the latter possibility, evidence has been presented for the presence of both cysteine and tyrosine groups in the ligand binding site of alpha-2 adrenergic receptors (Nakata et al., 1986).

2.4. ^{125}I-Rauwolscine-AzPC

A newer ligand has been described for photoaffinity labeling alpha-2 adrenergic receptors. This compound is a radioiodinated derivative of rauwolscine called ^{125}I-rau-AZPC (Lanier et al., 1986a). The affinity of ^{125}I-rau-AzPC was not reported, although the parent amino derivative had a K_i of 3 nM for rat renal alpha-2 adrenergic receptors and 10 nM for receptors from pig brain. When affinity-purified alpha-2 adrenergic receptors from pig brain were photolyzed in the presence of 1 nM ^{125}I-rau-AzPC, a peptide with an M_r of 62,000 was identified by autoradiography following SDS-PAGE. Competition studies established the alpha-2 adrenergic character of the labeled peptide. Thus yohimbine, phentolamine, and (–)-epinephrine blocked photolabeling of the M_r 62,000 peptide by ^{125}I-rau-AzPC, whereas (–)-isoproterenol and prazosin were ineffective. This result is in essential agreement with the results obtained from purification of the human platelet alpha-2 adrenergic receptor showing that the receptor migrates as a broad band on SDS-PAGE with an M_r in the range of 60,000 to 70,000 (Regan et al., 1986a). As compared with [^{3}H]SKF 102229, ^{125}I-rau-AzPC has the advantage of having much higher specific radioactivity and therefore greater sensitivity. On the other hand the yield of covalent insertion of ^{125}I-rau-AzPC may be significantly less than that of [^{3}H]SKF 102229, and it appears that labeling alpha-2 adrenergic receptors in membranes may be very difficult using ^{125}I-rau-AzPC. Given these uncertainties, and the limitations of [^{3}H]SKF 102229, there is ample room for improvement in the area of affinity ligands for alpha-2 adrenergic receptors.

3. Biochemical Isolation of alpha-2 Adrenergic Receptors

The development of specific, high-affinity radioligands has led to a thorough characterization of alpha-2 adrenergic receptors in membranes prepared from a variety of tissues. This characterization has improved our understanding of the distribution of alpha-2 adrenergic receptors between tissues and across species. It has also resulted in the elucidation of biochemical properties such as the kinetics of ligand binding, the number of receptors in various tissues, and the interaction of alpha-2 adrenergic receptors with guanine nucleotide binding proteins. Finally, the study of radioligand binding in membranes has greatly improved our knowledge of the pharmacology of alpha-2 adrenergic receptors. Nevertheless this knowledge probably only represents a fraction of the useful information that can be gleaned from the alpha-2 adrenergic receptor. To further our understanding will require acquisition of the techniques needed for the isolation of the receptor itself. This represents a formidable task since the amount of alpha-2 adrenergic receptor present in even the most abundant source is less than 0.001% of the total cellular protein. This latter reality also means that even with the ability to purify, the amount of structural information to be gained may be limited by the low abundance of the receptor. To this end the techniques of gene isolation, cloning, and protein expression may be the key to a detailed understanding of the alpha-2 adrenergic receptor.

3.1. Solubilization of alpha-2 Adrenergic Receptors from Membranes

In a procedure somewhat akin to taking a fish out of water, the alpha-2 adrenergic receptor must be removed from the environment of the membrane before it can be purified and studied. The problem in both cases is not so much with the removal process, but with the preservation of normal biological activity. Like other membrane proteins, the alpha-2 adrenergic receptor is quite hydrophobic, and by itself it is essentially insoluble in conventional aqueous buffer systems. Therefore the presence of a detergent is required both to solubilize the receptor and to maintain biological activity. Although most detergents are capable of solubilizing the alpha-2 adrenergic receptor, very few can do so without causing a loss of binding activity. The tissues, the detergents, and the radioligands that have been used successfully for the solubilization of alpha-2 adrenergic receptors are listed in Table 6. The list shows that in most tissues in which alpha-2 adrenergic receptors have been well characterized, such as in platelets, brain, and kidney, solubilization has been achieved. The yield of solubilized alpha-2 adrenergic receptors has ranged from approximately 15 to 90% with the best results, both in terms of yield and specific binding activity, being obtained with human platelets. As alluded to previously, the type and concentration of a particular detergent can have marked effects on the binding activity of the alpha-2 adrenergic receptor. In this regard only two detergents have provided acceptable results: they are digitonin and CHAPS.

The first of these, digitonin, is a nonionic glycoside isolated from the seeds of foxglove (*Digitalis purpurea*). Chemically it consists of a pentasaccharide attached to a steroidal aglycone. Although a detailed mechanism for the solubilization of intrinsic membrane proteins by digitonin is lacking, it is at present the best detergent for the solubilization of alpha-2 adrenergic receptors while still preserving binding activity. In the case of human platelet membranes, optimum solubilization was achieved with a digitonin-to-protein ratio of ~2 (Regan et al., 1986a). With lower digitonin-to-protein ratios, of ~0.5, it was possible to solubilize a significant fraction of the total membrane protein without appreciable solubilization of binding activity. By treating membranes first with a low concentration of digitonin, and then again (after centrifugation) with a higher concentration, it was possible to obtain a three- to fourfold purification of the alpha-2 adrenergic receptor during solubilization. In contrast, a single solubilization step with a high concentration of digitonin yielded a soluble preparation with approximately the same specific binding activity as the original membranes (~200 fmol/mg protein). The final yield of binding activity under both conditions was about the same (Regan et al., 1986a).

Following solubilization of human platelet alpha-2 adrenergic receptors with 1% digitonin, the concentration of digitonin needed to maintain binding activity is less. In fact as the digitonin concentration is decreased to a lower limit of ~0.02%, specific [^{3}H]yohimbine binding activity increases (Nakata and Regan, unpublished). This effect of the digitonin concentration should be kept in mind when other variables are the subject of investigation. For most purposes, after solubilization and an initial purification step, a final concentration of digitonin in the range of 0.05 to 0.1% maintains good binding activity, does not interfere with a variety of chromatographic procedures, and is economical.

Recently digitonin has been used in a two-step solubilization procedure in which membranes are first solubilized with sodium cholate (Repaske et al., 1987). Although binding activity could not be measured in this detergent, alpha-2 adrenergic receptors were solubilized since binding activity was restored following removal of the cholate and resolubilization with digitonin. Using this procedure, some enrichment of binding activity was achieved. Thus starting with a specific activity of ~75 fmol/mg protein in the membranes, a final specific activity of ~200 fmol/mg protein was obtained in the soluble preparation; however, the efficiency in terms of the yield of binding activity was not reported. Solubilization of alpha-2 adrenergic receptors by sodium cholate was also documented by Kremenetzky and Atlas (1984). In the latter studies, binding activity could only be measured following removal of the cholate in the presence of phospholipids.

The use of CHAPS to solubilize alpha-2 adrenergic receptors with retention of binding activity has been demonstrated in several studies. CHAPS is chemically related to cholic acid, but, unlike cholic acid, CHAPS is zwitterionic. This means that CHAPS has no formal charge in the physiologic range of pH. The yield of solubilized receptor is less than that obtained with digitonin; however, unlike digitonin, solubilization with CHAPS preserves agonist binding. Using the partial agonists [^{3}H]p-aminoclonidine and [^{3}H]clonidine, specific binding to alpha-2 adrenergic receptors has been obtained following solubilization from bovine brain (Sladeczek et al., 1984) and from rat brain membranes (Matsui et al., 1984; Kitamura et al., 1986). The fact that agonist binding is preserved suggests that CHAPS does not perturb functional interactions between the alpha-2 adrenergic receptor and the inhibitory guanine nucleotide binding protein, G_i, which couples agonist occupancy of the receptor to the inhibition of adenylate cyclase. This conclusion is supported by evidence that the addition of guanine nucleotides, such as GTP and Gpp(NH)p, caused a marked loss of agonist binding (Sladeczek et al., 1984; Matsui et al., 1984). These results are essentially the same as those obtained with membrane preparations of alpha-2 adrenergic receptors. In

Table 6
Radioligand Binding to Solubilized alpha-2 Adrenergic Receptors

Reference	Species/tissue	Ligand
Smith and Limbird, 1981	Human platelet	[3H]Yohimbine
Michel et al., 1981	Human platelet	[3H]Yohimbine
Regan et al., 1982	Human platelet	[3H]Yohimbine
Nambi et al., 1982	Rat adrenocortical carcinoma 494	[3H]Dihydroergocryptine
Limbird et al., 1982	Human platelet	[3H]Yohimbine
Regan et al., 1982	Human platelet	[3H]Yohimbine
Sladeczek et al., 1984	Bovine brain	[3H]p-Aminoclonidine
Matusi et al., 1985	Rat brain	[3H]Clonidine
Kawahara and Bylund, 1985	Human platelet	[3H]Yohimbine
Kawahara and Bylund, 1985	Rat cerebral cortex	[3H]Yohimbine
Kitamura et. al., 1986	Rat cerebral cortex	[3H]Clonidine
Limbird et al., 1985	Human platelet	[3H]Yohimbine
Jaiswal and Sharma, 1985	Rat adrenocortical carcinoma 494	[3H]p-Aminoclonidine
Regan et al., 1986a	Human platelet	[3H]Yohimbine
Cheung et al., 1986	Human platelet	[3H]Rauwolscine
Cheung et al., 1986	Rat kidney	[3H]Rauwolscine
Cheung et al., 1986	Rabbit kidney	[3H]Rauwolscine
Lanier et al., 1986a	Porcine brain	[3H]Yohimbine
McKernan et al., 1986	Human spleen	[3H]Yohimbine
	Rabbit spleen	[3H]Yohimbine

[a]Technique used for the separation of radioligand bound to the receptor (B) from the unbound or free (F) radioligand.
[b]The maximum specific binding of radioligand per mg of protein.
[c]Size exclusion chromatography: for example, by gel filtration over Sephadex G-50.
[d]Precipitation of protein-bound radioligand by polyethylene glycol (PEG) followed by filtration through glass fiber filters.
[e]Adsorption of protein-bound radioligand by filtration through polyethylenimine (PEI)-treated glass fiber filters.

membranes incubated in the absence of guanine nucleotides, the binding of agonists leads to the formation of a high-affinity state of the alpha-2 adrenergic receptor, which is believed to be caused by a physical interaction of the receptor with G_i. In the presence

Table 6 (*Continued*)

Separation of B and F[a]	Detergent	Solubili-zation, %	Specific activity[b]
Gel filtration[c]	Digitonin	75	265 fmol/mg
Gel filtration	Digitonin	—	55 fmol/mg
Gel filtration	Digitonin	49	95 fmol/mg
Gel filtration	CHAPS	—	35 fmol/mg
Gel filtration	Digitonin	—	333 fmol/mg
Gel filtration	Digitonin	—	100 pmol/mg
PEG precipitation-filtration[d]	CHAPS	—	150 fmol/mg
Filtration-PEI-treated filters[e]	CHAPS	40	35 fmol/mg (H) 195 fmol/mg (L)
Gel filtration	Digitonin	28	169 fmol/mg
Gel filtration	Digitonin	21	47 fmol/mg
PEG precipitation-filtration	CHAPS	30	16 fmol/mg (H) 21 fmol/mg (L)
Gel filtration	Digitonin	40	—
Gel filtration	Digitonin	—	4 nmol/mg
Gel filtration	Digitonin	65	14.5 nmol/mg
PEG precipitation-filtration	Digitonin	90	177 fmol/mg
PEG precipitation-filtration	Digitonin	40	63 fmol/mg
PEG precipitation-filtation	Digitonin	40	34 fmol/mg
Gel filtration	Cholate-digitonin	—	600 pmol/mg
Gel filtration	Digitonin	—	368 fmol/mg
Gel filtration	Digitonin	—	854 fmol/mg

of agonist and guanine nucleotides, however, the association of the receptor with G_i is disrupted, and the receptor reverts to a low-affinity state with respect to the binding of agonists. The binding of antagonists, on the other hand, is not markedly affected by this receptor–G_i interaction.

The absence of radiolabeled agonist binding and of guanine nucleotide effects on receptor–agonist interactions has led to the conclusion that the receptor and G_i are physically divorced from one another after solubilization with digitonin (Kawahara and Bylund, 1985; Cheung et al., 1986). There is mounting evidence, however, that this is not the case. Early studies with digitonin showed that it was possible to solubilize a GTP-sensitive form of the alpha-2 adrenergic receptor if agonist was present prior to solubilization (Smith and Limbird, 1981). Likewise, agonist binding that was GTP-sensitive could be obtained following solubilization if the digitonin was removed in the presence of phospholipid (Kremenetzky and Atlas, 1984). Although effectively reconstituting a functional interaction, these studies did not establish whether the receptor and G_i were solubilized independently or as a complex. More recent evidence indicates that even in the absence of added agonist a significant fraction of the alpha-2 adrenergic receptors is physically coupled with G_i after solubilization with digitonin. Thus, following solubilization and a 750-fold purification of the alpha-2 adrenergic receptor, G_i was still found to be present in stoichiometric amounts with the receptor (Cerione et al., 1986). Furthermore a functional interaction of the alpha-2 adrenergic receptor with this endogenous G_i was obtained following removal of the digitonin and reconstitution into phospholipid vesicles.

The previous results suggest a possible reinterpretation of earlier studies, which reported changes in the apparent size of alpha-2 adrenergic receptors as a consequence of agonist binding (Smith and Limbird, 1981; Michel et al., 1981). As determined by sucrose gradient centrifugation, these studies showed that agonist occupation of the solubilized receptor increased its rate of sedimentation. Although considered, hydrodynamic changes were rejected in favor of the hypothesis that the change in sedimentation reflected a change in molecular size caused by a physical association of the agonist occupied receptor with G_i. It now seems likely that the data could also be explained by hydrodynamic changes resulting from an agonist-induced conformational change in either the receptor, G_i, or both proteins in combination.

It is evident from the results obtained with digitonin and CHAPS that the choice of detergent and the concentration employed can have a significant impact on studies of solubilized alpha-2 adrenergic receptors. In addition to possible effects on protein–protein interactions, detergents may influence other types of interactions such as those with chromatographic supports. For example CHAPS-solubilized brain alpha-2 adrenergic receptors do not bind to DEAE-agarose, an anion exchange support, under conditions in which digitonin solubilized alpha-adrenergic receptors do (Matsui and Rouot, personal communications). Detergents are also characterized by their micellar size, which has obvious implications for size exclusion chromatography. Finally, if the detergent interferes with ligand binding, it is of no use for affinity chromatography. For the latter reason in particular, digitonin has been the only detergent used for studies involved with further purification of the alpha-2 adrenergic receptor.

3.2. Approaches to the Purification of alpha-2 Adrenergic Receptors

The first significant step toward the isolation of alpha-2 adrenergic receptors came with the development of a biospecific affinity chromatographic support (Regan et al., 1982). The ligand used for this support, SKF 101253, was the best of a series of compounds that were designed for the purpose of affinity chromatography (DeMarinis et al., 1984). Starting with a high-affinity alpha-2 adrenergic antagonist (SKF 86466), modifications were made so that the derivatized product could be covalently coupled, and thereby immobilized, to a suitable insoluble support such as agarose. The synthetic approach for covalent coupling was the same as that taken for the beta-adrenergic ligand, alprenolol, in which an allyl group was coupled to sulfhydryl-activated Sepharose by means of a free radical reaction (Caron et al., 1979). On an analytical scale, SKF 101253-Sepharose provided a 100-fold purification of human platelet alpha-2 adrenergic receptors when compared to the starting specific binding activity of the crude membrane preparation. Adsorption to and elution from this affinity matrix showed an alpha-2 adrenergic specificity.

Other matrices that were investigated in initial studies of solubilized alpha-2 adrenergic receptors included chromatography over DEAE-Sepharose (Limbird et al., 1985). Being an acidic protein at neutral pH, it was found that alpha-2 adrenergic receptors could be adsorbed to and eluted from this anion exchange support. Using DEAE-Sepharose it was possible to separate agonist-occupied receptor from agonist-free receptor; however, the extent of purification, if any, was not reported. On an analytical scale, using 1 mL columns and 2 mL of solubilized receptor, we have obtained an approximate fivefold purification and a 60% recovery using DEAE-Fractogel TSK (Raymond and Regan, unpublished). On a scale that was ~1000 times larger, however, the recovery was disastrous, and purification with this matrix was not pursued.

Another support that was found to give some purification of the alpha-2 adrenergic receptor was wheat germ agglutinin-agarose (Smith and Limbird, 1982). In this interaction, advantage was taken of the fact that the alpha-2 adrenergic receptor is a glycoprotein. Following adsorption, therefore, it was possible to elute the receptor with N-acetyl-D-glucosamine. Starting with a crude soluble preparation of the alpha-2 adrenergic receptor, yields on the order of 60% were found with a three- to sixfold purification.

alpha-2 Adrenergic receptor preparations that were over 500-fold purified could be obtained by repeating the initial affinity chromatographic step (Regan et al., 1984) or by combining it with chromatography over wheat germ agglutinin-agarose (Regan et al., 1985). Using such partially purified preparations it was possible to identify directly the alpha-2 adrenergic receptor by covalent labeling with the affinity ligand [^{3}H]phenoxybenzamine (see Fig. 1 and previous discussion of [^{3}H]phenoxybenzamine). The information provided by this labeling, i.e., that the alpha-2 adrenergic receptor was a protein with an M_r of ~64,000, led to a couple of important conclusions with regards to its purification. First it predicted that the theoretical specific binding activity of the purified receptor would be ~16 nmol/mg protein if there were one ligand binding site per receptor molecule. Second, it indicated that even with receptor preparations that were purified 500-fold from human platelets (~100 pmol/mg protein), the receptor itself only accounted for ~1% of the total protein.

Complete purification of the alpha-2 adrenergic receptor from rat adrenocortical carcinoma was reported (Jaiswal and Sharma, 1985); however, these studies can be criticized on several grounds. In particular there is a very large discrepancy between the results obtained by SDS-PAGE and the results obtained from the determination of the specific binding activity of the receptor. The reported specific activity was 4 nmol/mg protein; however, the methods used to arrive at this estimate were not detailed, and calculations based upon the published data would suggest a specific activity closer to 1 nmol/mg protein. In either case the purity of the alpha-2 adrenergic receptor as a percentage of the total protein would be expected to be in the range of 6 to 25%. These data are in conflict with the results of SDS-PAGE, which clearly show a single sharp band of radiolabeled protein with an M_r close to that of the bovine serum albumin (BSA) standard. The sharpness of this band of putative iodinated receptor is also at odds with results obtained from photoaffinity labeling and other purification studies, which show the alpha-2 adrenergic receptor migrating on SDS-PAGE as a rather broad band (Regan et al., 1986a; Lanier et al., 1986a). A possible explanation of this discrepancy between the specific activity data and the results of SDS-PAGE would be if the radioiodinated protein were actually BSA. The basis of this supposition is that the authors used a gel filtration column presaturated with 1% BSA (10 mg/mL) to separate the radioiodinated protein from the unreacted Na^{125}I. According to the procedure described by Greenwood et al. (1963), the presence of BSA during this gel filtration step resulted in the iodination of BSA even though the reaction had already been quenched by the addition of sodium metabisulfite. If radioiodination of BSA in fact took place, the results of the SDS-PAGE shown by Jaiswal and Sharma (1985) would appear as one band of protein, since the concentration of BSA would be many orders of magnitude greater than the concentration of protein(s) obtained from the purification. It should be stressed that as an alternative, it is possible that the alpha-2 adrenergic receptor from this cell line is particularly labile. Thus the difference between the specific activity data and the results of SDS-PAGE might be caused by a significant loss of binding activity without a concommitant loss of receptor protein. Some attempt to address these possibilities experimentally would be useful.

A purification of alpha-2 adrenergic receptors from human platelets has been described that employed a five-step procedure involving chromatography over columns of SKF 101253-Sepharose, WGA-agarose, and heparin-agarose (Regan et al., 1986a). The use of heparin-agarose was novel in the sense that it does not appear to have been used previously for the purification of hormone receptors. Although originally intended to remove contaminants, heparin-agarose was found to give a good purification of the alpha-2 adrenergic receptor itself. Specific binding activity obtained after four chromatographic steps was ~14 nmol/mg protein corresponding to a receptor purity of ~90%. Radioiodination of the latter material, followed by SDS-PAGE, revealed three bands of protein with M_r values of 80,000, 64,000, and 55,000. Identification of the 64,000 protein as the alpha-2 adrenergic receptor was made by specific covalent labeling of this protein using [^{3}H]phenoxybenzamine (Regan et al., 1986a) and by photoaffinity labeling using [^{3}H]SKF 102229 (Regan, unpublished data). Complete purification of the alpha-2 adrenergic receptor was achieved by a final chromatographic step over heparin-agarose. Figure 3 shows the results of SDS-PAGE following radioiodination of the final heparin eluate by either Na^{125}I and chloramine T or ^{125}I-Bolton Hunter reagent. The purified alpha-2 adrenergic receptor migrates as a broad band with an M_r centered at ~64,000. Radioiodination with chloramine T was quite variable and often led to both aggregation (seen at the top of the gel in Fig. 3) and degradation of the receptor to a species of M_r 41,000. The overall yield starting from the membrane preparation was only ~2% with an overall fold purification of ~80,000. The binding of ligands to the purified receptor protein

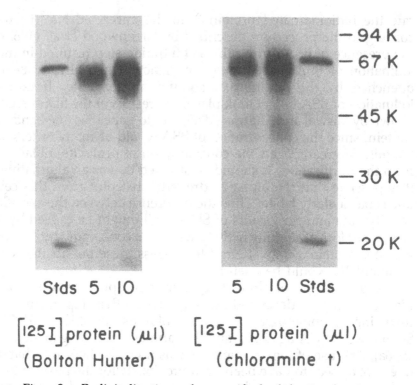

Fig. 3. Radioiodination of a purified alpha-2 adrenoreceptor preparation by the Bolton Hunter and chloramine T methods. alpha-2 Adrenergic receptors were purified from human platelet membranes as previously described (Regan et al., 1986a). A sample (100 μL) of the final heparin eluate in a buffer containing 300 mM KCl, 50 mM HEPES, 1 mM EDTA, and 0.1% digitonin (pH 7.2) was radiolabeled using ^{125}I-Bolton Hunter reagent (Regan et al., 1986a). Another sample (100 μL) was mixed with an additional 250 μL of buffer, and then 1 μL of Na[125]I (~350 μCi) was added followed by 75 μL of 1 mg/mL chloramine T. After 20 s, 75 μL of 8 mg/mL sodium metabisulfite was added, and then both samples were desalted by Sephadex G-50 chromatography into a buffer containing 100 mM Tris HCl, 1 mM EDTA, and 0.1% digitonin (pH 7.2). Aliquots of the radiolabeled material were mixed with SDS-PAGE sample buffer and were electrophoresed on 10% polyacrylamide gels. The gels were dried, and autoradiographs were obtained.

showed a classic alpha-2 adrenergic pattern, and stereospecificity was preserved.

The primary reason for the low overall yield in the purification of human platelet alpha-2 adrenergic receptors was the affinity chromatographic step (Regan et al., 1986a). Thus, using SKF 101253-Sepharose on a preparative scale gave a yield of 26% with respect to the material that was carried on for further purification. More problematic, however, was the fact that only a

30-fold purification was achieved, which necessitated further chromatographic steps. This situation may be relieved by the use of immobilized yohimbinic acid as an affinity support for alpha-2 adrenergic receptors (Repaske et al., 1987). Although the free acid has relatively low affinity for alpha-2 adrenergic receptors, the product, coupled to agarose by means of a carbodiimide reaction, was able to adsorb porcine brain alpha-2 adrenergic receptors, which were then specifically eluted with phentolamine. Based upon the total amount of solubilized receptor applied, a purification of ~2000-fold was obtained with a yield of ~35%. The purity of the receptor after this affinity chromatography was ~4% given that the specific binding activity was reported to be ~600 pmol/mg protein. In our hands yohimbinic acid-agarose has been used for the purification of alpha-2 adrenergic receptors from human platelets (Regan and Matsui, unpublished data). On a preparative scale, 40% yields have been obtained with a fold purification ranging from 100 to 200. The lower fold purification may be related to the source of the receptors, the synthesis of the yohimbinic acid-agarose, or in the conditions used for either washing or eluting of this affinity support. Nevertheless the fold purification is better than that obtained with SKF 101253-Sepharose, and yohimbinic acid-agarose is easier to make.

Yohimbinic acid-agarose has also been used in combination with other steps, for the complete purification of porcine brain alpha-2 adrenergic receptors (Repaske et al., 1987). The specific binding activity is reported to be 13.4 nmol/mg protein, which is consistent with an essentially homogenous preparation of the receptor protein. On SDS-PAGE the receptor migrates as a single band with an M_r of ~65,000, and it has been identified by specific photoaffinity labeling with ^{125}I-rau-AzPC. Interestingly, the binding of agonists to the purified receptor protein is modulated by agents that interact with the Na^+/H^+ antiporter system of cell membranes (Nunnari et al., 1987). These agents include sodium ions, hydrogen ions, and various 5-amino substituted analogs of amiloride. Thus both sodium ions and hydrogen ions can decrease agonist affinity for purified porcine brain alpha-2 adrenergic receptors by approximately 10-fold, whereas the amiloride analogs accelerate the dissociation of bound [^{3}H]-yohimbine. The modulation of agonist binding to the purified receptor by these agents may be related to observations in both platelet and NG 108-15 cells that the effects of alpha-2 adrenergic stimulation appear to involve Na^+/H^+ exchange (Sweatt et al., 1986; Isom et al., 1987). The intriguing possibility that the alpha-2 adrenergic receptor itself is involved in this exchange will need to be tested in reconstituted preparations of the purified receptor.

4. Structural Aspects of the alpha-2 Adrenergic Receptors

With the development of purification procedures and of affinity labeling techniques, structural aspects of the alpha-2 adrenergic receptor are beginning to emerge. Initial studies tried to establish structural similarities of the alpha-2 adrenergic receptor with alpha-1 and beta-2 adrenergic receptors by the techniques of peptide mapping. In this technique, purified preparations of the radioiodinated receptors are treated with various specific endopeptidases, and the digests are then electrophoresed on SDS-PAGE to create a map of the peptides that are generated. If extensive structural correspondence exists, one could expect that certain common peptides would be observed. Peptide mapping of purified alpha-2 and beta-2 adrenergic receptors with three different endopeptidases, however, produced no common peptides (Regan et al., 1986a). Likewise peptide mapping of alpha-2 and alpha-1 adrenergic receptors with four different proteinases produced no common peptides (Lomasney et al., 1986). Although these results suggest that structural similarities are not extensive, they do not rule out the possibility that these receptors may be homologous and share certain structural features. For example some proteins, like alpha- and beta-tubulin, do not give similar peptide maps even though they share considerable amino acid sequence homology (Cleveland et al., 1977).

Recent experiments with partially and completely purified preparations of alpha-2 adrenergic receptors indicate that cysteine and tyrosine residues may be involved with ligand binding to the receptor. Phenylmercuric chloride (PMC) is a reagent that reacts specifically with sulfhydryl groups to form stable mercaptide bonds. These bonds can be broken, however, by treatment with reducing agents such as dithiothreitol and 2-mercaptoethanol. Treatment of purified human platelet alpha-2 adrenergic receptors with PMC resulted in a loss of binding activity (Regan et al., 1986a). This loss, however, was prevented when treatment of the receptors with PMC was done in the presence of specific alpha-2 adrenergic agonists and antagonists. Furthermore, the decrease in binding activity caused by PMC treatment was reversed completely by subsequent treatment of the alpha-2 adrenergic receptors with dithiothreitol. These data indicate that a cysteine residue(s) is important for the binding activity of alpha-2 adrenergic receptors either directly, by participating in binding, or indirectly, by helping to maintain the proper conformation of the receptor. Similar results implicating sulfhydryl groups in ligand binding have also been obtained with membrane preparations of alpha-2 adrenergic receptors (Kawai and Nomura, 1983; Quenneday et al., 1984; Mattens et al., 1985).

Tyrosine residues also appear to be essential for the binding of ligands to alpha-2 adrenergic receptors (Nakata et al., 1986). Treatment of purified alpha-2 adrenergic receptors with tetranitromethane (TNM), a reagent that nitrates the phenolic ring of tyrosine residues, blocked subsequent binding of [^{3}H]yohimbine. Since TNM also has oxidative effects, the possible involvement of sulfhydryl groups in the blockade of binding was precluded by "protecting" the cysteine residues by pretreatment with PMC. Following exposure to TNM, dithiothreitol did not restore binding activity to the protected receptors, which is contrary to expectations if TNM were only reacting with cysteine residues. Additionally other reagents that also react with tyrosine residues, such as sodium iodide/chloramine T and p-nitrobenzenesulfonyl fluoride, caused a blockade of ligand binding (Nakata et al., 1986). As with PMC, the loss of binding activity caused by TNM could be prevented if alpha-2 adrenergic ligands were present during exposure to the reagent. The concentrations of ligands needed for protection, however, were almost three orders of magnitude less for experiments done with TNM than for experiments done with PMC. The latter result might be interpreted to signify a more intimate involvement of tyrosine with the mechanism of ligand binding to the alpha-2 adrenergic receptor.

As discussed previously, the alpha-2 adrenergic receptor is a glycoprotein as evidenced by its ability to bind to wheat germ agglutinin-agarose and to be eluted specifically with appropriate sugars, such as sialic acid and N-acetyl-D-glucosamine. alpha-2 Adrenergic receptors can also be adsorbed to lentil bean lectin-agarose and to Jimsom weed lectin-agarose, lectins that have sugar specificities for alpha-D-mannose and N,N'diacetyl-chitobiose, respectively (Regan, unpublished data). Elution of the receptor from these lectin columns, however, was quite variable. Further evidence of the glycoprotein nature of the alpha-2 adrenergic receptor is provided by Fig. 4, which shows the receptor following treatment with endoglycosidase F. Endoglycosidase F is an enzyme that cleaves both high-mannose and complex-type polysaccharides from their N-linked attachment sites to asparagine residues of the peptide chain. Treatment of purified alpha-2 adrenergic receptors with endoglycosidase F resulted in a change of the apparent molecular weight from ~64,000 to ~48,000 as judged by the results of SDS-PAGE. Essentially identical results were obtained with radioiodinated beta-adrenergic receptors purified from hamster lung (also see Stiles et al., 1984; Benovic et

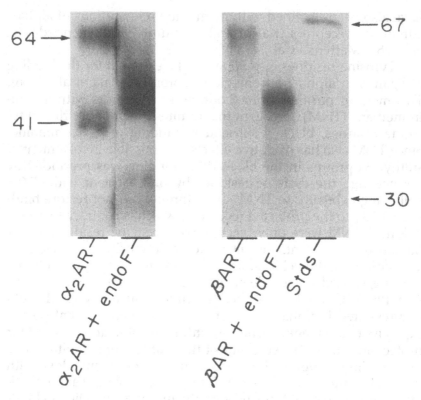

Fig. 4.   Treatment of radioiodinated beta-2 and alpha-2 adrenergic receptors with endoglycosidase F. beta-2 Adrenergic receptors were purified from hamster lung (Benovic et al., 1984), and alpha-2 adrenergic receptors were purified from human platelets (Regan et al., 1986a). Both receptor preparations were then treated with ^{125}I-Bolton Hunter reagent and the radiolabeled receptors were separated from unreacted reagent by gel filtration over Sephadex G-50. Samples of the radioiodinated receptor preparations were incubated for 2 h at 37°C with 1 U of endoglycosidase F (New England Nuclear-Dupont) in a final volume of 20 µL and in the presence of a final concentration of 25 mM EDTA. The samples were then mixed directly with SDS-PAGE sample buffer and were electrophoresed on an 8% polyacrylamide gel (Laemmli, 1970). The gel was dried and autoradiographs were obtained. The band at ~41,000 in the control alpha-2 adrenergic receptor preparation is a degradation product of the 64,000 band which arises during storage following purification of the receptor.

al., 1987). These results indicate that ~25% of the apparent molecular weight of the alpha-2 adrenergic receptor is from its carbohydrate content: whether this reflects the true mass of carbohydrate, or anomalous mobility on SDS-PAGE, cannot be answered by this experiment. If one assumes that the molecular weight of the average amino acid is 110, then the alpha-2 adrenergic receptor may consist of a peptide chain of ~440 amino acids. The gene for the hamster beta-2 adrenergic receptor has been cloned, and the deduced amino acid sequence is 418 amino acids, which is in reasonable agreement with the present results (Dixon et al., 1986). The previous study also identified two consensus sites for N-linked glycosylation near the amino terminus of the hamster lung beta-2 adrenergic receptor. Again this agrees with the results of the present experiment and may allow one to speculate that the glycosylation site(s) of the alpha-2 adrenergic receptor is near its amino terminus.

One final question with regard to the structure of the alpha-2 adrenergic receptor concerns the events that might take place following the binding of ligands. One unmistakable event that occurs as a consequence of agonist binding is that a change takes place that permits an interaction of the alpha-2 adrenergic receptor with the guanine nucleotide binding protein G_i. This has been shown conclusively by reconstitution experiments using partially purified alpha-2 adrenergic receptors and purified G_i (Cerione et al., 1986). In these studies alpha-2 adrenergic receptors that had been purified up to 5000-fold from human platelet membranes were inserted into phospholipid vesicles along with purified human erythrocyte G_i. In the presence of the agonist $(-)$-epinephrine there was a dramatic stimulation of the GTPase activity intrinsic to G_i. Additionally, this stimulation by $(-)$-epinephrine could be specifically blocked by co-incubation with the antagonist, phentolamine. It is assumed that conformational changes, arising from the binding of the agonist, are responsible for this stimulation of GTPase activity by the receptor.

More direct evidence for conformational changes in the alpha-2 adrenergic receptor, caused by agonist binding, comes from recent experiments done with the beta-adrenergic receptor kinase or β-ARK. β-ARK is a protein kinase that is capable of phosphorylating only the agonist-occupied form of the beta-adrenergic receptor (Benovic et al., 1986). β-ARK is also capable, however, of phosphorylating partially purified preparations of reconstituted human platelet alpha-2 adrenergic receptors (Regan et al., 1987; Benovic et al., 1988). This phosphorylation is totally agonist-dependent, and it can be specifically prevented by antagonists. Such results suggest the occurrence of agonist-induced conformational changes leading to the exposure of previously cryptic phosphorylation sites.

It may be wondered whether or not the binding of antagonists can lead to conformational changes as well. The results shown in Fig. 5 suggest that they do. In this experiment attempts were made to radioiodinate partially purified alpha-2 adrenergic

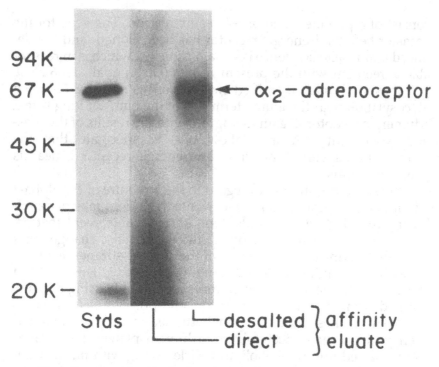

Fig. 5. Specific blockade of alpha-2 adrenergic receptor radioiodination in the presence of phentolamine. Human platelet alpha-2 adrenergic receptors were partially purified by a combination of affinity, heparin, and WGA chromatography (Regan et al., 1986a). The eluted receptors, in a buffer containing 200 μM phentolamine, were then radiolabeled either directly with ^{125}I-Bolton Hunter reagent, or following removal of the phentolamine by Sephadex G-50 chromatography. The samples were then electrophoresed (Laemmli, 1970) on 10% polyacrylamide gels and autoradiographs were obtained.

receptors either in the presence or in the absence of phentolamine. As shown by the autoradiograph obtained following SDS-PAGE, radiolabeling of alpha-2 adrenergic receptors did not take place in the presence of phentolamine. Removal of phentolamine, on the other hand, permitted radioiodination of the receptor. A nonspecific effect of the phentolamine on the iodination reaction seems unlikely since the iodination of other contaminating proteins was unaffected by the presence of phentolamine. One possible explanation of the data is that phentolamine induces or stabilizes a conformation of the receptor that is resistant to radioiodination. Another possibility for these findings is that failure to iodinate was caused by direct blockade of the binding site by the ligand itself. Since the labeling of proteins by ^{125}I-Bolton Hunter reagent is specific for primary amino groups, an inhibition of iodination by direct blockade would imply the presence of a lysine residue(s) in the ligand binding site of the alpha-2 adrenergic receptor. However, given that lysine often accounts for a significant fraction of the amino acid composition of a protein (4% in the beta-adrenergic receptor; Dixon et al., 1986), it would seem unlikely that antagonist binding could block directly all of these residues simultaneously.

5. Conclusions

The structure of the alpha-2 adrenergic receptor is beginning to reveal itself. Recent progress on the molecular cloning of opsins, beta-adrenergic receptors, and muscarinic cholinergic receptors is exposing a level of structural and functional similarity never imagined (Dohlman et al., 1987). It seems that most, if not all, of the membrane-bound receptors coupled to guanine nucleotide binding proteins (G-proteins) may share certain common features. At present these features would appear to include sites of N-linked glycosylation, multiple membrane spanning regions, consensus sites for the binding of G-proteins, regulatory phosphorylation sites, and finally, the ligand binding domains. Figure 6 presents a generic prototype with the label "alpha-2 adrenergic receptor": it could probably serve equally well for other receptors about which we know little. This model does, however, provide a framework for some of the structural features that we do know about alpha-2 adrenergic receptors. Thus, the alpha-2 adrenergic receptor is an integral membrane protein coupled to adenylate cyclase, and possibly to other effector systems, by means of a guanine nucleotide binding protein(s). The alpha-2 adrenergic receptor is a glycoprotein with an M_r of ~64,000, of which ~25% is accounted for by its carbohydrate content. The alpha-2 adrenergic receptor has been virtually defined by its ability to bind ligands, and whether or not it is regulated in vivo by phosphorylation, it certainly contains phosphorylation sites that are exposed by the binding of agonists. Finally the alpha-2 adrenergic receptor contains tyrosine and cysteine residues that might be involved directly in ligand binding or in the conformational stability of the receptor protein. The true involvement of these residues, their precise location in the receptor, and the answers to other questions, as well as the new questions themselves, await the results of further experimentation.

Several relevant papers, published after the completion of this review, are briefly described as follows. [³H]Idazoxan has been used in detailed com-

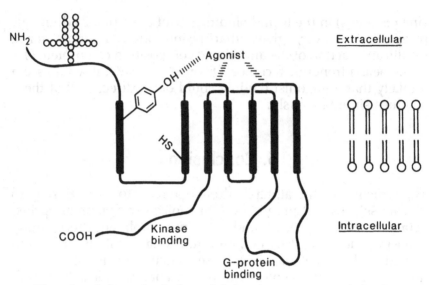

Fig. 6. A hypothetical model for the structure of the human platelet alpha-2 adrenergic receptor. The peptide backbone is defined by the alternating thin and thick line, which defines the presumed hydrophilic and hydrophobic domains, respectively. The chain of circles near the amino-terminus represents the glycosylation site of the receptor. "G-protein binding" and "kinase binding" indicate that the alpha-2 adrenergic receptor can be phosphorylated, and can activate G_i, in an agonist-dependent fashion. The interaction sites of these proteins with the receptor are assumed to be intracellular, but not necessarily in the regions given above.

parative studies with [^{3}H]rauwolscine, and the results support the conclusion of alpha-2 adrenergic receptor heterogeneity in rat brain (Boyajian et al., 1987; Boyajian and Leslie, 1987). The purification of alpha-2 adrenergic receptors from pig brain has been published, which includes a detailed protocol for the preparation of yohimbinic acid-agarose (Repaske et al., 1987). The gene coding for the human platelet alpha-2 adrenergic receptor has been cloned, sequenced, and expressed in Xenopus laevis oocytes (Kobilka et al., 1987). The deduced amino acid sequence supports the model shown in Fig. 6, and shows that the human platelet alpha-2 adrenergic receptor is a member of the family of G-protein coupled receptors, which presently includes the beta adrenergic receptors, the muscarinic cholinergic receptors, and rhodopsin.

Acknowledgments

The excellent secretarial assistance of Donna Addison is sincerely appreciated. Thanks also go to Dr. Robert J. Lefkowitz and Dr. Marc G. Caron for their support and to Dr. Mark Hnatowich for helpful comments. Finally I would like to thank the other members of the alpha-2 team, including Kiefer Daniel, Dr. John R. Raymond, Dr. Hiroyasu Nakata, and Dr. Hiroaki Matsui for their hard work.

REFERENCES

Agrawal, D. K., and Daniel, E. E. (1985) Two distinct populations of [^{3}H]prazosin and [^{3}H]yohimbine binding sites in the plasma membranes of rat mesenteric artery. *J. Pharmacol. Exp. Ther.* **233**, 195–203.

Alabaster, V. A., and Brett, J. M. (1983) Different affinities of alpha-2 adrenergic receptor antagonists for [^{3}H]rauwolscine binding sites in brain and spleen membranes. *Br. J. Pharmacol.* **79**, 314P.

Arnett, C. D., and Davis, J. N. (1979) Denervation-induced changes in alpha and beta adrenergic receptors of the rat submandibular gland. *J. Pharmacol. Exp. Ther.* **211**, 394–400.

Asakura, M., Tsukamoto, T., Imafuku, J., Matsui, H., Ino, M., and Hasegawa, K. (1985) Quantitative analysis of rat brain alpha-2 receptors discriminated by [^{3}H]clonidine and [^{3}H]rauwolscine. *Eur. J. Pharmacol.* **106**, 141–147.

Barnes, P. J., Skoogh, B. E., Nadel, J. A., and Roberts, J. M. (1983) Postsynaptic alpha$_2$-adrenergic receptors predominate over alpha$_1$-adrenergic receptors in canine tracheal smooth muscle and mediate neuronal and hormonal alpha-adrenergic contraction. *Mol. Pharmacol.* **23**, 570–575.

Benovic, J. L., Regan, J. W., Matsui, H., Mayor, F., Cotecchia, S., Leeb-Lundberg, L. M. F., Caron, M. G., and Lefkowitz, R. J. (1988) Agonist-dependent phosphorylation of the alpha-2 adrenergic receptor by the beta adrenergic receptor kinase. *J. Biol. Chem.* **262**, 17521–17253.

Benovic, J. L., Shorr, R. G. L., Caron, M. G., and Lefkowitz, R. J. (1984) The mammalian beta-2 adrenergic receptor: Purification and characterization. *Biochemistry* **23**, 4510–4518.

Benovic, J. L., Strasser, R. H., Caron, M. G., and Lefkowitz, R. J. (1986) Beta adrenergic receptor kinase: Identification of a novel protein kinase that phosphorylates the agonist-occupied form of the receptor. *Proc. Natl. Acad. Sci. USA* **83**, 2797–2801.

Benovic, J. L., Staniszewski, C., Cerione, R. A., Codina, J., Lefkowitz, R. J., and Caron, M. G. (1987) The mammalian beta adrenergic receptor: Structural and functional characterization of the carbohydrate moiety. *J. Receptor Res.* **7**, 257–281.

Berthelsen, S., and Pettinger, W. A. (1977) A functional basis for classification of alpha adrenergic receptors. *Life Sci.* **21**, 595–606.

Bobik, A. (1982) Identification of alpha adrenergic receptor subtypes in dog arteries by [^{3}H]yohimbine and [^{3}H]prazosin. *Life Sci.* **30**, 219–228.

Bottari, S. P., Vokaer, A., Kaivez, E., Lescrainier, J. P., and Vauquelin, G. (1983) Identification and characterization of alpha-2 adrenergic receptors in human myometrium by [^{3}H]rauwolscine binding. *Am. J. Obstet. Gynecol.* **146**, 639–643.

Boyajian, C. L., and Leslie, F. M. (1987) Pharmacological evidence for alpha-2 adrenoceptor heterogeneity: Differential binding properties of [^{3}H]rauwolscine and [^{3}H]idazoxan in rat brain. *J. Pharmacol. Exp. Ther.* **241**, 1092–1098.

Boyajian, C. L., Loughlin, S. E., and Leslie, F. M. (1987) Anatomical evidence for alpha-2 adrenoceptor heterogeneity: Differential auto-radiographic distributions of [^{3}H]rauwolscine and [^{3}H]idazoxan in rat brain. *J. Pharmacol. Exp. Ther.* **241**, 1079–1091.

Broadhurst, A. M., and Wyllie, M. G. (1986) A reassessment of the binding of [^{3}H]rauwolscine to membranes from rat cortex. *Neuropharmacology* **25**, 287–295.

Brodde, O. E., Hardung, A., Ebel, H., and Bock, K. D. (1982) GTP regulates binding of agonists to alpha-2 adrenergic receptors in human platelets. *Arch. Int. Pharmacodyn.* **258**, 193–207.

Brodde, O. E., Eymer, T., and Arroyo, J. (1983) [^{3}H]Yohimbine binding to guinea pig kidney and calf cerebral cortex membranes: Comparison with human platelets. *Arch. Int. Pharmacodyn.* **266**, 208–220.

Bylund, D. B. (1985) Heterogeneity of alpha-2 adrenergic receptors. *Pharmacol. Biochem. Behav.* **22**, 835–843.

Bylund, D. B., and Martinez, J. R. (1980) Alpha-2 adrenergic receptors appear in rat salivary glands after reserpine treatment. *Nature* **285**, 229–230.

Bylund, D. B., and Martinez, J. R. (1981) Postsynaptic localization of alpha-2 adrenergic receptors in rat submandibular gland. *J. Neurosci.* **1**, 1003–1007.

Bylund, D. B., and U'Prichard, D. C. (1983) Characterization of alpha-1 and alpha-2 adrenergic receptors. *Internat. Rev. Neurobiol.* **24**, 343–431.

Bylund, D. B., Martinez, J. R., and Pierce, D. L. (1982) Regulation of autonomic receptors in rat submandibular gland. *Mol. Pharmacol.* **21**, 27–35.

Cambridge, D. (1981) UK-14,304, a potent and selective alpha-2 agonist for the characterization of alpha-2 adrenergic receptor subtypes. *Eur. J. Pharmacol.* **72**, 413–415.

Caron, M. G., Srinivasan, Y., Pitha, J., Kociolek, K., and Lefkowitz, R. J. (1979) Affinity chromatography of the beta adrenergic receptor. *J. Biol. Chelm.* **254**, 2923–2927.

Carter, R. J., and Shuster, S. (1982) The association between the melanocyte-stimulating hormone receptor and the alpha-2 adrenergic receptor on the *Anolis* melanophore. *Br. J. Pharmacol.* **75**, 169–176.

Cerione, R. A., Regan, J. W., Nakata, H., Codina, J., Benovic, J. B., Gierschik, P., Somers, R. L., Spiegel, A. M., Birnbaumer, L., Lefkowitz, R. J., and Caron, M. G. (1986) Functional reconstitution of the alpha-2 adrenergic receptor with guanine nucleotide regulatory proteins in phospholipid vesicles. *J. Biol. Chem.* **261**, 3901–3909.

Chang, E. B., Field, M., and Miller, R. J. (1983) Enterocyte alpha-2 adrenergic receptors: Yohimbine and p-aminoclonidine binding relative to ion transport. *Am. J. Physiol.* **244**, G76–G82.

Chapleo, C. B., Doxey, J. C., Meyers, P. L., and Roach, A. G. (1981) RX 781094, a new potent, selective antagonist of alpha-2 adrenergic receptors. *Br. J. Pharmacol.* **74**, 842P.

Cheung, Y. D., Barnett, D. B., and Nahorski, S. R. (1982) [^{3}H]Rauwolscine and [^{3}H]yohimbine binding to rat cerebral and human platelet membranes: Possible heterogeneity of alpha-2 adrenergic receptors. *Eur. J. Pharmacol.* **84**, 79–85.

Cheung, Y. D., Barnett, D. B., and Nahorski, S. R. (1984) Interactions of endogenous and exogenous norepinephrine with alpha$_2$ adrenergic receptor binding sites in rat cerebral cortex. *Biochem. Pharmacol.* **33**, 1293–1298.

Cheung, Y. D., Barnett, D. B., and Nahorski, S. R. (1986) Heterogeneous properties of alpha$_2$ adrenergic receptors in particulate and soluble preparations of human platelet and rat and rabbit kidney. *Biochem. Pharmacol.* **35**, 3767–3775.

Cleveland, D. W., Fischer, S. G., Kirschner, M. W., and Laemmli, U. K. (1977) Peptide mapping by limited proteolysis in sodium dodecyl sulfate and analysis by gel electrophoresis. *J. Biol. Chem.* **252**, 1102–1106.

Cornett, L. E., and Norris, J. S. (1986) Affinity labeling of the DDT$_1$ MF-2 cell alpha-1 adrenergic receptor with [^{3}H]phenoxybenzamine. *Biochem. Pharmacol.* **35**, 1663–1669.

Dabire, H., Dausse, J. P., Mouille, P., Schmitt, H., and Meyer, P. (1983) In vitro studies with (imidazolinyl-2)-2-benzodioxane-1-4 (170 150), a new potent alpha-2 adrenergic receptor blocking agent. *Eur. J. Pharmacol.* **86**, 87–90.

Dabire, H., Mouille, P., Andrejak, M., Fournier, B., and Schmitt, H. (1981) Pre- and postsynaptic alpha adrenergic receptor blockade by (imidazolinyl-2)-2-benzodioxane 1-4(170 150): Antagonist action on central effects of clonidine. *Arch. Int. Pharmacodyn.* **254**, 252–270.

Daiguji, M., Meltzer, H. Y., and U'Prichard, D. C. (1981) Human platelet alpha-2 adrenergic receptors: Labeling with [^{3}H]yohimbine, a selective antagonist ligand. *Life Sci.* **28**, 2705–2717.

DeMarinis, R. M., Krog, A. J., Shah, D. H., Lafferty, J., Holden, K. G., Hieble, J. P., Matthews, W. D., Regan, J. W., Lefkowitz, R. J., and Caron, M. G. (1984) Development of an affinity ligand for purification of alpha-2 adrenergic receptors from human platelet membranes. *J. Med. Chem.* **27**, 918–921.

Dickinson, K. E. J., McKernan, R. M., Miles, C. M. M., Leys, K. S., and Sever, P. S. (1986) Heterogeneity of mammalian alpha-2 adrenergic receptors delineated by [^{3}H]yohimbine binding. *Eur. J. Pharmacol.* **120**, 285–293.

Diop, L., Dausse, J. P., and Meyer, P. (1983) Specific binding of [^{3}H]rauwolscine to alpha-2 adrenergic receptors in rat cerebral cortex: Comparison between crude and synaptosomal plasma membranes. *J. Neurochem.* **41**, 710–715.

Dixon, R. A. F., Kobilka, B. K., Strader, D. J., Benovic, J. L., Dohlman, H. G., Frielle, T., Bolanowski, M. A., Bennett, C. D., Rands, E., Diehl, R. E., Mumford, R. A., Slater, E. E., Sigal, I. S., Caron, M. G., Lefkowitz, R. J., and Strader, C. D. (1986) Cloning of the gene and cDNA for mammalian beta adrenergic receptor and homology with rhodopsin. *Nature* **321**, 75–79.

Dohlman, H. G., Caron, M. G., and Lefkowitz, R. J. (1987) A family of receptors coupled to guanine nucleotide regulatory proteins. *Biochemistry* **27**, 2657–2664.

Doxey, J. C., Gadie, B., Lane, A. C., and Tullock, I. F. (1983a) Evidence for pharmacological similarity between alpha-2 adrenergic receptors in the vas deferens and central nervous system of the rat. *Br. J. Pharmacol.* **80**, 155–161.

Doxey, J. C., Roach, A. G., and Smith, C. F. C. (1983b) Studies on RX 781094: A selective, potent and specific antagonist of alpha-2 adrenergic receptors. *Br. J. Pharmacol.* **78**, 489–505.

Doxey, J. C., Roach, A. C., Strachan, D. A., and Virdee, N. K. (1984) Selectivity and potency of 2-alkyl analogues of the alpha-2 adrenergic receptor antagonist idazoxan (RX 781094) in peripheral systems. *Br. J. Pharmacol.* **83**, 713–722.

Elliot, J. M., and Rutherford, M. G. (1983) Binding characteristics of [^{3}H]RX 781094 on human intact platelets. *Br. J. Pharmacol.* **79**, 313P.

Fedan, J. S., Hogaboom, G. K., and O'Donnell, J. P. (1984) Photoaffinity labels as pharmacological tools. *Biochem. Pharmac.* **33**, 1167–1180.

Feller, D. J., and Bylund, D. B. (1984) Comparison of alpha-2 adrenergic receptors and their regulation in rodent and porcine species. *J. Pharmacol. Exp. Ther.* **228**, 275–281.

Gadie, B., Lane, A. C., McCarthy, P. S., Tulloch, I. F., and Walter, D. S. (1984) 2-Alkyl analogues of idazoxan (RX 781094) with enhanced antagonist potency and selectivity at central alpha-2 adrenergic receptors in the rat. *Br. J. Pharmacol.* **83**, 707–712.

Garcia-Sevilla, J. A., and Fuster, M. J. (1986) Labelling of human platelet alpha-2–adrenoceptors with the full agonist [^{3}H](−)adrenaline. *Eur. J. Pharmacol.* **124**, 31–41.

Garcia-Sevilla, J. A., Hollingsworth, P. J., and Smith, C. B. (1981) Alpha-2 adrenergic receptors on human platelets: Selective labeling by [^{3}H]clonidine, [^{3}H]yohimbine and competitive inhibition by antidepressant drugs. *Eur. J. Pharmacol.* **74**, 329–341.

Glusa, E., and, Markwardt, F. (1983) Characterization of alpha-2 adrenergic receptors on blood platelets from various species using [^{3}H]yohimbine. *Haemostasis* **13**, 96–101.

Grant, J. A., and Scrutton, M. C. (1980) Interaction of selective alpha adrenergic receptor agonists and antagonists with human and rabbit blood platelets. *Br. J. Pharmacol.* **71**, 121–134.

Greenwood, F. C., Hunter, W. M., and Glover, J. S. (1963) The preparation of ^{131}I-labelled human growth hormone of high specific radioactivity. *Biochem. J.* **89**, 114–123.

Haga, T. and Haga, K. (1980) Characterization of alpha-adrenergic receptor subtypes in rat brain: Estimation of ability of adrenergic ligands to displace [^{3}H]dihydroergocryptine from the receptor subtypes. *Life Sci.* **26**, 211–218.

Hannah, J. A. M., Hamilton, C. A., and Reid, J. L. (1983) RX 781094, a new potent alpha-2 adrenergic receptor antagonist. *Naunyn Schmiedebergs Arch. Pharmacol.* **322**, 221–227.

Hoffman, B. B., DeLean, A., Wood, C. L., Schocken, D. D., and Lefkowitz, R. J. (1979) Alpha-adrenergic receptor subtypes: Quantitative assessment by ligand binding. *Life Sci.* **24**, 1739–1746.

Hoffman, B. B., Michel, T., Brenneman, T. B., and Lefkowitz, R. J. (1982) Interactions of agonists with platelet alpha-2 adrenergic receptors. *Endocrinology* **110**, 926–932.

Howlett, D. R., Taylor, P., and Walter, D. S. (1982) Alpha adrenoceptor selectivity studies with RX 781094 using radioligand binding to cerebral membranes. *Br. J. Pharmacol.* **76**, 294P.

Isom, L. L., Cragoe, E. J., and Limbird, L. E. (1987) Alpha-2 adrenergic receptors accelerate Na$^+$/H$^+$ exchange in neuroblastoma X glioma cells. *J. Biol. Chem.* **262**, 6750–6757.

Jaiswal, R. K., and Sharma, R. K. (1985) Purification and biochemical characterization of alpha-2 adrenergic receptor from the rat adrenocortical carcinoma. *Biochem. Biophys. Res. Comm.* **130**, 58–64.

Jakobs, K. H., and Rauschek, R. (1978) [^{3}H]Dihydroergonine binding to alpha adrenergic receptors in human platelets. *Klin. Wochenschr.* **56** (suppl. 1), 139–145.

Jarrott, B., Louis, W. J., and Summers, R. J. (1982) [^{3}H]Guanfacine: A radioligand that selectively labels high affinity alpha-2 adrenergic receptor sites in homogenates of rat brain. *Br. J. Pharmacol.* **75**, 401–408.

Karlsson, J. O. G., Grundstrom, N., Wikberg, J. E. S., Friedman, R., and Anderson, R. G. G. (1985) The effect of pertussis toxin on alpha-2 adrenergic receptor-mediated pigment migration in fish melanophores. *Life Sci.* **37**, 1043–1049.

Kawahara, R. S. and Bylund, D. B. (1985) Solubilization and characterization of putative alpha-2 adrenergic isoreceptors from the human platelet and the rat cerebral cortex. *J. Pharmacol. Exp. Ther.* **233**, 603–610.

Kawahara, R. S., Byington, K. H. and Bylund, D. B. (1985) p-Azidoclonidine: A photoaffinity label for the alpha-2 adrenergic receptor. *Eur. J. Pharmacol.* **117**, 43–50.

Kawai, M., and Nomura, Y. (1983) Involvement of sulfhydryl groups in cerebral cortical [^{3}H]clonidine binding in developing rats. *Eur. J. Pharmacol.* **91**, 449–454.

Kerry, R., Scrutton, M. C., and Wallis, R. B. (1984) Mammalian platelet adrenergic receptors. *Br. J. Pharmacol.* **81**, 91–102.

Kitamura, Y., Tanaka, H., and Nomura, Y. (1986) [^{3}H]Clonidine and [^{3}H]yohimbine binding to solubilized alpha-2 adrenergic receptor from rat cerebral cortex. *Eur. J. Pharmacol.* **123**, 263–270.

Kobilka, B. K., Matsui, H., Kobilka, T. S., Yang-Feng, T. L., Francke, U., Caron, M. G., Lefkowitz, R. J., and Regan, J. W. (1987) Cloning, sequencing, and expression of the gene coding for the human platelet alpha-2 adrenergic receptor. *Science* **238,** 650–656.

Kremenetzky, R., and Atlas, D. (1984) Solubilization and reconstitution of alpha-2 adrenergic receptors from rat and calf brain. *Eur. J. Biochem.* **138**, 573–577.

Kunos, G., Kan, W. H., Greguski, R., and Venter, J. C. (1983) Selective affinity labeling and molecular characterization of hepatic alpha-1 adrenergic receptors with [^{3}H]phenoxybenzamine. *J. Biol. Chem.* **258**, 326–332.

Laemmli, U. K. (1970) Cleavage of structural proteins during the assembly of the head of bacteriophage T4. *Nature* **227**, 680–685.

Lane, A. C., Howlett, D. R., and Walter, D. S. (1983) The effects of metal ions on the binding of a new alpha-2 adrenergic receptor antagonist radioligand [^{3}H]RX 781094 in rat cerebral cortex. *Biochem. Pharmacol.* **32**, 3122–3125.

Langer, S. Z. (1974) Presynaptic regulation of catecholamine release. *Biochem. Pharmacol.* **23**, 1793–1800.

Langer, S. Z., Pimoule, C., and Scatton, B. (1983) [^{3}H]RX 781094, a preferential alpha-2 adrenergic receptor antagonist radioligand, labels alpha-2 adrenergic receptors in the rat brain cortex. *Br. J. Pharmacol.* **78**, 109P.

Lanier, S. M., Graham, R. M., Hess, H. J., Grodski, A., Repaske, M. G., Nunnari, J. M., Limbird, L. E., and Homcy, C. J. (1986a) Photoaffinity labeling of the porcine brain alpha-2 adrenergic receptor using a radioiodinated arylazide derivative of rauwolscine: Identification of the hormone-binding subunit. *Proc. Natl. Acad. Sci. USA* **83**, 9358–9362.

Lanier, S. M., Hess, H. J., Grodski, A., Graham, R. M., and Homcy, C. J.

(1986b) Synthesis and characterization of a high affinity radioiodinated probe for the alpha-2 adrenergic receptor. *Mol. Pharmacol.* **29**, 219–227.

Latifpour, J., Jones, S. B., and Bylund, D. B. (1982) Characterization of [^{3}H]yohimbine binding to putative alpha-2 adrenergic receptors in neonatal rat lung. *J. Pharmacol. Exp. Ther.* **223**, 606–611.

Lattimer, N., and Rhodes, K. F. (1985) A difference in the affinity of some selective, alpha-2 adrenergic receptor antagonists when compared on isolated vasa deferentia of rat and rabbit. *Naunyn Schmiedebergs Arch. Pharmacol.* **329**, 278–281.

Limbird, L. E., Speck, J. L., and Smith, S. K. (1982) Sodium ion modulates agonist and antagonist interactions with the human platelet alpha$_2$-adrenergic receptor in membrane and solubilized preparations. *Mol. Pharmacol.* **21**, 609–617.

Limbird, L. E., MacMillan, S. T., and Kalinoski, D. L. (1985) The resolution of agonist alpha-2 adrenergic receptor complexes from unoccupied receptors or antagonist-alpha-2 receptor complexes using DEAE chromatography. *J. Cyclic Nucl. Protein Phosphor. Res.* **10**, 75–82.

Loftus, D. J., Stolk, J. M., and U'Prichard, D. C. (1984) Binding of the imidazoline UK-14,304, a putative full alpha-2 adrenergic receptor agonist, to rat cerebral cortex membranes. *Life Sci.* **35**, 61–69.

Lomasney, J. W., Leeb-Lundberg, L. M. F., Cotecchia, S., Regan, J. W., DeBarnardis, J. F., Caron, M. G., and Lefkowitz, R. J. (1986) Mammalian alpha-1 adrenergic receptor: Purification and characterization of the native receptor ligand binding subunit. *J. Biol. Chem.* **261**, 7710–7716.

Lynch, C. J., and Steer, M. L., (1981) Evidence for high and low affinity alpha-2 adrenergic receptors. *J. Biol. Chem.* **256**, 3298–3303.

Macfarlane, D. E., Wright, B. L., and Stump, D. C. (1981) Use of [methyl-^{3}H]yohimbine as a radioligand for alpha-$_2$-adrenoreceptors on intact platelets. *Throm. Res.* **24**, 31–43.

Matsui, H., Imafuku, J., Asakura, M., Tsukamoto, T., Ino, M., Saitoh, N., Miyamura, S., and Hasegawa, K. (1984) Solubilization of active alpha-2 adrenergic receptor from rat brain: Regulation by cations and GTP. *Biochem. Pharmacol.* **33**, 3311–3314.

Matsui, H., Asakura, M., Tsukamoto, T., Imafuku, J., Ino, M., Saitoh, N., Miyamura, S., and Hasegawa, K. (1985) Solubilization and characterization of rat brain alpha-2 adrenergic receptor. *J. Neurochem.* **44**, 1625–1632.

Matsushima, Y., Akabane, S., and Ito, K. (1986) Characterization of alpha-1 and alpha-2 adrenergic receptors directly associated with basolateral membranes from rat kidney proximal tubules. *Biochem. Pharmacol.* **35**, 2593–2600.

Mattens, E., Bottari, S., Vokaer, A., and Vauquelin, G. (1985) Arginine and cysteine residues in the ligand binding site of alpha 2-adrenergic receptors. *Life Sci.* **36**, 355–362.

McGrath, J. C., and Reid, J. L. (1985) Commentary on workshop on alpha adrenergic receptors. *Clin. Sci.* **68** (suppl. 10), 1–7.

McKernan, R. M., Dickinson, K. E. J., Miles, C. M. M., Sever, P. S. (1986) Heterogeneity between soluble human and rabbit splenic alpha-2 adrenergic receptor. *Biochem. Pharmacol.* **35**, 3517–3523.

McLauglin, N. J., and Collins, G. G. S. (1986) Binding characteristics of the selective α-2-adrenoceptor antagonist [^{3}H]idazoxan to rat olfactory cortex membranes. *Eur. J. Pharmacol.* **121**, 91–96.

McPherson, G. A., and Summers, R. J. (1983) Evidence from binding studies for alpha-2 adrenergic receptors directly associated with glomeruli from rat kidney. *Eur. J. Pharmacol.* **90**, 333–341.

Miach, P. J., Dausse, J. P., and Meyer, P. (1978) Direct biochemical demonstration of two types of alpha adrenergic receptor in rat brain. *Nature* **274**, 492–494.

Michel, T., Hoffman, B. B., Lefkowitz, R. J., and Caron, M. G. (1981) Different sedimentation properties of agonist and antagonist-labelled platelet alpha$_2$ adrenergic receptors. *Biochem. Biophys. Res. Comm.* **100**, 1131–1136.

Motulsky, H. J., and Insel, P. A. (1982) [^{3}H]Dihydroergocryptine binding to alpha-adrenergic receptors of human platelets. *Biochem. Pharmacol.* **31**, 2591–2597.

Motulsky, H. J., Shattil, S. J., and Insel, P. A. (1980) Characterization of alpha-2 adrenergic receptors on human platelets using [^{3}H]yohimbine. *Biochem. Biophys. Res. Commun.* **97**, 1562–1570.

Mouille, P., Dabire, H., Fournier, B., and Schmitt, H. (1981) A further attempt to characterize the alpha-2 adrenergic receptor blocking properties of (imidazolyl-2)-2-benzodioxane 1-4 (170 150) in pithed rats. *Eur. J. Pharmacol.* **73**, 367–370.

Mukherjee, A. (1981) Characterization of alpha-2 adrenergic receptors in human platelets by binding of a radioactive ligand [^{3}H]yohimbine. *Biochim. Biophys. Acta* **676**, 148–154.

Nakaki, T., Nakadate, T., Ishii, K. and Kato, R. (1981) Postsynaptic alpha-2 adrenergic receptors in isolated rat islets of Langerhans: Inhibition of insulin release and cyclic 3′:5′-adenosine monophosphate accumulation. *J. Pharmacol. Exp. Ther.* **216**, 607–612.

Nakaki, T., Nakadate, T., Yamamoto, S., and Kato, R. (1983) Alpha$_2$-adrenergic receptor in intestinal epithelial cells. Identification by [^{3}H]yohimbine and failure to inhibit cyclic AMP accumulation. *Mol. Pharmacol.* **23**, 228–234.

Nakata, H., Regan, J. W., and Lefkowitz, R. J. (1986) Chemical modification of alpha-2 adrenergic receptors: Possible role for tyrosine in the ligand binding site. *Biochem. Pharmacol.* **35**, 4089–4094.

Nambi, P., Aiyar, N. V., and Sharma, R. K. (1982) Solubilization of epinephrine-specific alpha-2 adrenergic receptors from adrenocortical carcinoma. *FEBS Lett.* **140**, 98–102.

Neubig, R. R., Gantzos, R. D., and Brasier, R. S. (1985) Agonist and antagonist binding to alpha-2 adrenergic receptors in purified membranes from human platelets. *Mol. Pharmacol.* **28**, 475–486.

Neylon, C. B., and Summers, R. J. (1985) [^{3}H]-Rauwolscine binding to alpha-2 adrenergic receptors in the mammalian kidney: Apparent receptor heterogeneity between species. *Br. J. Pharmacol.* **85**, 349–359.

Nunnari, J. M., Repaske, M. G., Brandon, S., Cragoe, E. J. Jr., and Limbird, L. E. (1987) Regulation of porcine brain alpha$_2$-adrenergic receptors by Na$^+$, H$^+$ and inhibitors of Na$^+$/H$^+$ exchange. *J. Biol. Chem.* **262**, 12387–12392.

Perry, B. D., and U'Prichard, D. (1981) [^{3}H]Rauwolscine (alpha-yohimbine): A specific antagonist radioligand for brain alpha-2 adrenergic receptors. *Eur. J. Pharmacol.* **76**, 461–464.

Petrash, A. C., and Bylund, D. B. (1986) Alpha-2 adrenergic receptor subtypes indicated by [^{3}H]yohimbine binding in human brain. *Life Sci.* **38**, 2129–2137.

Pimoule, C., Briley, M. S. and Langer, S. Z. (1980) Short-term surgical denervation increases [^{3}H]clonidine binding in rat salivary gland. *Eur. J. Pharmacol.* **63**, 85–87.

Pushpendran, C. K., and Garcia-Sainz, J. A. (1984) RX 781094 a potent and selective alpha-2 adrenergic antagonist. Effects in adipocytes and hepatocytes. *Eur. J. Pharmacol.* **99**, 337–339.

Quennedey, M. C., Bockaert, J., and Rouot B. (1984) Direct and indirect effects of sulfhydryl blocking agents on agonist and antagonist binding to central alpha-1 and alpha-2 adrenergic receptors. *Biochem. Pharmacol.* **33**, 3923–3928.

Regan, J. W., Barden, N., Lefkowitz, R. J., Caron, M. G., DeMarinis, R. M., Krog, A. J., Holden, K. G., Matthews, W. D., and Hieble, J. P. (1982) Affinity chromatography of human platelet alpha$_2$-adrenergic receptors. *Proc. Natl. Acad. Sci. USA* **79**, 7223–7227.

Regan, J. W., Benovic, J. L., Matsui, H., Mayor, F., Caron, M. G., and Lefkowitz, R. J. (1987) Agonist-dependent phosphorylation of the alpha-2 adrenergic receptor by the beta adrenergic receptor kinase. *Clin. Res.* **35**, 648A.

Regan, J. W., DeMarinis, R. M., Caron, M. G., and Lefkowitz, R. J. (1984) Identification of the subunit binding site of alpha-2 adrenergic receptors using [^{3}H]phenoxybenzamine. *J. Biol. Chem.* **259**, 7864–7869.

Regan, J. W., DeMarinis, R. M., and Lefkowitz, R. J. (1985) Arylazide photoaffinity probe for alpha-2 adrenergic receptors. *Biochem. Pharmacol.* **34**, 3667–3672.

Regan, J. W., Nakata, H., DeMarinis, R. M., Caron, M. G., and Lefkowitz, R. J. (1986a) Purification and characterization of the human platelet alpha-2 adrenergic receptor. *J. Biol. Chem.* **261**, 3894–3900.

Regan, J. W., Raymond, J. R., Lefkowitz, R. J., and DeMarinis, R. M. (1986b) Photoaffinity labeling of human platelet and rabbit kidney alpha-2 adrenergic receptors with [^{3}H]SKF 102229. *Biochem. Biophys. Res. Commun.* **137**, 606–613.

Repaske, M. G., Nunnari, J. M., and Limbird, L. E. (1987) Purification of the alpha-2 adrenergic receptor from porcine brain using a yohimbine-agarose affinity matrix. *J. Biol. Chem.* **262**, 12381–12386.

Rouot, B. R., and Snyder, S. H. (1979) [^{3}H]Para-amino-clonidine: A novel ligand which binds with high affinity to alpha adrenergic receptors. *Life Sci.* **25**, 769–774.

Rouot, B., Quennedey, M. C., and Schwartz, J. (1982) Characteristics of the [^{3}H]yohimbine binding on rat brain alpha-2 adrenergic receptors. *Naunyn Schmiedebergs Arch. Pharmacol.* **321**, 253–259.

Schmitz, J. M., Graham, R. M., Sagalowsky, A., and Pettinger, W. A. (1981) Renal alpha-1 and alpha-2 adrenergic receptors: Biochemical and pharmacological correlations. *J. Pharmacol. Exp. Ther.* **219**, 400–406.

Shreeve, S. M., Fraser, C. M., and Venter, J. C. (1985) Molecular comparison of alpha-1 and alpha-2 adrenergic receptors suggests that these proteins are structurally related isoreceptors. *Proc. Natl. Acad. Sci. USA* **82**, 4842–4846.

Sladeczek, F., Bockaert, J., and Rouot, B. (1984) Solubilization of brain alpha-2 adrenergic receptor with a zwitterionic detergent: Preservation of agonist binding and its sensitivity to GTP. *Biochem. Biophys. Res. Comm.* **119**, 1116–1121.

Smith, S. K., and Limbird, L. E. (1981) Solubilization of human platelet alpha adrenergic receptors: Evidence that agonist occupancy of the receptor stabilizes receptor-effector interactions. *Proc. Natl. Acad. Sci. USA* **78**, 4026–4030.

Smith, S. K. and Limbird, L. E. (1982) Evidence that human platelet alpha adrenergic receptors coupled to inhibition of adenylate cyclase are not associated with the subunit of adenylate cyclase ADP-ribosylated by cholera toxin. *J. Biol. Chem.* **257**, 10471–10478.

Snavely, M. D. and Insel, P. A. (1982) Characterization of alpha-adrenergic receptor subtypes in the rat renal cortex. *Mol. Pharmacol.* **22**, 532–546.

Starke, K. (1981) Alpha adrenoceptor subclassification. *Rev. Physiol. Biochem. Pharmacol.* **88**, 199–236.

Steer, M. J., Khorana, J., and Galgoci, B. (1979) Quantitation and characterization of human platelet alpha-adrenergic receptors using [^{3}H]phentolamine. *Mol. Pharmacol.* **16**, 719–728.

Stiles, G. L., Benovic, J. L., Caron, M. G., and Lefkowitz, R. J. (1984) Mammalian beta adrenergic receptors: Distinct glycoprotein populations containing high mannose or complex type carbohydrate chains. *J. Biol. Chem.* **259**, 8655–8663.

Summers, R. J., Barnett, D. B., and Nahorski, S. R. (1983) The characteristics of adrenergic receptors in homogenates of human cerebral cortex labelled by [^{3}H]rauwolscine. *Life Sci.* **33**, 1105–1112.

Sweatt, J. D., Blair, I. A., Cragoe, E. J., and Limbird, L. E. (1986) Inhibitors of Na^+/H^+ exchange block epinephrine- and ADP-induced stimulation of human platelet phospholipase C by blockade of arachidonic acid release at a prior step. *J. Biol. Chem.* **261**, 8660–8666.

Tanaka, T. and Starke, K. (1979) Binding of ^{3}H-clonidine to an alpha adrenergic receptor in membranes of guinea-pig ileum. *Naunyn Schmiedebergs Arch. Pharmacol.* **309**, 207–215.

Tanaka, T., Ashida, T. Deguchi, F., and Ikeda, M. (1983) [^{3}H]Clonidine and [^{3}H]rauwolscine binding to membranes from rat cerebral cortex and kidney. *Japan J. Pharmacol.* **33**, 713–716.

Tharp, M. D., Hoffman, B. B., and Lefkowitz, R. J. (1981) Alpha adrenergic receptors in human adipocyte membranes: Direct determination by [^{3}H]yohimbine binding. *J. Clin. Endocrinol. Metab.* **52**, 709–714.

Timmermans, P. B. M. W. M., Schoop, A. M. C., and van Zwieten, P. (1982) Binding characteristics of [^{3}H]guanfacine to rat brain alpha adrenergic receptors. *Biochem. Pharmacol.* **31**, 899–905.

Timmermans, P. B. M. W. M., Qian, J. Q., Ruffolo, R. R., and van Zwieten, P. A. (1984) A study of the selectivity and potency of rauwolscine, RX 781094 and RS 21361 as antagonists of alpha-1 and alpha-2 adrenergic receptors. *J. Pharmacol. Exp. Ther.* **228**, 739–748.

Tsukahara, T., Taniguchi, T., Fujiwara, and Handa, H. (1983) Characterization of alpha adrenergic receptors in pial arteries of the bovine brain. *Naunyn Schmiedebergs Arch. Pharmacol.* **324**, 88–93.

Turner, J. T., Ray-Prenger, C., and Bylund, D. B. (1985) Alpha-2 Adrenergic receptors in the human cell line, HT29. *Mol. Pharmacol.* **28**, 422–430.

Turtle, J. R., and Kipnis, D. M. (1967) An adrenergic receptor mechanism for the control of cyclic 3'5' adenosine monophosphate synthesis in tissues. *Biochem. Biophys. Res. Commun.* **28**, 797–802.

U'Prichard, D. C., and Ernsberger, P. (1983) p-Azidoclonidine: A potential photoaffinity ligand for the alpha-2 receptor. *Soc. Neurosci. Abst.* **9**, 1117.

U'Prichard, D. C., and Snyder, S.H. (1977) Binding of [^{3}H]catecholamines to alpha-noradrenergic receptor sites in calf brain. *J. Biol. Chem.* **252**, 6450–6463.

U'Prichard, D. C., Greenberg, D. A., and Snyder, S. H. (1977). Binding characteristics of a radiolabeled agonist and antagonist at central nervous system alpha-noradrenergic receptors. *Mol. Pharmacol.* **13**, 454–473.

U'Prichard, D. C., Mitrius, J. C., Kahn, D. J., and Perry, B. D. (1983) The alpha-2 adrenergic receptor: Multiple affinity states and regulation of a receptor inversely coupled to adenylate cyclase. *Adv. in Biochem. Psychopharmacol.* **36**, 53–72.

Walter, D. S., Flockhart, I. R., Haynes, M. J., Howlett, D. R., Lane, A. C., Burton, R., Johnson, J., and Dettmar, P. W. (1984) Effects of idazoxan on catecholamine systems in rat brain. *Biochem. Pharmacol.* **33**, 2553–2557.

Wikberg, J. E. S. (1978) Pharmacological classification of adrenergic alpha receptors in the guinea pig. *Nature* **273**, 164–166.

Williams, L. T., and Lefkowitz, R. J. (1976) Alpha-Adrenergic receptor identification by [^{3}H]dihydroergocryptine binding. *Science* **192**, 791–793.

Wood, C. L., Arnett, C. D., Clarke, W. R., Tsai, B. S., and Lefkowitz, R. J. (1979) Subclassification of alpha-adrenergic receptors by direct binding studies. *Biochem. Pharmacol.* **28**, 1277–1282.

Woodcock, E. A., and Johnston, C. I. (1982) Characterization of adenylate cyclase-coupled alpha$_2$-adrenergic receptors in rat renal cortex using [^{3}H]yohimbine. *Mol. Pharmacol.* **22**, 589–594.

Woodcock, E. A., and Murley, B. (1982) Increased central alpha-2 adrenergic receptors measured with [^{3}H]yohimbine in the presence of sodium ion and guanyl-nucleotides. *Biochem. Biophys. Res. Commun.* **105**, 252–258.

Yamazaki, S., Katada, T., and Ui, M. (1982) Alpha$_2$-adrenergic inhibition of insulin secretion via interference with cyclic AMP generation in rat pancreatic islets. *Mol. Pharmacol.* **21**, 648–653.

SECTION 3
BIOCHEMICAL MECHANISMS OF
RECEPTOR ACTION

Chapter 3

Mechanisms for Inhibition of Adenylate Cyclase by alpha-2 Adrenergic Receptors

Peter Gierschik and Karl H. Jakobs

1. Introduction

Regulation of cellular functions by hydrophilic transmitters, e.g., catecholamines, requires efficient mechanisms for signal transduction across the plasma membrane. Generally speaking, these mechanisms function by translating the primary message, i.e., presence or absence of a hormone or neurotransmitter at the outer surface of the plasma membrane, into one or more second messages inside the cell, e.g., changes in cytosolic concentrations of cyclic nucleotides or calcium ions, modulation of ion fluxes across the cell membrane, altered metabolism of phosphoinositides, or phosphorylation of cellular proteins by receptor-driven protein kinases.

Substances that are able to bind to and activate adrenergic receptors of the pharmacologically defined alpha-2 subtype have been demonstrated to decrease cellular cyclic AMP levels by inhibiting the activity of the membrane-bound enzyme complex adenylate cyclase. Although this effect is well documented in many systems (*see below*), it is important to note that it is not known if inhibition of adenylate cyclase is the most important or even the only mechanism by which alpha-2 adrenergic substances express their physiologic effects. It is clear, however, that mechanisms by which these agents inhibit adenylate cyclase need to be integrated into any concept of their action.

The scope of this review is to analyze the current knowledge about structure and function of components involved in inhibition of adenylate cyclase by alpha-2 adrenergic agonists. We will not consider inhibition of adenylate cyclase by other hormones or neurotransmitters nor the properties of the receptor moiety responsible for recognition of alpha-2 adrenergic substances. These issues have been reviewed recently elsewhere (Jakobs et al., 1983a, 1984b, 1985a) and are discussed in other chapters of this volume as well.

2. Components and Mechanisms Involved in Stimulation of Adenylate Cyclase

Before discussing mechanisms involved in alpha-2 adrenergic inhibition of adenylate cyclase, it is necessary to present a brief outline of the current knowledge about the enzyme's stimulation by hormones and neurotransmitters. At least three different proteins, all of them intrinsic plasma membrane proteins, are necessary for stimulation of intracellular cyclic AMP formation by extracellular messengers: receptor proteins (R_s) facing the extracellular space, the stimulatory guanine nucleotide binding protein (G_s) at the inside of the plasma membrane, and the catalytic unit (C), which possibly spans the plasma membrane and catalyzes cyclic-AMP formation at its cytoplasmatic pole. The individual components have been purified to homogeneity and characterized (Shorr et al., 1981; Northup et al., 1980; Pfeuffer et al., 1985), and their reconstitution into phospholipid vesicles has been accomplished (May et al., 1985).

As illustrated in Fig. 1, activation of the catalytic unit is thought to result from a series of sequential interactions of these protein components. First, the activated, i.e., hormone-bound, receptor interacts with the heterotrimeric G_s holoprotein and facilitates an exchange of GTP for GDP bound to its 45-kDa α subunit. Activation of the GTP-bound form of G_s appears to coincide with its dissociation from the receptor as well as its own dissociation into a free α subunit and a complex of a 35-kDa β subunit and a 8-kDa γ subunit. The GTP-bound free α subunit has been shown to be fully capable of stimulating cyclic AMP formation by the catalytic unit in vitro (Northup et al., 1983b).

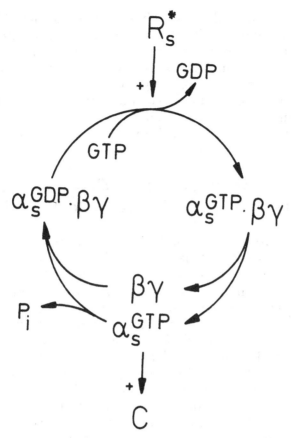

Fig. 1. Model for stimulation of adenylate cyclase activity by stimulatory hormones and neurotransmitters. The activated, i.e., agonist-bound stimulatory receptor (R_s^*) activates the heterotrimeric stimulatory G-protein (G_s, $\alpha_s \cdot \beta/\gamma$) by facilitating an exchange of GTP for GDP bound to α_s. Activation of G_s is thought to result in a dissociation of the GTP-bound form of α_s from the β/γ complex. Activated, free α_s is capable of stimulating the activity of the catalytic unit of adenylate cyclase (C). Deactivation of α_s results in hydrolysis by α_s of GTP to GDP and/or reassociation of α_s with the β/γ complex.

Deactivation of the enzyme is thought to mainly result from two mechanisms: hydrolysis by α_s of bound GTP to GDP and reassociation of the α subunit with β/γ complexes. As will be discussed below, it is not clear which of these mechanisms is more important for the deactivation process. The latter, however, appears to contribute, at least partially, to the molecular basis for the inhibition of adenylate cyclase for alpha-2 adrenergic agonists.

In vitro studies of adenylate cyclase stimulation have utilized four classes of agents in addition to hormones and naturally occurring guanine nucleotides. Hydrolysis-resistant GTP analogs (e.g., GTP-γ-S and GMPPNHP) and fluoride ions (complexed by Al^{3+} to AlF_4^-) persistently activate G_s and lead to increased cyclic-AMP formation. Cholera toxin (CT) has been found to activate G_s permanently by catalyzing the covalent linkage of α_s with an ADP-ribose moiety from NAD (Moss and Vaughan, 1979). Finally, the diterpene forskolin appears to stimulate the catalytic unit directly (*see* Seamon and Daly, 1981, for review).

3. Components Involved in Inhibition of Adenylate Cyclase by alpha-2 Adrenergic Agonists

Many studies focusing on inhibition of adenylate cyclase activity by alpha-2 adrenergic agonists have utilized membranes prepared from human platelets as a test system, mainly for two reasons: First, the alpha-adrenergic receptors present on human platelets have been demonstrated to exclusively be of the alpha-2 subtype (Hoffman et al., 1980). Second, the inhibitory effect of alpha-2 adrenergic substances on both basal and hormone [e.g., prostaglandin (PGE_1)]-stimulated adenylate cyclase is well defined in this system (Jakobs et al., 1976) and appears to be correlated with their ability to lower cellular cyclic AMP levels (Salzman and Neri, 1969). The relative ease of membrane preparation together with the availability of platelets in reasonable quantities are additional reasons for using this system. There are, however, several other cell types in which alpha-2 adrenergic inhibition of adenylate cyclase has been studied, e.g., rabbit platelets (Tsai and Lefkowitz, 1978), rat pancreatic islets (Katada and Ui, 1981), NG108-15 neuroblastoma × glioma hybrid cells (Sabol and Nirenberg, 1979), and hamster adipocytes (Aktories et al., 1979, 1980).

3.1. A Guanine Nucleotide-Binding Protein Is Involved in Inhibition of Adenylate Cyclase by alpha-2 Adrenergic Agonists

In retrospect, the observation that GTP is required for inhibition of human platelet adenylate cyclase by epinephrine was clearly one of the first steps toward understanding the coupling of alpha-2 adrenergic receptors to adenylate cyclase (Jakobs et al., 1978). Specifically, there is an absolute requirement for GTP for epinephrine-induced inhibition of both basal and hormone (e.g., prostaglandin PGE_1)-stimulated adenylate cyclase activity. Interestingly, as shown in Fig. 2, the concentrations of GTP required for half-maximal and maximal expression of epinephrine-induced inhibition of the enzyme (about 0.3 and 10 μM, respectively) are higher by about one order of magnitude than the concentrations of GTP required to support stimulation of the enzyme by PGE$_1$.

Another important observation indicating that GTP is in fact involved in epinephrine-induced inhibition of adenylate cyclase was the stimulation of a high-affinity GTPase by epinephrine in human platelet membranes (Aktories and Jakobs, 1981). Of importance, both inhibition of adenylate cyclase and stimulation of GTP hydrolysis occur at the same concentrations of epinephrine,

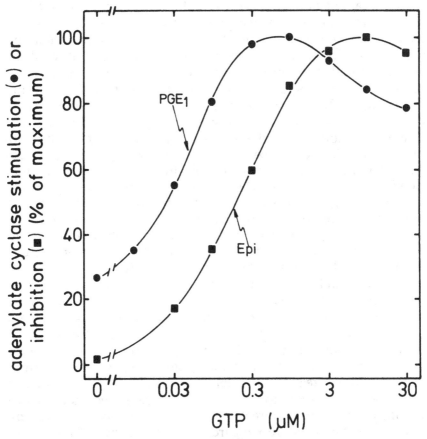

Fig. 2. GTP requirements of hormonal stimulation and inhibition of human platelet adenylate cyclase. Adenylate cyclase was determined with either PGE$_1$ (10 μM) or epinephrine (Epi, 30 μM) at increasing GTP concentrations. Stimulation by PGE$_1$ and inhibition by epinephrine are given on the ordinate as % of maximal hormonal effect.

i.e., half-maximal effects are typically observed at about 0.3 μM epinephrine.

Interactions between alpha-2 adrenergic agonists and guanine nucleotides are not a unidirectional process. Rather, GTP has effects on the binding properties of alpha-2 adrenergic receptors for agonists as well. This effect of GTP shows striking similarity to the modulation by guanine nucleotides of agonist affinity of receptors linked to stimulation of adenylate cyclase, e.g., beta-adrenergic receptors (*see* Lefkowitz et al., 1983, for review). In this case, the receptor shows a high affinity for agonists upon interaction with G_s. Exchange of GTP (or one of its nonhydrolyzable analogs) for GDP and concomitant dissociation of the receptor and G_s provides the receptor with a low affinity for agonists. Typically, antagonists do not discriminate between the two affinity states. GTP regulation of receptor affinity for agonists is usually analyzed by comparing agonist competition for radiolabeled antagonist binding in the absence or presence of GTP or its analogs and quantitative analysis of the competition binding profiles using computer programs for nonlinear least-squares curve fitting (Munson, 1983). This approach has been considered in detail in Chapter 2 of this volume. Using this approach, Hoffman et al. (1980) have found the alpha-2 adrenergic receptors of human platelet membranes to exist in two affinity states for agonists, termed alpha-2(H) and alpha-2(L), in the absence of guanine nucleotides. In the presence of guanine nucleotides, however, analysis of binding data revealed only one affinity state of the alpha-2 adrenergic receptor for epinephrine, which is identical to alpha-2(L). Disappearance of high-affinity sites is best explained by their conversion to low-affinity sites, since total receptor number is unchanged by addition of guanine nucleotides. The authors hypothesized that the high-affinity state of the receptor reflects a ternary complex of agonist, receptor, and guanine nucleotide binding protein. Studies performed with alpha-2 adrenergic receptors solubilized from human platelets proved to be very much in support of such an hypothesis (Smith and Limbird, 1981). If the receptors were treated with agonist prior to solubilization, their sedimentation rate in sucrose density gradients was clearly higher than the rate obtained for antagonist-pretreated receptors. In addition, only agonist-pretreated receptors could be regulated by guanine nucleotides with regards to their agonist affinity. Pretreatment of the receptor with agonists, therefore, appears to stabilize its interaction with an additional component, most likely a guanine nucleotide-binding protein.

In summary, four observations strongly suggested that a guanine nucleotide-binding protein is involved in mediation of the inhibitory response to alpha-2 adrenergic agonists: (1) GTP is required for inhibition of adenylate cyclase by epinephrine. (2) Epinephrine stimulates a high affinity GTPase in plasma membranes. (3) The affinities for agonists of alpha-2 adrenergic receptors are regulated by guanine nucleotides. (4) Solubilized preparations of alpha-2 receptors that have retained guanine nucleotide regulation of agonist affinity are larger in size than those that have lost this property. Since the guanine nucleotide-binding protein G_s had already been shown to be involved in stimulation of adenylate cyclase, it now remains to be clarified whether the protein relevant to the enzyme's inhibition was identical to G_s or represented a distinct entity.

3.2. Identification of the Inhibitory Guanine Nucleotide-Binding Protein G_i by Pertussis Toxin

A major breakthrough in studying the mechanisms involved in inhibition of adenylate cyclase was the observation that one of the toxins found in the supernatant of *Bordetella pertussis* cultures was able to prevent inhibition of cyclic AMP accumulation by epinephrine in rat pancreatic islet cells, and consequently, activate insulin release from these cells (Katada and Ui, 1979, 1980). The specific toxin is variably referred to as "islet-activating protein (IAP)" or simply "pertussis toxin (PT)." Typically, PT-treated cells showed increased stimulation of cyclic AMP accumulation by stimulatory hormones, e.g., glucagon, and decreased inhibition by epinephrine. However, cellular cyclic AMP content was not changed by PT treatment in the absence of stimulatory or inhibitory hormones. Similar actions of PT have been reported for other systems as well, including NG108-15 cells (Kurose et al., 1983) and dog thyroid slices (Cochaux et al., 1985). In addition, prevention by PT of receptor-mediated decreases in cyclic AMP content is not confined to alpha-2 adrenergic agonists, but rather is a general property of agonists acting at receptors linked to inhibition of adenylate cyclase, e.g., those acting via muscarinic cholinergic (Hazeki and Ui, 1981), prostaglandin (PGE_1), and adenosine (A_1) (Murayama and Ui, 1983) and opioid receptors (Kurose et al., 1983). Experiments performed with membranes prepared from rat islets showed that pretreatment with PT does not change the kinetic properties of adenylate cyclase when stimulated by GTP or its nonhydrolyzable analogs, by AlF_4^-, or by cholera toxin, thus making it unlikely that either G_s or C are modified by exposure to PT. Instead, PT pretreatment renders adenylate cyclase slightly more sensitive to stimulation by glucagon and markedly less sensitive to inhibition by epinephrine. There were no changes, however, in the potency of epinephrine and the GTP requirement for epinephrine-induced inhibition of the enzyme (Katada and Ui, 1981). Therefore, neither the components necessary for stimulation of adenylate cyclase nor the alpha-2 receptor appeared to be modified by the toxin.

Treatment of intact cells with PT had initially been the only means by which effects of this toxin could be elicited on cellular cyclic AMP content or on adenylate cyclase activity in plasma membranes. Maximal effects were observed after a definite lag phase, suggesting that the toxin had to be activated and/or internalized by the cells (Katada and Ui, 1980). In addition, whereas toxin-dependent effects could be seen with most cell types, a few appeared to be resistant to toxin treatment, e.g., platelets and reticulocytes. Most of these features are caused by the molecular properties of PT, which have been analyzed in great detail by Ui and coworkers and reviewed recently (Ui, 1984).

PT is a hexamer with five of its subunits (B-oligomer) being responsible for its binding to specific receptors at the plasma membrane of cells to be modified. Subsequent to binding, the sixth subunit is thought to be introduced into the cytoplasm and activated by disulfide reduction (A protomer). The molecular mechanism responsible for PT action has been identified as the transfer of an ADP-ribose moiety from NAD to a 41-kDa plasma membrane protein (Katada and Ui, 1982a). The relevant PT-substrate, although clearly different from the proteins ADP-ribosylated by CT, appears to be a guanine nucleotide-binding protein. ADP-ribosylation by PT of the 41-kDa protein correlates well with the functional changes observed with PT treatment (Katada and Ui, 1982b). These findings by Katada and Ui represent the first physical demonstration of a distinct guanine nucleotide-binding protein, termed G_i, responsible for inhibition of adenylate cyclase and clearly paved the way for the purification of this protein, which will be discussed below.

Certain requirements for PT action are implicit in its molecular properties and need to be considered in practice. They are summarized as follows: (1) If intact cells are to be treated with PT, receptors for the B-oligomer need to be present on their plasma membrane. (2) For ADP-ribosylation of G_i in membranes, the toxin is usually activated by DTT reduction, since enzymes for endogenous activation do not exist in all membrane systems. (3)

NAD is needed as a cosubstrate for PT treatment of membrane preparations. When radioactively labeled NAD is used for covalent identification of the PT substrate, G_i, the investigator must be cognitant that some membrane preparations, e.g., brain and liver membranes, do contain relatively high amounts of NAD-degrading activities. Their action can be counteracted, at least in part, by substituting NAD several times during incubation and adding products of NAD degradation, e.g. nicotinamide (Bokoch and Gilman, 1984). (4) ATP is needed as a cofactor for maximal expression of the toxin's activity (Lim et al., 1985). GTP has been found to be beneficial in some systems, whereas GTP-γ-S, magnesium ions, and phosphate buffers inhibit the ADP-ribosylation of G_i by pertussis toxin (Bokoch et al., 1983, 1984; Ribeiro-Neto et al., 1985).

Knowing that PT very likely ADP-ribosylates the inhibitory guanine nucleotide-binding protein, it was obviously of great interest to identify the exact mechanism by which this covalent modification leads to diminished response to agents inhibitory to adenylate cyclase, e.g., alpha-2 adrenergic agonists. In analyzing hormone-stimulated GTPase activity in membranes prepared from human platelets, Aktories et al. (1983) found that PT treatment inhibits epinephrine-induced GTPase activity at concentrations of toxin similar to those necessary to observe abolished inhibition of adenylate cyclase by epinephrine. In contrast, prostaglandin (PGE_1)-stimulated GTPase activity remained unchanged by PT treatment. Murayama and Ui (1984) conducted further studies in membranes prepared from hamster adipocytes to determine which of the partial reactions contributing to overall GTPase activity was most likely the one impaired by PT treatment. They showed that epinephrine stimulated the binding of tritiated GTP to these membranes under appropriate conditions. GTP bound this way could be released subsequent to hydrolysis to GDP only in the presence of epinephrine or other agonists that evoke inhibition of adenylate cyclase. Thus, one of the actions of inhibitory receptors appears to be to facilitate the release of GDP bound to the guanine nucleotide-binding protein coupled to the receptor. If, however, membranes are treated with PT either before or after agonist-dependent guanine nucleotide binding, epinephrine-dependent GDP release was markedly diminished, suggesting that the ADP-ribosylated inhibitory guanine nucleotide-binding protein was "uncoupled" from inhibitory receptors. Decreased GTPase activity in response to epinephrine could be explained very well with this mechanism of toxin action, since GTP hydrolysis could not be detected if hydrolyzed GDP could not be removed from G_i subsequent to PT-catalyzed ADP-ribosylation. In addition, studies focusing on the effect of PT on agonist affinity of alpha-2 adrenergic receptors in hamster adipocyte membranes appear to support the "uncoupling hypothesis." In this system, PT treatment drastically reduces the number of receptors in the high-affinity state, typically found in the absence of guanine nucleotides and believed to reflect receptors coupled to guanine nucleotide-binding proteins, without changing total receptor number (Garcia-Sainz et al., 1984).

In summary, pertussis toxin appears to modify inhibition of adenylate cyclase via ADP-ribosylation of G_i, a guanine nucleotide-binding protein clearly different from G_s. This covalent modification apparently perturbs the interaction of G_i with inhibitory receptors, i.e., prevents the receptor mediated activation of G_i.

3.3. Purification of G_i and Characterization of the Purified Protein

As indicated above, ADP-ribosylation by PT of G_i using $[^{32}P]NAD$ as a substrate provided a good assay for the G_i protein. Its purification from rabbit liver (Bokoch et al., 1983) and human erythrocytes (Codina et al., 1983) has been accomplished. As illustrated in Fig. 3, the relevant protein appears to be a heterotrimer like G_s. The 41-kDa α subunit,* although clearly being distinct from α_s, is the protein that binds guanine nucleotides and is ADP-ribosylated by PT. In its nonactivated, i.e., GDP-bound form, the α subunit is associated with a β/γ complex with molecular weights of 35 and 8 kDa, respectively. The β subunit of G_i appears to be very similar, if not identical, to the β subunit of G_s by comparison of amino acid composition and peptide maps after proteolytic digestion (Manning and Gilman, 1983). In addition, the two proteins could also be shown to be indistinguishable by immunochemical criteria (Gierschik et al., 1985). Recent evidence (Hildebrandt et al., 1985) suggests that the γ subunits of G_i and G_s may be very similar as well.

Heterotrimeric G_i, as shown in sucrose density gradients, dissociates into free α subunits and β/γ complexes upon binding of GTP-γ-S or in the presence of AlF_4^- (Northup et al., 1983b). Interestingly, the free α subunit, but not the β/γ complex or the nonactivated holoprotein, appears to be soluble in the absence of detergents, suggesting that the β/γ complex may be the compo-

---

*Some laboratories report 40 kDa as the apparent molecular weight of α_i on SDS gels.

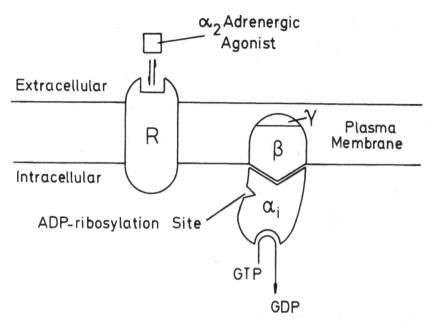

Fig. 3. Components involved in alpha-2 adrenergic inhibition of adenylate cyclase. Binding of the alpha-2 adrenergic agonist to the alpha-2 adrenergic receptor (R) at the outside of the plasma membrane results in the activation of the receptor protein. The activated receptor is thought to interact with the heterotrimeric inhibitory guanine nucleotide-binding protein (G_i, $\alpha_i \cdot \beta/\gamma$) in the lipid bilayer and activate G_i by an exchange of GTP for GDP bound to α_i. For subsequent steps involved in inhibition of adenylate cyclase, *see* Fig. 5. ADP-ribosylation of α_i by pertussis toxin uncouples G_i from the receptor and leads to a decreased or abolished inhibitory response to alpha-2 adrenergic agonists.

nent responsible for anchoring G_i in the plasma membrane (Sternweis, 1986). Binding of guanine nucleotides to purified preparations of G_i is complex and has only been studied with the nonhydrolyzable GTP analog GTP-γ-S (Bokoch et al., 1984). Both increasing the concentration of magnesium and decreasing the protein concentration increased the rate of high-affinity binding. On the other hand, dissociation of bound GTP-γ-S was found only in the absence of free magnesium and was stimulated by increasing amounts of β/γ complexes. The authors speculated that increasing the magnesium concentration might provoke dissociation of G_i and thereby increase its affinity for GTP-γ-S. In contrast, Birnbaumer and colleagues have suggested, based on hydrodynamic data, that Mg^{2+} promotes the formation of a preactivated G_i trimer, which has a high affinity for guanine nucleotide triphosphates even before it dissociates (Codina et al., 1984).

Studies performed with purified G_i also provided the final proof that the protein has GTPase activity (Sunyer et al., 1984). This activity was found to be a property of the 41-kDa α subunit and be independent of the amount of β/γ subunits present during the assay. Interestingly, relatively low concentrations of Mg^{2+} (apparent K_m = 5–15 nM) were required to support GTP hydrolysis. It is important to note at this point that GTPase activity of a given guanine nucleotide-binding protein reflects the result of GDP release, GTP binding, activation of the binding protein, actual hydrolysis of GTP, and deactivation of the binding protein. The exact mechanism(s) by which this complex enzyme activity is regulated by activated receptors and modified by pertussis toxin-dependent ADP-ribosylation remains to be determined.

3.4. Analogous G-Proteins

Receptor-effector coupling via guanine nucleotide-binding proteins appears to be a mechanism for signal transduction that is not limited to regulation of adenylate cyclase activity (for a recent review, see Spiegel et al., 1985). Light-dependent stimulation of cyclic-GMP phosphodiesterase in retinal rod outer segments is mediated by transducin (G_t), a guanine nucleotide-binding heterotrimer with β subunits similar or identical to, and α and γ subunits clearly distinct from, those of G_s and G_i (Manning and Gilman, 1983; Gierschik et al., 1985; Pines et al., 1985). Yet another heterotrimeric GTP-binding protein, termed G_o, has recently been identified in bovine brain (Sternweis and Robishaw, 1984; Neer et al., 1984), although no function has yet been ascribed to the protein except for its ability to couple to muscarinic cholinergic receptors in vitro (Florio and Sternweis, 1985). Recently, however, the α subunit of G_o has been shown to be effective in restoring opiate- and alpha-2 adrenergic receptor-induced decreases in voltage-sensitive Ca^{2+} channel activity to differentiated NG108-15 cells pretreated with PT to eliminate the effectiveness of endogenous G_i or G_o (Hescheler et al., 1987a,b). The α subunits of G_i, G_o, and G_t seem to be closely related proteins, since all three of them are substrates for ADP-ribosylation by PT (Sternweis and Robishaw, 1984; Neer et al., 1984; Van Dop et al., 1984), and certain antisera raised against α_t cross-react either with α_i (Gierschik et al., 1985; Pines et al., 1985) or with α_o (Gierschik and Spiegel, unpublished observations).

   Numa and colleagues recently have cloned the cDNA for β_t from a retinal cDNA library and found that it hybridizes with cDNA from bovine brain poly(A)$^+$ RNA that has also been cloned by this group (Sugimoto et al., 1985). Preliminary cDNA sequence analysis revealed no differences in the coding region of those clones. There were, however, differences in the 5'-noncoding region of those clones, which may be important for differences in regulation of synthesis of β_t versus other G-protein β subunits. In addition, a cDNA clone for the α subunit of G_i (Nukada et al., 1986) has been isolated, and the primary structure has been compared to that of α_s or α_t. A much higher degree of amino acid sequence homology exists for α_i and α_t (68%) than for α_i and α_s (43%).

Therefore, signal transducing G-proteins (the known G-proteins are summarized in Table 1) represent a rapidly growing family of highly related proteins, some of which may in fact be unknown today. For this reason, the regulation of signal transfer by these proteins may be more complex than initially expected. The involvement of more than one G-protein in one function, as well as one G-protein in several functions, has to be considered.

4. Mechanisms Involved in Inhibition of Adenylate Cyclase in the Plasma Membrane

In the foregoing discussion we have presented information about the structure of the inhibitory guanine nucleotide-binding protein. We will now turn our attention to the one question, which has been most intriguing or even tantalizing for many workers in the field and still remains unanswered today: How does this protein bring about inhibition of adenylate cyclase in the plasma

Table 1
Signal Transducing G-Proteins

G-protein	MW α-subunit[a]	ADP-ribo-sylation	Function
G_s	45 kDa	CT[b]	Stimulation of adenylate cyclase
G_i	41 kDa	PT[c]	Inhibition of adenylate cyclase
G_o	39 kDa	PT	Inhibition of voltage-dependent Ca^{2+} channels
G_t	39 kDa	PT,CT	Stimulation of retinal cGMP-phosphodiesterase

[a]Apparent molecular weight on SDS polyacrylamide gels.
[b]CT, cholera toxin.
[c]PT, pertussis toxin.

membrane? Two main mechanisms have been considered, and there is ample experimental support for each of them, much of which is based on observations of the behavior of G_s and G_i in detergent-containing solutions. It is not clear, however, which of these mechanisms is more important for the function of G_i in the plasma membrane. We will, therefore, separately discuss the pros and cons of both mechanisms and unfortunately have to leave it to future experimentation to prove or disprove the relevance of one or the other (or both) mechanisms for inhibition of adenylate cyclase in the bilayer.

4.1. Inhibition of Adenylate Cyclase by β/γ Complexes Mimics Hormonal Inhibition of the Enzyme

As indicated above, G_s was found to dissociate (in detergent-containing solutions) upon activation with GTP-γ-S or AlF_4^-. Based on this observation, Northup et al. (1983a) succeeded in separating α subunits and β/γ complexes of G_s and were thus able to separately study their functions. Two important findings emerged from these studies. First, the resolved, activated α subunit was shown to be fully capable of activating the catalytic unit. Second, the β/γ complexes could be shown to inhibit the activation of G_s by GTP-γ-S (Northup et al., 1982) and facilitate the deactivation of either isolated, activated α_s or activated G_s-holoprotein (Northup et al., 1983a,b). The observation that G_i was also apt to dissociate into its subunits upon activation (Bokoch et al., 1983) in context with the demonstration of a very high degree of similarity (if not identity) between the β subunits of G_s and G_i (Manning and Gilman, 1983) led Gilman and coworkers to suggest that the β subunit plays a major role in inhibition of adenylate cyclase (Gilman, 1984a,b). According to their theory, β subunits would be released by activated G_i in the lipid bilayer and subsequently be able to reduce the level of active G_s molecules in the membrane. Several experimental findings are very much in support of such a mechanism. (1) Inhibition of adenylate cyclase activity in human platelet membranes by epinephrine plus GTP-γ-S and addition to the membranes of purified β/γ complexes are nonadditive processes (Katada et al., 1984b). (2) Inhibition of adenylate cyclase by added β/γ complexes closely resembles inhibition of the enzyme by epinephrine and/or guanine nucleotides. For example, inhibition is best observed at low magnesium concentration and if forskolin is used as activating agent. In contrast, effects of β/γ complexes are less impressive or even absent if PGE_1 is used as the stimulating agent or if the enzyme is assayed at higher Mg^{2+} concentrations or in the presence of millimolar concentrations of Mn^{2+} ions (Katada et al., 1984a). (3) Addition of purified, nonactivated α_i to platelet membranes counteracts the inhibition caused by epinephrine plus GTP-γ-S treatment, presumably by acting as a sink for free β/γ complexes (Katada et al., 1984b). (4) GTP-γ-S-activated α_i inhibits the enzyme with lower apparent affinity than do β/γ complexes (Katada et al., 1984a). (5) The amount of β/γ subunits present in the plasma membrane exceeds the amount associated with G_s by at least 5–10-fold (Northup et al., 1983a). Provided these complexes are associated with α_i, the amounts released with activation of G_i are high enough to cause inhibition of G_s function. (6) The retinal G-protein G_t was found to inhibit G_s-stimulated adenylate cyclase in a reconstituted system (Cerione et al., 1985). The β/γ complexes were found to be primarily responsible for this inhibition, when assayed under physiologic conditions, i.e., in the presence of agonist stimulated beta-adrenergic receptors and GTP (Cerione et al., 1986).

In addition to their interaction with α_s, β/γ complexes have recently been suggested to directly inhibit the activity of the catalytic unit itself (Katada et al., 1986; Enomoto and Asakawa, 1986). However, differences in the degree of inhibition have been reported for different preparations of purified C (Smigel, 1986), and an indirect effect of β/γ subunits on the C activity via α_s contaminating certain preparations can not be excluded with certainty.

4.2. Release of β/γ Complexes by Activated G_i Cannot Be the Sole Mechanism Responsible for Inhibition of Adenylate Cyclase

Inhibition of adenylate cyclase via β/γ subunits, although certainly attractive for the reasons mentioned above, has been a matter of considerable dispute over the past several years. There are several lines of evidence, which either argue strongly against a mechanism of adenylate cyclase inhibition exclusively depending on β/γ release or are not compatible with such a mechanism at all. (1) Inhibition of adenylate cyclase by hormones and/or guanine nucleotides can clearly be seen in membranes derived from S-49 cyc⁻ cells (Hildebrandt et al., 1982, 1983, 1984; Jakobs et al., 1983d,e; Jakobs and Schultz, 1983), which appear to not only functionally (Ross et al., 1978; Johnson et al., 1978), but also structurally, be deficient of α_s (Harris et al., 1985). (2) Purified, GTP-γ-S-activated α_i has been shown to be capable to inhibit adenylate cyclase activity in several membrane systems (Katada et al., 1984a,c) and partially purified preparations of C in the presence of activated α_s (Katada et al., 1986). (3) Treatment of membranes with cholera toxin results in markedly decreased inhibitory potency of exogenous β/γ complexes, presumably by decreasing the affinity of α_s for β/γ complexes (Kahn and Gilman, 1984). Yet, as shown in Fig. 4, inhibition of adenylate cyclase by epinephrine in human platelet membranes is completely unaltered by pretreatment of membranes with cholera toxin, which at the same time leads to a several-fold increase in stimulation of the enzyme by GTP (not shown). (4) Recent evidence suggests that a pertussis toxin-sensitive GTP-binding protein may be involved in regulating phosphoinositide turnover in neutrophils, mast cells,

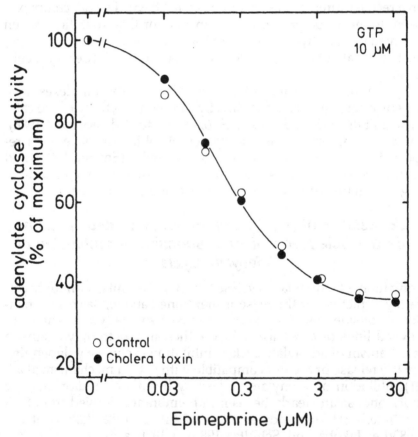

Fig. 4. Inhibition of basal- and cholera toxin-stimulated human platelet adenylate cyclase by epinephrine. Adenylate cyclase activity is given on the ordinate as % of activity measured in the absence of epinephrine, which was 35 and 250 pmol of cyclic AMP/min/mg protein without and with cholera toxin (50 µg/mL) present, respectively.

and several other cell types (*see* Litosch and Fain, 1986, for a recent review). Immunochemical studies suggest that the major pertussis toxin substrate in membranes prepared from human neutrophils may be different from previously identified PT substrates (Gierschik et al., 1986), and several observations indicate that this protein is associated with β subunits similar to those of other G-proteins in the native membrane (Gierschik et al., 1987). However, conditions that activate phosphoinositide turnover in the cell types mentioned (most likely via activation of the GTP-binding protein) do not decrease intracellular levels of cyclic-AMP and fail to result in any alteration of adenylate cyclase activity in plasma membrane preparations (Verghese et al., 1985). Clearly, more direct studies are required to prove the existence of unique PT substrates in these systems and analyze their interaction with components of the adenylate cyclase system, but the evidence presented is very suggestive of a specific role of α_i in inhibition of adenylate cyclase.

In summary, several hypotheses exist as to how activated G_i could inhibit adenylate cyclase. These mechanisms are illustrated in Fig. 5. (1) G_i inhibits adenylate cyclase via release of β/γ subunits, which are able to deactivate α_s. (2) α_i directly inhibits the activity of the catalytic unit. (3) Active α_i competes with active α_s for binding sites on C, but has no (or less) intrinsic activity. (4) β/γ complexes directly inhibit C.

5. Modulation of G_i-Mediated Adenylate Cyclase Inhibition

The following section of this review will focus on modification of adenylate cyclase inhibition by agents other than inhibitory hormones and neurotransmitters. At the outset it needs to be pointed out that it is difficult to resolve actions of individual agents from those of others, since these are interconnected in a complex fashion. For example, inhibitory action of GTP in the absence of inhibitory agonists may only be seen at appropriate Mg^{2+} concentrations and may be markedly diminished with increasing Na^+ concentrations. In addition, modification of the inhibitory pathway by a particular agent may easily be demonstrated in one system, while being absent from another one. For example, the biphasic GTP response is clearly evident in rat fat cell membranes (Murayama and Ui, 1983), whereas membranes prepared from rat C6 glioma cells only show a stimulatory GTP response (Katada et al., 1981). Clearly, no hard and safe rule can

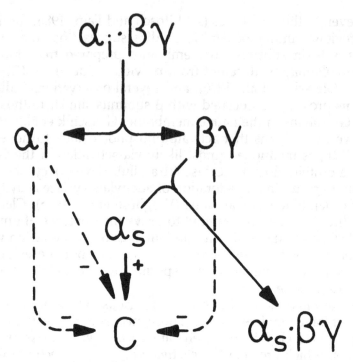

Fig. 5. Model for inhibition of adenylate cyclase by the inhibitory guanine nucleotide binding protein (G_i). Activation of heterotrimeric G_i ($\alpha_i \cdot \beta/\gamma$) is thought to result in the dissociation of α_i and β/γ complex. The β/γ complex of G_i is highly homologous or even identical to the β/γ complex of G_s. Thus, β/γ complexes released by activated G_i could deactivate α_s by forming the inactive complex $\alpha_s \cdot \beta/\gamma$. In addition, the β/γ complex may inhibit the activity of the catalytic unit of adenylate cyclase (C) directly. The activated, GTP-bound form of α_i may inhibit the activity of C either directly or via competition with activated α_s for binding sites on C.

be presented. We will, nevertheless, try to carve out characteristic effects of several agents known to influence adenylate cyclase inhibition by alpha-2 adrenergic agonists.

5.1. Modulation by GTP and Stable GTP Analogs

As mentioned above, hormonal stimulation as well as inhibition of adenylate cyclase requires the presence of GTP. In several cases, however, the enzyme's activity may be regulated by the nucleotide itself, even in the absence of either stimulatory or inhibitory hormones, although there is considerable variation between different systems in terms of the phenomenology of the GTP regulation, as outlined below.

Certain systems, e.g., rat fat cell membranes, show a biphasic GTP response, i.e., GTP stimulates the enzyme's activity at low concentrations (0.1 μM) and becomes inhibitory at higher concentrations (e.g., 30 μM) (Cooper et al., 1979). Other systems, e.g., hamster fat cell membranes (Aktories et al., 1981), do not appear to be regulated bimodally by GTP itself; instead, the nucleotide only decreases the enzyme's activity. It is not clear, however, whether the lack of stimulatory GTP action in these cases is intrinsic to the systems examined or is a consequence of a masking of stimulatory effects caused by the presence of GTP, which may contaminate the membrane preparations. As mentioned above, in some model systems, e.g., C6 glioma cells and rabbit myocardium (Katada et al., 1982; Jakobs et al., 1979), GTP appears to only stimulate the enzyme's activity, despite the fact that there is good evidence for the presence of inhibitory components (i.e., G_i) in these preparations. Finally, some systems, e.g., human platelet membranes, reveal only minimal or even no response to GTP. There is as yet no simple explanation for the different patterns for guanine nucleotide regulation of the adenylate cyclase enzyme in different tissues. Certainly, differences in composition of the various components of the adenylate cyclase system may be responsible for these functional differences, but additional, as yet unknown, factors may contribute as well.

The inhibitory action of GTP, if present in a particular system, may be modified by several agents. Stimulatory hormones generally amplify both the stimulatory and inhibitory action of GTP, whereas inhibitory hormones enhance the inhibitory GTP response. Sodium ions have been shown to diminish the effects of GTP alone, while preserving the inhibition by hormones plus GTP. On the other hand, Mg^{2+} and Mn^{2+} are able to block inhibition by both GTP alone and inhibitory hormones plus GTP (see below).

As mentioned above, nonhydrolyzable analogs of GTP are known to persistently activate adenylate cyclase. There are situations, however, in which these analogs will also inhibit the enzyme. In certain systems, e.g., adipocyte membranes (Rodbell, 1975), GTP-γ-S has been shown to inhibit basal enzyme activity. In most other membrane systems evaluated to date, expression of the inhibitory effects of GTP-γ-S requires prior stimulation of the enzyme by stimulatory hormones, cholera toxin, or forskolin. Surprisingly enough, the concentration of GTP-γ-S necessary for inhibition is substantially lower than concentrations found to be stimulatory on basal activity. In platelet membranes, for example, maximal inhibition and stimulation are observed at about 30 and 1000 nM, respectively (Jakobs and Aktories, 1983). Apparent K_d

values for GTP-γ-S binding for purified G_i and G_s of 12 and 710 nM (Bokoch et al., 1984; Northup et al., 1982) may explain this phenomenon. In contrast, apparent affinities of the purified proteins for GTP (25 and 1300 nM for G_i and G_s, respectively), found by the same authors, cannot account for the ability of GTP to stimulate adenylate cyclase at concentrations by far lower than those necessary for inhibition. Presumably, the behavior of G_s and G_i in detergent-containing solutions differs from that in their physiologic lipid environment. In addition, regulation of the enzyme by GTP may be more complex than regulation by its stable analogs. Clearly, more direct studies will be necessary to fully understand the regulation of adenylate cyclase by physiologic ligands.

Several characteristics of GTP-γ-S-induced adenylate cyclase inhibition can be derived from time course experiments, as shown in Fig. 6. Forskolin-stimulated cyclic AMP formation of human platelet membranes is linear within the assay period and is not altered by epinephrine alone. Addition of GTP-γ-S alone, however, drastically reduces the rate of cyclic-AMP formation. Note that GTP-γ-S added at zero time expresses its full activity only after a lag phase of about 5 min. Simultaneous addition of epinephrine and GTP-γ-S has no effect on steady-state cyclic-AMP formation compared to the inhibition by GTP-γ-S alone, but reduces the lag period by a factor of about three, presumably by facilitating the exchange of GTP-γ-S for GDP bound to G_i. Both GTP-γ-S and GTP-γ-S plus epinephrine inhibit adenylate cyclase to the same degree as epinephrine in the presence of GTP (Jakobs and Aktories, 1983), suggesting that inhibition of adenylate cyclase by GTP-γ-S utilizes mechanisms similar to those responsible for receptor-mediated inhibition, i.e., activation of G_i by a guanine nucleotide triphosphate.

5.2. Modulation by Sodium Ions

Adenylate cyclase inhibition may be profoundly altered by monovalent cations, in particular by sodium ions. Basically, sodium effects can be divided into a generalized inhibition of many functions of this multicomponent enzyme complex observed at concentrations above 200 mM and more specific effects on regulation of enzyme activity by extracellular mediators or guanine nucleotides at lower concentrations. Former effects are presumably caused by chaotropic actions of the highly concentrated cations and will not be considered here, whereas the latter will be discussed below. Unfortunately, it is again not possible to provide general rules for specific sodium effects on adenylate cyclase ac-

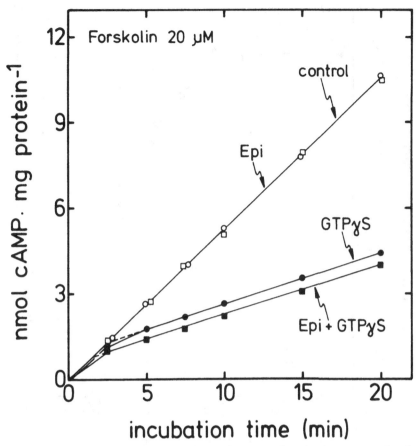

Fig. 6. Inhibition of forskolin-stimulated human platelet adenylate cyclase by GTP-γ-S and epinephrine. The assay conditions were as described in Jakobs and Aktories (1983). When present, GTP-γ-S was 100 n*M* and epinephrine (Epi), 30 μ*M*.

tivity, since certain effects are best observed in, or may even be limited to, certain cell types or membrane preparations. For example, modification by sodium ions of alpha-2 adrenergic receptor agonist affinities are well documented for platelet membranes, whereas effects of sodium ions on coupling of alpha-2 adrenergic receptors to adenylate cyclase inhibition are best defined in adipocyte membranes. Furthermore, these effects of sodium are not limited to alpha-2 adrenergic receptors; they can rather be observed with other receptors inhibitory to adenylate cyclase as well (*see* Jakobs et al., 1981, for review).

As shown in Fig. 7, dose–response curves for epinephrine-induced inhibition of forskolin-stimulated adenylate cyclase obtained in membranes prepared from human platelets are shifted

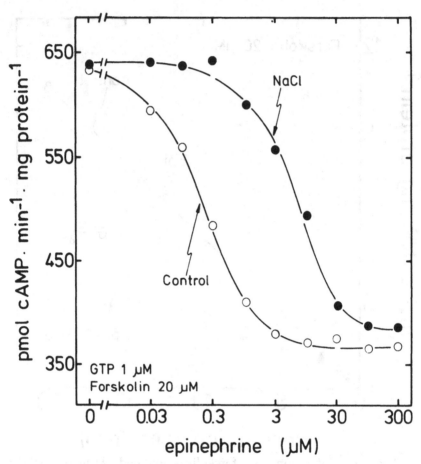

Fig. 7.   Influence of sodium on epinephrine-induced inhibition of human platelet adenylate cyclase. Adenylate cyclase activity was determined as described (Jakobs et al., 1984a) with the additions indicated in the absence (control) and the presence of 100 mM NaCl.

to the right in the presence of 100 mM sodium ions. Therefore, inhibition of adenylate cyclase by suboptimal concentrations of epinephrine may be substantially diminished by increasing the concentration of sodium ions (Fig. 8). Binding studies performed with membranes prepared from rabbit platelets (Michel et al., 1980) provided important insights into regulation by sodium ions of alpha-2 adrenergic receptor agonist affinities, which may in fact be responsible for the decreased potency of epinephrine mentioned above. In this system, sodium ions have been found to decrease the affinity of alpha-2 adrenergic receptors for agonists, but not for antagonists. More specifically, K_d values for high- and low-affinity states of agonist binding increased by about two and

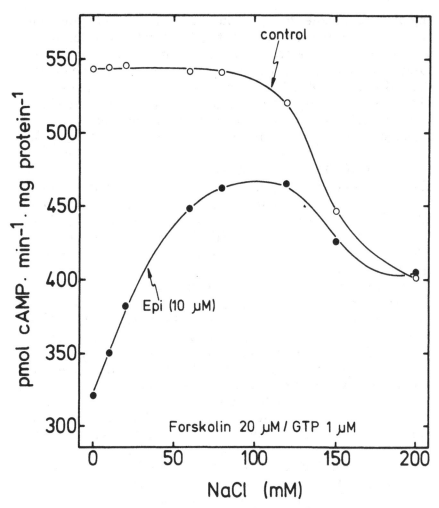

Fig. 8. Interference of sodium with epinephrine-induced inhibition of human platelet adenylate cyclase. The assay conditions were as described in Fig. 7. Epi, epinephrine.

one order of magnitude, respectively. Addition of GMPP(NH)P on top of sodium ions resulted in a complete conversion of high-affinity sites into the low-affinity type. Thus, modification of agonist affinity by monovalent cations clearly differs from guanine nucleotide-induced changes in terms of expression of receptor affinity states for agonists. This raises the distinct possibility that the structural components responsible for these regulatory mechanisms may indeed be different as well.

In addition to regulating the binding properties of alpha-2 adrenergic receptors, sodium ions do modify the inhibitory coupling of these moieties to adenylate cyclase as well. In membranes prepared from hamster adipocyctes, the inhibitory action of epinephrine (in the presence of GTP) was found to be minimal in the absence of sodium ions, whereas the guanine nucleotide by itself showed striking inhibition of adenylate cyclase (Aktories et al., 1979). Increasing sodium concentrations, however, were shown to dramatically enhance the inhibitory hormone effect and simultaneously abolish guanine nucleotide-dependent inhibition of the enzyme, with half-maximal and maximal effects of sodium at concentrations of 30 and 200 mM NaCl, respectively. A similar sodium "requirement" for epinephrine-induced inhibition of adenylate cyclase has been found in other membrane systems, e.g., rat renal cortex (Woodcock et al., 1980), rat liver (Jard et al., 1981), and pancreas (Katada and Ui, 1981), as well as for other hormones inhibitory to adenylate cyclase in numerous systems (see Jakobs et al., 1981, for review).

Distinct effects of sodium ions and guanine nucleotides on agonist affinity had already suggested that distinct protein components or domains may be targets for these agents. This hypothesis was further strengthened by studies of human platelet alpha-2 adrenergic receptors performed by Limbird and coworkers. Several perturbants, including N-ethylmaleimide and high temperature, were found to abolish effects of guanine nucleotides on the receptor's affinity for agonists, whereas effects of sodium ions appeared to be perfectly preserved (Limbird and Speck, 1983). In addition, alpha-2 adrenergic receptors when solubilized in the absence of agonists are smaller in size than those with agonists present during solubilization and appear to have lost guanine nucleotide sensitivity (see above). Regulation of agonist affinity by sodium ions, however, can still be observed in these preparations (Limbird et al., 1982). Sodium and guanine nucleotides, therefore, very likely exert their effects by acting on different protein moieties rather than on distinct domains of the α subunit of G_i.

As shown recently in our laboratory (Jakobs et al., 1984a), modification by sodium ions of hormonal regulation of adenylate cyclase is not limited to inhibition of adenylate cyclase, but may also be seen with stimulation of the enzyme. In membranes prepared from hamster adipocytes, human platelets, and S-49 lymphoma wild-type cells, the potencies of both stimulatory and inhibitory hormones were found to be decreased in the presence of sodium ions. In addition, lag phases required for activation by GTP-γ-S of both G_s and G_i are prolonged by sodium ions and may be shortened by addition of appropriate hormones.

In summary, several properties of both the alpha-2 receptor and G_i appear to be modified by sodium ions at similar concentrations: (1) Sodium ions decrease the potency of alpha-2 agonists to activate G_i, presumably by decreasing the agonist affinity of alpha-2 adrenergic receptors. (2) Sodium ions are required in certain systems for maximal inhibition of adenylate cyclase by alpha-2 adrenergic agonists. (3) The ability of GTP to inhibit the enzyme by itself is reduced. (4) The target for sodium ions is likely to be different from α_i. It remains an intriguing question, however, whether sodium acts on a single protein among the other components necessary for adenylate cyclase inhibition or simultaneously modifies several proteins. Interestingly, the modification by sodium ions of the coupling of alpha-2 adrenergic receptors to adenylate cyclase closely resembles the action of sodium ions on the coupling of other inhibitory (see above) and even stimulatory receptors (Heidenreich et al., 1980; Minuth and Jakobs, 1986) to adenylate cyclase. Therefore, it is tempting to speculate that a component common to the inhibitory and the stimulatory pathway may be the target. Clearly, the β subunit of G_s and G_i is an attractive candidate for this role.

5.3. Divalent Cations Regulate Adenylate Cyclase at Several Sites

Divalent cations, in particular magnesium ions, fulfill many functions within the adenylate cyclase system. On the one hand, they are necessary as complexed ions for ATP to serve as a substrate for the catalytic unit. On the other hand, they may by themselves profoundly alter adenylate cyclase by acting on allosteric sites. For example, magnesium ions were shown to facilitate markedly the activation by guanine nucleotides of G_s and G_i. Interestingly, concentrations of magnesium required for activation of purified G_s by guanine nucleotides (Sternweis et al., 1981; Hanski et al., 1981) are substantially higher than those required for activation of purified G_i (Bokoch et al., 1984, and see below). However, only micromolar concentrations of magnesium appear to be required to support activation of G_s by stimulatory receptors, as seen for the beta-adrenergic receptor and G_s in a reconstituted system (Brandt and Ross, 1986). In addition to the guanine nucleotide regulatory proteins, the catalytic unit itself has been suggested to be activated by magnesium ions (Garbers and Johnson, 1975), although this effect remains to be demonstrated for the purified protein.

Therefore, magnesium ions may modulate adenylate cyclase activity by acting at several sites, which, nevertheless, leads to characteristic effects of magnesium ions on hormonal regulation of the enzyme, as shown in Fig. 9 for human platelet membranes. Three of these effects are clearly evident from this data: (1) Basal activity is clearly enhanced by Mg^{2+} at half-maximal and maximal concentrations of 5 and 20 mM, respectively. This effect could be caused by a direct interaction of Mg^{2+} with the catalytic unit or, alternatively, by an effect of Mg^{2+} on the basal activity of G_s. The observation of a similar effect of magnesium on basal activity of the enzyme in membranes prepared from cyc⁻ cells (which do

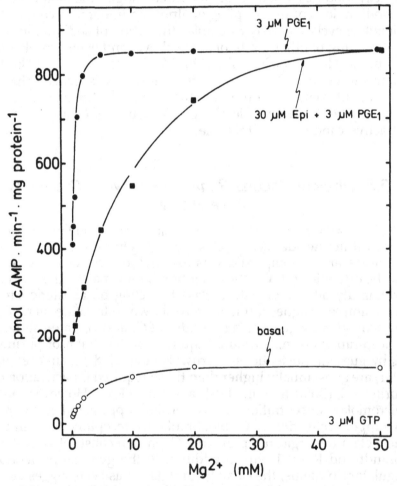

Fig. 9. Influence of stimulatory and inhibitory hormones on Mg^{2+} kinetics of human platelet adenylate cyclase. Adenylate cyclase activity was determined as described (Jakobs et al., 1983b) without (basal) and with 3 μM PGE₁ or 3 μM PGE₁ plus 30 μM epinephrine (Epi).

the enzyme in membranes prepared from cyc$^-$ cells (which do not contain functional G_s) argues against the latter possibility (Jakobs et al., 1983d). (2) In the presence of PGE_1, magnesium ions stimulate the enzyme at far lower concentrations than those required for Mg^{2+}-dependent stimulation of basal activity. This could conceivably reflect the ability of activated stimulatory receptors to activate G_s at micromolar concentrations of magnesium (*see above*) or an increase in affinity for Mg^{2+} of the catalyst when stimulated by G_s. The observation that stimulation of adenylate cyclase by GMPP(NH)P preactivated G_s shifts the magnesium sensitivity of the enzyme to values identical to those seen with stimulation by PGE_1 plus GTP is in support of the latter mechanism (Jakobs et al., 1983b). (3) Inhibition by epinephrine of PGE_1-stimulated activity is markedly diminished by increasing concentrations of magnesium ions, at half maximal and maximal concentrations of 10 and 50 mM, respectively. Unfortunately, it is again not possible to unequivocally correlate these effects of magnesium ions with their interaction with specific components. Certainly, they could reflect the lower tendency of G_s to be activated (*see above*) and possibly higher tendency to be deactivated at low magnesium concentrations; additional effects, however, may contribute as well.

For practical purposes, it is important to note that a low magnesium concentration (1–2 mM Mg^{2+} in excess of chelating agents, e.g., EDTA) should be maintained in experiments analyzing inhibition of adenylate cyclase, because detection of inhibition of cyclase is optimal under these circumstances.

Aside from magnesium ions, which we have focused on in the foregoing discussion because of their possible involvement on in vivo regulation of adenylate cyclase (see Maguire, 1984, for review), manganese ions have been shown to elicit effects qualitatively similar to those of magnesium ions. However, the concentrations of manganese ions required are about 30-fold lower than those of magnesium ions, i.e., hormonal inhibition of adenylate cyclase, is usually abolished in the presence of 1 mM manganese ions (Jakobs et al., 1981; Jakobs and Aktories, 1981; Hoffman et al., 1981).

5.4. Inhibitory Coupling of alpha-2 Adrenergic Receptors Is Highly Susceptible to Alkylation by N-Ethylmaleimide

N-Ethylmaleimide (NEM), a lipid-soluble agent that irreversibly alkylates sulfhydryl groups, has long been known to inhibit epinephrine-induced aggregation of platelets (Robinson et al., 1963) and simultaneously abolish the epinephrine-induced decrease in platelet cyclic-AMP (Mills and Smith, 1972). Studies performed with membranes prepared from NEM-treated human platelets or with membranes of the same type treated directly with NEM showed that the alkylating agent indeed perturbs the inhibitory coupling of alpha-2 adrenergic receptors to adenylate cyclase in this system (Jakobs et al., 1982). A component or domain distal to the actual epinephrine binding site appears to be the NEM target, since the binding properties of alpha-2 adrenergic receptors were not altered by NEM treatment. In contrast, epinephrine stimulation of GTPase was abolished after treatment with NEM. Interestingly, stimulation of adenylate cyclase by PGE_1, NaF, or GTP-γ-S appeared to be unaltered or even enhanced under the same conditions. In a more detailed analysis of the binding properties of alpha-2 adrenergic receptors, Limbird and Speck (1983) found that the ability of guanine nucleotides to regulate agonist affinities (which presumably depends on interaction between the receptor and G_i) was abolished after NEM treatment. Similar or identical effects of NEM treatment on the inhibitory coupling to adenylate cyclase have been reported for other receptors, e.g., for muscarinic cholinergic (Wei and Sulakhe, 1980; Harden et al., 1982; Korn et al., 1983; Smith and Harden, 1984), adenosine (Yeung and Green, 1983), and dopamine receptors (Kilpatrick et al., 1982). Thus, alkylation by NEM inactivates a component commonly involved in adenylate cyclase inhibition, presumably one of the subunits of G_i. This hypothesis is supported by the observation that NEM treatment completely abolishes the ability of G_i to be ADP-ribosylated by PT (Martin et al., 1985). In light of the recent suggestion that a cystein may be the ADP-ribose acceptor on α_i (Hsia et al., 1985), it is tempting to speculate that the same amino acid may be modified by NEM as well. A common target amino acid could explain the very similar functional consequences of both modifications.

5.5. alpha-2 Adrenergic Inhibition of Adenylate Cyclase Can Be Specifically Blocked by Limited Proteolysis

Activation of adenylate cyclase by mild proteolysis is a well-established phenomenon and has been observed with several proteases in many adenylate cyclase systems (for a recent review, see Hanoune et al., 1982). Since different sites on the various components of adenylate cyclase may serve as targets for different proteases, and multiple targets may exist for one specific proteolytic enzyme, no unique molecular mechanism may exist for proteolytic activation of the enzyme. In certain cases, however, e.g., human platelets, components involved in inhibition of adenylate cyclase appear to be specifically inactivated by proteolysis. Hence, proteases can be used to analyze mechanisms of alpha-2 adrenergic regulation of the enzyme. Specifically, three proteases, trypsin, chymotrypsin, and a protease found in bovine sperm, have been found to block alpha-2 adrenergic inhibition of human platelet adenylate cyclase without altering stimulation of the enzyme by PGE_1 under certain conditions (Stiles and Lefkowitz, 1982; Ferry et al., 1982; Jakobs et al., 1983c). In all cases, a component other than the receptor, most likely G_i or one of its subunits, appears to be the target, since antagonist binding parameters were not changed by protease treatment. Interestingly, regulation of agonist affinity by guanine nucleotides is abolished by chymotrypsin, but still preserved after treatment with trypsin and sperm protease. Presumably, different components or domains of G_i are targets for these proteases. Clearly, exact localization of these sites will provide important insights into structure function relationships of adenylate cyclase inhibition by alpha-2 adrenergic agonists.

5.6. Protein Kinase C Interferes with alpha-2 Adrenergic Inhibition of Adenylate Cyclase in Human Platelet Membranes

As mentioned at the outset of this chapter, there are several mechanisms by which cellular functions are regulated by extracellular messengers. Most substances are studied as if they act exclusively through one particular mechanism. Similarly, investigators often study receptor-provoked signal transduction mechanisms as if they function independently from each other. However, recent experience tells us that assuming that each receptor activates a sole signaling mechanism that modulates cell function without influence from other signaling mechanisms has been a myopic perspective. First, several hormones and neurotransmitters are now known to simultaneously activate two or more signaling systems. For example both epinephrine and norepinephrine are known to be capable of regulating adenylate cyclase activity via beta- and alpha-2 adrenergic receptors and stimulate the hydrolysis of phosphoinositides via alpha-1 adrenergic receptors. Second, recent evidence suggests that various components of different signaling systems may in fact interact with each other in a complex fashion and thus integrate the consensus of cellular activation by a variety of extracellular signals. For example, it has been shown in our laboratory that activation of protein kinase C, generally believed to result from increased formation of diacylglycerol from phosphoinositolphosphates (*see* Nishizuka, 1986, for review), greatly interferes with inhibition of adenylate cyclase by GTP or inhibitory hormones plus GTP (Jakobs et al., 1985b). Membranes prepared from human platelets pretreated with the phorbol ester 12-O-tetradecanoylphorbol-13-acetate (TPA) (a synthetic analog of diacylglycerol) showed no inhibition of PGE_1-stimulated adenylate cyclase activity by GTP and inhibition of basal-, PGE_1-, or forskolin-stimulated enzyme activity by epinephrine and GTP was either markedly reduced or abolished. In contrast, stimulation of the enzyme by various agents showed either minimal or no changes after TPA treatment. Similar results were obtained when treatment of platelet membranes with TPA-activated, partially purified protein kinase C was substituted for pretreatment with TPA of intact platelets (Watanabe et al., 1985). Effects of phorbol esters on inhibition of adenylate cyclase are not limited to epinephrine and human platelet membranes, but are also seen with other inhibitory hormones in the same and other systems (Jakobs et al., 1985b; Katada et al., 1985).

In analyzing the possible targets of protein kinase C in platelet membranes and detergent extracts prepared therefrom, Katada et al. (1985) found a 41-kDa protein specifically phosphorylated by the enzyme, raising the distinct possibility that the α subunit of G_i may be the modified entity. This hypothesis was further strengthened by the finding that purified preparations of α_i were indeed substrates for activated, partially purified protein kinase C. Interestingly, phosphorylation of α_i by C kinase was markedly reduced in the presence of β/γ complexes or if GTP-γ-S-pretreated α_i was used as a substrate. These findings suggest that α_i may serve as a substrate only in a specific configuration. If only a GDP-liganded form of α_i were the C kinase substrate, for example, then this would have important implications for the role of activated protein kinase C in regulation of adenylate cyclase inhibition in vivo. However, these are only a few of the questions that still exist and need to be addressed in the future. For example, a clear-cut physical demonstration of the phosphorylation by protein kinase C of α_i in vivo is still missing. It is clear, however, that the link between inhibition of adenylate cyclase and other signaling systems may be closer than initially suspected, and their regulatory relationship has to be taken into account to fully understand alpha-2 adrenergic regulation of cellular functions.

6. Conclusion

Substantial progress has been made over the past several years toward understanding the mechanisms involved in inhibition of adenylate cyclase by alpha-2 adrenergic agonists, a large amount of which is certainly due to the discovery of new tools and development of new techniques. However, as indicated above, many important questions still exist and need to be addressed in the future. For example, it is not known how the various components necessary for adenylate cyclase inhibition behave in the native membrane. We will also have to determine how the adenylate cyclase system is integrated into the complex network of transmembrane signaling. Finally, it will be fascinating to learn more about the dynamics of the system; for example, the regulation of the synthesis of the various components, their posttranslational processing and transfer to the plasma membrane, and the mechanisms involved in their functional or physical removal from the membrane. There is no question that molecular biology techniques, which have become available recently, will now allow researchers to address or even answer many of these questions.

REFERENCES

Aktories, K. and Jakobs, K. H. (1981) Epinephrine inhibits adenylate cyclase and stimulates a GTPase in human platelet membranes via α-adrenoceptors. *FEBS Lett.* **130,** 235–238.

Aktories, K., Schultz, G., and Jakobs, K. H. (1979) Inhibition of hamster fat cell adenylate cyclase by prostaglandin E_1 and epinephrine: Requirement for GTP and sodium ions. *FEBS Lett.* **107,** 100–104.

Aktories, K., Schultz, G., and Jakobs, K. H. (1980) Regulation of adenylate cyclase activity in hamster adipocytes. *Naunyn Schmiedebergs Arch. Pharmacol.* **312,** 167–173.

Aktories, K., Schultz, G., and Jakobs, K. H. (1981) The hamster adipocyte adenylate cyclase system. II. Regulation of enzyme stimulation and inhibition by monovalent cations. *Biochim. Biophys. Acta.* **676,** 59–67.

Aktories, K., Schultz, G., and Jakobs, K. H. (1983) Islet-activating protein impairs α_2-adrenoceptor-mediated inhibitory regulation of human platelet adenylate cyclase. *Naunyn Schmiedebergs Arch. Pharmacol.* **324,** 196–200.

Bokoch, G. M. and Gilman, A. G. (1984) Inhibition of receptor mediated release of arachidonic acid by pertussis toxin. *Cell* **39,** 301–308.

Bokoch, G. M., Katada, T., Northup, J. K., Hewlett, E. L., and Gilman, A. G. (1983) Identification of the predominant substrate for ADP-ribosylation by islet activating protein. *J. Biol. Chem.* **258,** 2072–2075.

Bokoch, G. M., Katada, T., Northup, J. K., Ui, M., and Gilman, A. G. (1984) Purification and properties of the inhibitory guanine nucleotide regulatory component of adenylate cyclase. *J. Biol. Chem.* **259,** 3560–3567.

Brandt, D. R. and Ross, E. (1986) Catecholamine-stimulated GTPase cycle. Multiple sites of regulation by β-adrenergic receptor and Mg^{2+} studied in reconstituted receptor-G_s vesicles. *J. Biol. Chem.* **261,** 1656–1664.

Cerione, R. A., Codina, J., Kilpatrick, B. F., Staniszewski, C., Gierschik, P., Somers, R. L., Spiegel, A. M., Birnbaumer, L., Caron, M. G., and Lefkowitz, R. J. (1985) Transducin and the inhibitory nucleotide regulatory protein inhibit the stimulatory nucleotide regulatory protein mediated stimulation of adenylate cyclase in phospholipid vesicle systems. *Biochemistry* **24,** 4499–4503.

Cerione, R. A., Staniszewski, C., Gierschik, P., Codina, J., Somers, R. L., Birnbaumer, L., Spiegel, A. M., Caron, M. G., and Lefkowitz, R. J. (1986) Mechanism of guanine nucleotide regulatory protein-mediated inhibition of adenylate cyclase. Studies with isolated subunits of transducin in a reconstituted system. *J. Biol. Chem.* **261,** 9414–9520.

Cochaux, P., Van Sande, J., and Dumont, J. E. (1985) Islet-activating protein discriminates between different inhibitors of thyroidal cyclic AMP system. *FEBS Lett.* **179,** 303–306.

Codina, J., Hildebrandt, J., Iyengar, R., Birnbaumer, L., Sekura, R. D., and Manclark, C. R. (1983) Pertussis toxin substrate, the putative N_i component of adenylyl cyclases, in an αβ heterodimer regulated by guanine nucleotide and magnesium. *Proc. Natl. Acad. Sci. USA* **80,** 4276–4280.

Codina, J., Hildebrandt, J. D., Birnbaumer, L., and Sekura, R. D. (1984) Effects of guanine nucleotides and Mg on human erythrocyte N_i and N_s, the regulatory components of adenylyl cyclase. *J. Biol. Chem.* **259,** 11408–11418.

Cooper, D. M. F., Schlegel, W., Lin, M. C., and Rodbell, M. (1979) The fat cell adenylate cyclase system. Characterization and manipulation of its bimodal regulation by GTP. *J. Biol. Chem.* **254,** 8927–8931.

Enomoto, K. and Asakawa, T. (1986) Inhibition of the catalytic unit of adenylate cyclase and activation of GTPase of N_i protein by βγ-subunits of GTP-binding proteins. *FEBS Lett.* **202,** 63–68.

Ferry, N., Adnot, S., Borsodi, A., Lacombe, M. L., Guellaen, G., and Hanoune, J. (1982) Uncoupling by proteolysis of α-adrenergic receptor-mediated inhibition of adenylate cyclase in human platelets. *Biochem. Biophys. Res. Comm.* **108,** 708–714.

Florio, V. A. and Sternweis, P. C. (1985) Reconstitution of muscarinic cholinergic receptors with purified GTP-binding proteins. *J. Biol. Chem.* **260,** 3477–3483.

Garbers, D. L. and Johnson, R. A. (1975) Metal and metal-ATP interactions with brain and cardiac adenylate cyclases. *J. Biol. Chem.* **250,** 8449–8456.

Garcia-Sàinz, J. A., Boyer, J. L., Michel T., Sawyer, D., Stiles, G. L., Dohlman, H., and Lefkowitz, R. J. (1984) Effect of pertussis toxin on $α_2$-receptors: Decreased formation of the high affinity state for agonists. *FEBS Lett.* **172,** 95–98.

Gierschik, P., Codina, J., Simons, C., Birnbaumer, L., and Spiegel, A. (1985) Antisera against a guanine nucleotide binding protein from retina cross react with the β subunit of the adenylyl cyclase-associated guanine nucleotide binding proteins, N_s and N_i. *Proc. Natl. Acad. Sci. USA* **82,** 727–731.

Gierschik, P., Falloon, J., Milligan, G., Pines, M., Gallin, J. I., and Spiegel, A. M. (1986) Immunochemical evidence for a novel pertussis toxin substrate in human neutrophils. *J. Biol. Chem.* **261,** 8058–8062.

Gierschik, P., Sidiropoulos, D., Spiegel, A., and Jakobs, K. H. (1987) Purification and immunochemical characterization of the major pertussis toxin substrate of bovine neutrophil membranes. *Eur. J. Biochem.* **165,** 185–194.

Gilman, A. G. (1984a) Guanine nucleotide-binding regulatory proteins and dual control of adenylate cyclase. *J. Clin. Invest.* **73,** 1–4.

Gilman, A. G. (1984b) G proteins and dual control of adenylate cyclase. *Cell* **36,** 577–579.

Hanoune, J., Stengel, D., and Lacombe, M. L. (1982) Proteolytic activation and solubilization of adenylate and guanylate cyclases. *Mol. Cell. Endocrinol.* **31,** 21–41.

Hanski, E., Sternweis, P. C., Northup, J. K., Dromerick, A. W., and Gilman, A. G. (1981) The regulatory component of adenylate cyclase. Purifications and properties of the turkey erythrocyte protein. *J. Biol. Chem.* **256,** 12911–12919.

Harden, T. K., Scheer, A. G., and Smith, M. M. (1982) Differential modification of the interaction of cardiac muscarinic cholinergic and β-adrenergic receptors with a guanine nucleotide binding component(s). *Mol. Pharmacol.* **21,** 570–580.

Harris, B. A., Robishaw, J. D., Mumby, S. M., and Gilman, A. G. (1985) Molecular cloning of complementary DNA for the α subunit of the G protein that stimulates adenylate cyclase. *Science* **229,** 1274–1277.

Hazeki, O. and Ui, M. (1981) Modification by islet-activating protein of receptor-mediated regulation of cyclic AMP accumulation in isolated rat heart cells. *J. Biol. Chem.* **256,** 2856–2862.

Heidenreich, K. A., Weiland, G. A., and Molinoff, P. B. (1980) Characterization of radiolabeled agonist binding to β-adrenergic receptors in mammalian tissues. *J. Cyclic Nucl. Res.* **6,** 217–230.

Hescheler, J., Rosenthal, W., Trautwein, W., and Schultz, G. (1987a) The GTP-binding protein, G_o, regulates neuronal calcium channels. *Nature* **325,** 445–447.

Hescheler, J., Rosenthal, W., Wulfern, M., Tang, M., Motoyuji, Y., Trautwein, W., and Schultz, G. (1987b) Involvment of the guanine nucleotide-binding protein N_o, in the inhibitory regulation of neuronal calcium channels. *Adv. Cyclic Nucleotide Prot. Phosp. Res.* **21,** in press.

Hildebrandt, J. D., Hanoune, J., and Birnbaumer, L. (1982) Guanine nucleotide inhibition of cyc⁻ S49 mouse lymphoma cell membrane adenylyl cyclase. *J. Biol. Chem.* **257,** 14723–14725.

Hildebrandt, J. D., Sekura, R. D., Codina, J., Iyengar, R., Manclark, C. R., and Birnbaumer, L. (1983) Stimulation and inhibition of adenylyl cyclases mediated by distinct regulatory proteins. *Nature* **302,** 706–709.

Hildebrandt, J. D., Codina, J., and Birnbaumer, L. (1984) Interaction of the stimulatory and inhibitory regulatory proteins of the adenylyl cyclase system with the catalytic component of cyc⁻ S49 cell membranes. *J. Biol. Chem.* **259,** 13178–13185.

Hildebrandt, J. D., Codina, J., Rosenthal, W., Birnbaumer, L., Neer, E.,

Yamazaki, A., and Bitensky, M. W. (1985) Characterisation by two-dimensional peptide mapping of the γ subunits of N_s and N_i, the regulatory proteins of adenylyl cyclase, and of transducin, the guanine nucleotide-binding protein of rod outer segments of the eye. *J. Biol. Chem.* **260**, 14867–14872.

Hoffman, B. B., Mullikin-Kilpatrick, D., and Lefkowitz, R. J. (1980) Heterogeneity of radioligand binding to α-adrenergic receptors. *J. Biol. Chem.* **255**, 4645–4652.

Hoffman, B. B., Yim, S., Tsai, B. S., and Lefkowitz, R. J. (1981) Preferential uncoupling by manganese of α-adrenergic receptor mediated inhibition of adenylate cyclase in human platelets. *Biochem. Biophys. Res. Comm.* **100**, 724–731.

Hsia, J. A., Tsai, S. C., Adamik, R. A., Yost, D. A., Hewlett, E. L., and Moss, J. (1985) Amino acid-specific ADP-ribosylation. Sensitivity to hydroxylamine of (cysteine(ADP-ribose))protein and (arginine(ADP-ribose)protein linkages. *J. Biol. Chem.* **260**, 16187–16191.

Jakobs, K. H. and Aktories, K. (1981) The hamster adipocyte adenylate cyclase system. I. Regulation of enzyme stimulation and inhibition by manganese and magnesium ions. *Biochim. Biophys. Acta* **676**, 51–58.

Jakobs, K. H. and Aktories, K. (1983) Synergistic inhibition of human platelet adenylate cyclase by stable GTP analogues and epinephrine. *Biochim. Biophys. Acta* **732**, 352–358.

Jakobs, K. H. and Schultz, G. (1983) Occurrence of a hormone-sensitive inhibitory coupling component of the adenylate cyclase in S49 lymphoma cyc⁻ variants. *Proc. Natl. Acad. Sci. USA* **80**, 3899–3902.

Jakobs, K. H., Aktories, K., and Schultz, G. (1981) Inhibition of adenylate cyclase by hormones and neurotransmitters. *Adv. Cyclic Nucl. Res.* **14**, 173–187.

Jakobs, K. H., Saur, W., and Schultz, G. (1976) Reduction of adenylate cyclase activity in lysates of human platelets by the α-adrenergic component of epinephrine. *J. Cyclic Nucl. Res.* **2**, 381–392.

Jakobs, K. H., Saur, W., and Schultz, G. (1978) Inhibition of platelet adenylate cyclase by epinephrine requires GTP. *FEBS Lett.* **85**, 167–170.

Jakobs, K. H., Aktories, K., and Schultz, G. (1979) GTP-dependent inhibition of cardiac adenylate cyclase by muscarinic cholinergic agonists. *Naunyn Schmiedebergs Arch. Pharmacol.* **310**, 113–119.

Jakobs, K. H., Lasch, P., Minuth, M., Aktories, K., and Schultz, G. (1982) Uncoupling of α-adrenoceptor-mediated inhibition of human platelet adenylate cyclase by N-ethylmaleimide. *J. Biol. Chem.* **257**, 2829–2833.

Jakobs, K. H., Aktories, K., and Schultz, G. (1983a) Inhibitory coupling of hormone and neurotransmitter receptors to adenylate cyclase. *J. Rec. Res.* **3**, 137–149.

Jakobs, K. H., Schultz, G., Gaugler, B., and Pfeuffer, T. (1983b) Inhibition of N_s-protein-stimulated human platelet adenylate cyclase by epinephrine and stable GTP analogs. *Eur. J. Biochem.* **134**, 351–354.

Jakobs, K. H., Johnson, R. A., and Schultz, G. (1983c) Activation of human platelet adenylate cyclase by a bovine sperm component. *Biochim. Biophys. Acta* **756**, 369–375.

Jakobs, K. H., Gehring, U., Gaugler, B., Pfeuffer, T., and Schultz, G. (1983d) Occurrence of an inhibitory guanine nucleotide-binding regulatory component of the adenylate cyclase system in cyc⁻ variants of S49 lymphoma cells. *Eur. J. Biochem.* **130**, 605–611.

Jakobs, K. H., Aktories, K., and Schultz, G. (1983e) A nucleotide regulatory site for somatostatin inhibition of adenylate cyclase in S49 lymphoma cells. *Nature* **303,** 177–178.

Jakobs, K. H., Minuth, M., and Aktories, K. (1984a) Sodium regulation of hormone-sensitive adenylate cyclase. *J. Recept. Res.* **4,** 443–458.

Jakobs, K. H., Aktories, K., and Schultz, G. (1984b) Mechanisms and components involved in adenylate cyclase inhibition by hormones. *Adv. Cyclic Nucleotide Res.* **17,** 135–143.

Jakobs, K. H., Aktories, K., Minuth, M., and Schultz, G. (1985a) Inhibition of adenylate cyclase. *Adv. Cyclic Nucleotide Res.* **19,** 137–150.

Jakobs, K. H., Bauer, S., and Watanabe, Y. (1985b) Modulation of adenylate cyclase of human platelets by phorbol esters: Impairment of the hormone sensitive inhibitory pathway. *Eur. J. Biochem.* **151,** 425–430.

Jard, S., Cantau, B., and Jakobs, K. H. (1981) Angiotensin II and α-adrenergic agonists inhibit rat liver adenylate cyclase. *J. Biol. Chem.* **256,** 2603–2606.

Johnson, G. L., Kaslow, H. R., and Bourne, H. R. (1978) Genetic evidence that cholera toxin substrates are regulatory components of adenylate cyclase. *J. Biol. Chem.* **253,** 7120–7123.

Kahn, R. and Gilman, A. G. (1984). ADP-ribosylation of G_s promotes the dissociation of its α and β subunits. *J. Biol. Chem.* **259,** 6235–6240.

Katada, T. and Ui, M. (1979) Islet activating protein. Enhanced insulin secretion and cyclic AMP accumulation in pancreatic islets due to activation of native calcium ionophores. *J. Biol. Chem.* **254,** 469–479.

Katada, T. and Ui, M. (1980) Slow interaction of islet activating protein with pancreatic islets during primary culture to cause reversal of α-adrenergic inhibition of insulin secretion. *J. Biol. Chem.* **255,** 9580–9588.

Katada, T. and Ui, M. (1981) Islet-activating protein. A modifier of receptor-mediated regulation of rat islet adenylate cyclase. *J. Biol. Chem.* **256,** 8310–8317.

Katada, T. and Ui, M. (1982a) Direct modification of the membrane adenylate cyclase system by islet-activating protein due to ADP-ribosylation of a membrane protein. *Proc. Natl. Acad. Sci. USA* **79,** 3129–3133.

Katada, T. and Ui, M. (1982b) ADP-ribosylation of the specific membrane protein of C6 cells by islet-activating protein associated with modification of adenylate cyclase activity. *J. Biol. Chem.* **257,** 7210–7215.

Katada, T., Amano, T., and Ui, M. (1982) Modulation by islet activating protein of adenylate cyclase activity in C6 glioma cells. *J. Biol. Chem.* **257,** 3739–3746.

Katada, T., Bokoch, G. M., Northup, J. K., Ui, M., and Gilman, A. G. (1984a) The inhibitory guanine nucleotide-binding regulatory component of adenylate cyclase. Properties and function of the purified protein. *J. Biol. Chem.* **259,** 3568–3577.

Katada, T., Northup, J. K., Bokoch, G. M., Ui, M., and Gilman, A. G. (1984b) The inhibitory guanine nucleotide-binding regulatory component of adenylate cyclase. Subunit dissociation and guanine nucleotide-dependent hormonal inhibition. *J. Biol. Chem.* **259,** 3578–3585.

Katada, T., Bokoch, G. M., Smigel, M. D., Ui, M., and Gilman, A. G. (1984c) The inhibitory guanine nucleotide-binding regulatory component of adenylate cyclase. Subunit dissociation and the inhibition of adenylate cyclase in S49 lymphoma cyc⁻ and wild type membranes. *J. Biol. Chem.* **259**, 3586–3595.

Katada, T., Gilman, A. G., Watanabe, Y., Bauer, S., and Jakobs, K. H. (1985) Protein kinase C phosphorylates the inhibitory guanine-nucleotide-binding regulatory component and apparently suppresses its function in hormonal inhibition of adenylate cyclase. *Eur. J. Biochem.* **151**, 431–437.

Katada, T., Oinuma, M., and Ui, M. (1986) Mechanisms for inhibition of the catalytic activity of adenylate cyclase by the guanine nucleotide binding proteins serving as the substrate of islet activating protein, pertussis toxin. *J. Biol. Chem.* **261**, 5215–5221.

Kilpatrick, B. F., De Lean, A., and Caron, M. (1982) Dopamine receptor of the porcine anterior pituitary gland. Effects of N-ethylmaleimide and heat on ligand binding mimic the effects of guanine nucleotides. *Mol. Pharmacol.* **22**, 298–303.

Korn, S. J., Martin, M. W., and Harden, T. K. (1983) N-ethylmaleimide-in-duced alteration in the interaction of agonists with muscarinic cholinergic receptors in rat brain. *J. Pharmacol. Exp. Ther.* **224**, 118–126.

Kurose, H., Katada, T., Amano, T., and Ui, M. (1983) Specific uncoupling by islet-activating protein, pertussis toxin, of negative signal transduction via α-adrenergic, cholinergic, and opiate receptors in neuroblastoma glioma hybrid cells. *J. Biol. Chem.* **258**, 4870–4875.

Lefkowitz, R. J., Stadel, J. M., and Caron, M. (1983) Adenylate cyclase-coupled β-adrenergic receptors: Structure and mechanisms of activation and desensitation. *Ann. Rev. Biochem.* **52**, 159–186.

Lim, L. K., Sekura, R. D., and Kaslow, H. R. (1985) Adenine nucleotides directly stimulate pertussis toxin. *J. Biol. Chem.* **269**, 2585–2588.

Limbird, L. E. and Speck, J. L. (1983) N-ethylmaleimide, elevated temperature, and digitonin solubilization eliminate guanine nucleotide but not sodium effects on human platelet α₂-adrenergic receptor-agonist interaction. *J. Cyclic Nucleic Res.* **9**, 183–202.

Limbird, L. E., Speck, J. L., and Smith, S. K. (1982) Sodium ion modulates agonist and antagonist interactions with the human platelet α₂-adrenergic receptor in membrane and solubilized preparations. *Mol. Pharmacol.* **21**, 609–617.

Litosch, I. and Fain, J. N. (1986) Regulation of phosphoinositide breakdown by guanine nucleotides. *Life Sci.* **39**, 187–194.

Maguire, M. E. (1984) Hormone-sensitive magnesium transport and magnesium regulation of adenylate cyclase. *TIPS* **2**, 73–77.

Manning, D. R. and Gilman, A. G. (1983) The regulatory components of adenylate cyclase and transducin. A family of structurally homologous guanine nucleotide-binding proteins. *J. Biol. Chem.* **258**, 7059–7063.

Martin, M. W., Evans, T., and Harden, T. K. (1985) Further evidence that muscarinic cholinergic receptors of 1321N1 astrocytoma cells couple to a guanine nucleotide regulatory protein that is not N$_i$. *Biochem. J.* **229**, 539–544.

May, D. C., Ross, E. M., Gilman, A. G., and Smigel, M. D. (1985) Reconstitution of catecholamine-stimulated adenylate cyclase activity using three purified proteins. *J. Biol. Chem.* **260**, 15829–15833.

Michel, T., Hoffman, B. B., and Lefkowitz, R. J. (1980) Differential regulation of the α-adrenergic receptor by Na⁺ and guanine nucleotides. *Nature* **288**, 709–711.

Mills, D. C. B. and Smith, J. B. (1972) The control of platelet responsiveness by agents that influence cyclic AMP metabolism. *Ann. NY Acad. Sci.* **201,** 391–399.

Minuth, M. and Jakobs, K. H. (1986) Sodium regulation of agonist and antagonist binding to β-adrenoceptors in intact and N$_s$-deficient membranes. *Naunyn Schmiedebergs Arch. Pharmacol.* **333,** 124–129.

Moss, J. and Vaughan, M. (1979) Activation of adenylate cyclase by choleragen. *Ann. Rev. Biochem.* **48,** 581–600.

Munson, P. J. (1983) LIGAND: A computerized analysis of ligand binding data. *Meth. Enzymol.* **92,** 543–576.

Murayama, T. and Ui, M. (1983) Loss of the inhibitory function of the guanine nucleotide regulatory component of adenylate cyclase due to its ADP-ribosylation by islet activating protein, pertussis toxin, in adipocyte membranes. *J. Biol. Chem.* **258,** 3319–3226.

Murayama T. and Ui, M. (1984) [^{3}H]GTP release from rat and hamster adipocyte membranes independently linked to receptors involved in activation or inhibition of adenylate cyclase. *J. Biol. Chem.* **259,** 761–769.

Neer, E. J., Lok, J. L., and Wolf, L. G. (1984) Purification and properties of the inhibitory guanine nucleotide regulatory unit of brain adenylate cyclase. *J. Biol. Chem.* **259,** 14222–14229.

Nishizuka, Y. (1986) Studies and perspectives of protein kinase C. *Science* **233,** 305–312.

Northup, J. K., Sternweis, P. C., Smigel, M. D., Schleifer, L. S., Ross, E. M., and Gilman, A. G. (1980) Purification of the regulatory component of adenylate cyclase. *Proc. Natl. Acad. Sci. USA* **77,** 6516–6520.

Northup, J. K., Smigel, M. D., and Gilman, A. G. (1982) The guanine nucleotide activating site of the regulatory component of adenylate cyclase. *J. Biol. Chem.* **257,** 11416–11423.

Northup, J. K., Sternweis, P. C., and Gilman, A. G. (1983a) The subunits of the stimulatory regulatory component of adenylate cyclase. Resolution, activity, and properties of the 35,000-dalton β subunit. *J. Biol. Chem.* **258,** 11361–11368.

Northup, J. K., Smigel, M. D., Sternweis, P. C., and Gilman, A. G. (1983b) The subunits of the stimulatory regulatory component of adenylate cyclase. Resolution of the activated 45,000-dalton α subunit. *J. Biol. Chem.* **258,** 11369–11376.

Nukuda, T., Tanabe, T., Takahashi, H., Noda, M. Haga, K., Haga, T., Ichiyama, A., Kangawa, K., Hiranaga, M., Matsuo, H., and Numa, S. (1986) Primary structure of the α subunit of bovine adenylate cyclase-inhibiting G-protein deduced from the cDNA sequence. *FEBS Lett.* **197,** 305–310.

Pfeuffer, E., Mollner, S., and Pfeuffer, T. (1985) Adenylate cyclase from bovine brain cortex: Purification and characterization of the catalytic unit. *EMBO J.* **4,** 3675–3679.

Pines, M., Gierschik, P., Milligan, G., Klee, W., and Spiegel, A. (1985) Antibodies against the carboxy-terminal 5-kDa peptide of the α subunit of transducin crossreact with the 40 kDa, but not the 39 kDa guanine nucleotide binding protein from brain. *Proc. Natl. Acad. Sci. USA* **82,** 4095–4099.

Ribeiro-Neto, F. A. P., Mattera, R., Hildebrandt, J. D., Codina, J., Field, J. B., Birnbaumer, L., and Sekura, R. D. (1985) ADP-ribosylation of membrane components by pertussis and cholera toxin. *Meth. Enzymol.* **109,** 566–572.

Robinson, Jr., C. W., Mason, R. G., and Wagner, R. H. (1963) Effect of sulfhydryl inhibitors on platelet agglutinability. *Proc. Soc. Exp. Biol. Med.* **113**, 857–861.

Rodbell, M. (1975) On the mechanism of activation of fat cell adenylate cyclase by guanine nucleotides. An explanation for the biphasic inhibitory and stimulatory effects of the nucleotides and the role of hormones. *J. Biol. Chem.* **250**, 5826–5834.

Ross, E. M., Howlett, A. C., Ferguson, K. M., and Gilman, A. G. (1978) Reconstitution of hormone-sensitive adenylate cyclase activity with resolved components of the enzyme. *J. Biol. Chem.* **253**, 6401–6412.

Sabol, S. L. and Nirenberg, M. (1979) Regulation of adenylate cyclase of neuroblastoma × glioma hybrid cells by α-adrenergic receptors. *J. Biol. Chem.* **254**, 1913–1920.

Salzman, E. W. and Neri, L. L. (1969) Cyclic 3′,5′-adenosine monophosphate in human blood platelets. *Nature* **224**, 609–610.

Seamon, K. B. and Daly, J. W. (1981) Forskolin: A unique diterpene activator of cyclic AMP generating systems. *J. Cyclic Nucleotide Res.* **7**, 201–224.

Shorr, R. G. L., Lefkowitz, R. J., and Caron, M. G. (1981) Purification of the β-adrenergic receptor. *J. Biol. Chem.* **256**, 5820–5826.

Smigel, M. D. (1986) Purification of the catalyst adenylate cyclase. *J. Biol. Chem.* **261**, 1976–1982.

Smith, M. M. and Harden, T. K. (1984) Modification of receptor mediated inhibition of adenylate cyclase in NG 108-15 neuroblastoma × glioma cells by N-ethylmaleimide. *J. Pharmacol. Exp. Ther.* **228**, 425–433.

Smith, S. K. and Limbird, L. E. (1981) Solubilization of human platelet α-adrenergic receptors: Evidence that agonist occupancy of the receptor stabilizes receptor-effector interactions. *Proc. Natl. Acad. Sci. USA* **78**, 4026–4030.

Spiegel, A. M., Gierschik, P., Levine, M. A., and Downs, R. W., Jr. (1985) Clinical implications of guanine nucleotide-binding proteins as receptor-effector couplers. *N. Engl. J. Med.* **312**, 26–33.

Sternweis, P. C. (1986) The purified α subunits of G_o and G_i from bovine brain require βγ for association with phospholipid vesicles. *J. Biol. Chem.* **261**, 631–637.

Sternweis, P. C. and Robishaw, J. D. (1984) Isolation of two proteins with high affinity for guanine nucleotides from membranes of bovine brain. *J. Biol. Chem.* **259**, 13806–13813.

Sternweis, P. C., Northup, J. K., Smigel, M. D., and Gilman, A. G. (1981) The regulatory component of adenylate cyclase. Purification and properties. *J. Biol. Chem.* **256**, 11517–11526.

Stiles, G. L. and Lefkowitz, R. J. (1982) Hormone-sensitive adenylate cyclase. Delineation of a trypsin-sensitive site in the pathway of receptor-mediated inhibition. *J. Biol. Chem.* **257**, 6287–6291.

Sugimoto, K., Nukada, T., Tanabe, T., Takahashi, H., Noda, M., Minamino, N., Kangawa, K., Matsuo, H., Hirose, T., Inayama, S., and Numa, S. (1985) Primary structure of the β subunit of bovine transducin deduced from the cDNA sequence. *FEBS Lett.* **191**, 235–240.

Sunyer, T., Codina, J., and Birnbaumer, L. (1984) GTP hydrolysis by pure N_i, the inhibitory regulatory component of adenylyl cyclase. *J. Biol. Chem.* **259**, 15447–15451.

Tsai, B. S. and Lefkowitz, R. J. (1978) Agonist-specific effects of monovalent and divalent cations on adenylate cyclase coupled α-adrenergic receptors in rabbit platelets. *Mol. Pharmacol.* **14**, 540–548.

Ui, M. (1984) Islet-activating protein, pertussis toxin: A probe for functions of the inhibitory guanine nucleotide regulatory component of adenylate cyclase. *TIPS* **5**, 277–279.

Van Dop, C., Yamanaka, G., Steinberg, F., Sekura, R. D., Manclark, C. R., Stryer, L., and Bourne, H. R. (1984) ADP-ribosylation of transducin by pertussis toxin blocks light-stimulated hydrolysis of GTP and cGMP in retinal photoreceptors. *J. Biol. Chem.* **259**, 23–26.

Verghese, M. W., Fox, K., McPhail, L. C., and Snyderman, R. (1985) Chemoattractant-elicited alterations of cAMP levels in human polymorphonuclear leukocytes require a Ca^{2+}-dependent mechanism which is independent of transmembrane activation of adenylate cyclase. *J. Biol. Chem.* **260**, 6769–6775.

Watanabe, Y., Horn, F., Bauer, S., and Jakobs, K. H. (1985) Protein kinase C interferes with N_i-mediated inhibition of human platelet adenylate cyclase. *FEBS Lett.* **192**, 23–27.

Wei, J. W. and Sulakhe, P. V. (1980) Requirement for sulfhydryl groups in the differential effects of magnesium ion and GTP on agonist binding of muscarinic cholinergic receptor sites in rat atrial membrane fraction. *Naunyn Schmiedebergs Arch. Pharmacol.* **314**, 51–59.

Woodcock, E. A., Johnston, C. I., and Olson, C. A. (1980) α-Adrenergic inhibition of renal cortical adenylate cyclase. *J. Cyclic Nucleotide Res.* **6**, 261–269.

Yeung, S. M. H. and Green, R. D. (1983) Agonist and antagonist affinities for inhibitory adenosine receptors are reciprocally affected by 5'-guanylylimidodiphosphate or N-ethylmaleimide. *J. Biol. Chem.* **258**, 2334–2339.

SECTION 4
CORRELATION OF
RECEPTOR BINDING AND
FUNCTION

Chapter 4

Structure–Activity Relationships for alpha-2 Adrenergic Receptor Agonists and Antagonists

Robert R. Ruffolo, Jr., Robert M. DeMarinis, Margaret Wise, and J. Paul Hieble

1. Introduction

This chapter extends the discussion provided in Chapter 1 by providing an overview of the development of criteria for subclassification of alpha-adrenergic receptors into alpha-1 vs alpha-2 adrenergic receptor sybtypes and of more recent data suggesting the existence of subtypes of alpha-2 adrenergic receptors. A compendium of structure–activity data is then provided that ultimately permits a description of the structural requirements for the ligand-binding site of an alpha-2 adrenergic receptor.

2. alpha-Adrenergic Receptor Subclassification

2.1. Methods of Subclassification

2.1.1. ANATOMICAL SUBCLASSIFICATION

It has been known for many years that alpha-adrenergic receptor antagonists increase the overflow of norepinephrine evoked by sympathetic nerve stimulation. Brown and Gillespie (1957) suggested that this effect was caused by blockade of postjunctional alpha-adrenergic receptors, thus preventing the combination of released norepinephrine with alpha-adrenergic receptors on effector cells, which serve as site of loss, leading to an increase in norepinephrine overflow. Thoenen et al. (1964) offered the alternative explanation of inhibition of neuronal uptake of released norepinephrine unrelated to alpha-adrenergic receptor blockade, whereas Langer (1970) proposed that blockade of extraneuronal uptake may be responsible. Subsequently, Starke et al. (1971a,b) and Langer et al. (1971) postulated, independently, that this effect of alpha-adrenergic receptor antagonists to enhance norepinephrine release was a prejunctional phenomenon mediated by alpha-adrenergic receptors. Further studies showed that alpha-adrenergic receptor agonists could inhibit the neurogenic release of norepinephrine via a prejunctional alpha-adrenergic receptor system, and that norepinephrine inhibits its own release by this negative feedback mechanism (Starke, 1972a,b). It was later demonstrated that certain agonists, such as clonidine, and antagonists, such as phenoxybenzamine, could discriminate between pre- and postjunctional alpha-adrenergic receptors (Starke et al., 1974,1975; Dubocovich and Langer, 1974), thus providing evidence that these receptors are distinct entities. As a result of this discrimination between pre- and postjunctional alpha-adrenergic receptors in vitro, Langer (1974) proposed that the postjunctional alpha-adrenergic receptor that mediates the response in an effector organ be termed alpha-1, whereas the prejunctional alpha-adrenergic receptor that regulates neurotransmitter release be termed alpha-2. Support for the subclassification of pre- and postjunctional alpha-adrenergic receptors in vivo came from Drew (1976), who provided evidence that pre- and postjunctional alpha-adrenergic receptors in the pithed rat could be differentiated.

2.1.2. PHARMACOLOGICAL SUBCLASSIFICATION

In general, the anatomical subclassification of presynaptic alpha-2 and postsynaptic alpha-1 adrenergic receptors holds true. However, the inhibition of alpha-melanocyte-stimulating hormone-induced melanin granule dispersion in frog skin (Pettinger, 1977), and the inhibition of isoproterenol-induced glycolysis and lipolysis in hamster isolated epididymal adipocytes (Schimmel, 1976), both of which are postjunctional effects, are mediated by agonists that are highly selective for alpha-2 adrenergic receptors. Likewise, Drew and Whiting (1979) demonstrated that the vasoconstrictor responses to norepinephrine in the rat and the cat were inhibited not only by the selective alpha-1 adrenergic receptor antagonist, prazosin, but also by the selective alpha-2 adrenergic receptor antagonist, yohimbine, thereby showing that alpha-1 and alpha-2 adrenergic receptors can exist postjunctionally in the vascular. At the same time, Timmermans et al. (1979), using a pharmacologic approach, also found that vasoconstriction in the pithed rat could be mediated via postjunctional alpha-2 adrenergic receptors as well as by alpha-1 adrenergic receptors. This pharmacologic demonstration of the postjunctional localization of alpha-2 adrenergic receptors was confirmed in many laboratories (Docherty and McGrath, 1980; Kobinger and Pichler, 1980; Timmermans and van Zwieten, 1980a,b; reviewed by McGrath, 1982; Timmermans and van Zwieten, 1982; Ruffolo, 1984a,b,1985a,b), and casts serious doubt on the validity of the anatomic subclassification of alpha-adrenergic receptors. This led Berthelsen and Pettinger (1977) to propose that classification of alpha-adrenergic receptors be based on relative potencies of selective agonists and antagonists. For example, an alpha-adrenergic receptor that is activated by either methoxamine, cirazoline, or phenylephrine, and is blocked in a competitive manner by low concentrations of prazosin, WB4101, or corynanthine, is classified as an alpha-1 adrenergic receptor. Conversely, responses elicited by UK 14,304, B-HT 920, or B-HT 933, which are competitively blocked by either yohimbine, rauwolscine, or idazoxan in low concentrations, are classified as being mediated via alpha-2 adrenergic receptors. The natural sympathetic neurotransmitter, norepinephrine, is a relatively nonselective alpha-adrenergic receptor agonist, whereas the adrenal hormone, epinephrine, shows a very slight selectivity for alpha-2 adrenergic receptors (Berthelsen and Pettinger, 1977). The receptor subtype selectivity of several alpha-adrenergic receptor agonists and antagonists, commonly used in characterizing alpha-adrenergic receptor effects, is shown in Fig. 1. These drugs serve as important chemical tools to subclassify, on a pharmacological basis, alpha-adrenergic receptors.

2.2. alpha-2 Adrenergic Receptor Heterogeneity

There is now increasing evidence to suggest that alpha-2 adrenergic receptors may not represent one homogeneous population of receptors (Bylund, 1985; 1987). In a comprehensive review of the literature in which pharmacologically determined affinities of yohimbine are surveyed, Drew (1985) concludes that there were

118
Ruffolo et al.

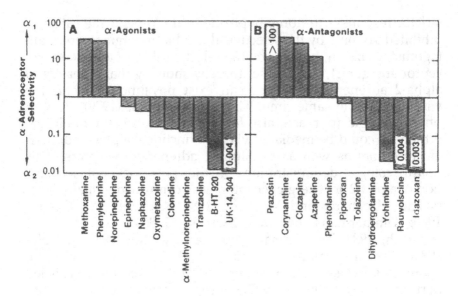

Fig. 1. Relative selectivities for alpha-1 and alpha-2 adrenergic receptors of a series of alpha-adrenergic receptor agonists (A) and antagonists (B) commonly used in the pharmacologic subclassification of alpha-adrenergic receptors.

at least two different populations of alpha-2 adrenergic receptors that were not related to their synaptic location. The major division appeared to be based on species variation, such as rodent versus nonrodent, although there were examples of intertissue variation within the same species. Consistent with this interspecies heterogeneity of alpha-2 adrenergic receptors, Waterfall et al. (1985) have shown that although the affinities of rauwolscine, yohimbine, and a series of substituted benzoquinoxolizines are roughly equivalent at presynaptic alpha-2 adrenergic receptors in the rat vas deferens, the affinities of yohimbine and rauwolscine are approximately 100-fold greater than those of the benzoquinoxolizines in the rabbit vas deferens.

Radioligand binding studies also suggest the existence of species differences in alpha-2 adrenergic receptors. Summers et al. (1983) have demonstrated that alpha-2 adrenergic receptors in human cerebral cortex are pharmacologically distinct from those present on human platelets (Cheung et al., 1982), based on the significantly higher affinities for rauwolscine and yohimbine and lower affinity for prazosin observed in the cerebral cortex. Both human alpha-2 adrenergic receptors on platelet and in cerebral cortex, however, appear to be different from those alpha-2 adrenergic receptors in rat cerebral cortex, where yohimbine and rauwolscine have significantly lower affinities and prazosin has a relatively higher affinity. In addition to marked differences in the affinities of antagonists, species variation in guanine nucleotide regulation of agonist binding are also apparent. p-Aminoclonidine binding to porcine submandibular gland and lung is reduced in a dose-dependent manner by GTP as a result of a decrease in agonist affinity, whereas in the rat submandibular gland, GTP increases agonist binding by elevating the number of specific binding sites (Feller and Bylund, 1984).

The evidence now strongly supports the concept of species-dependent heterogeneity in alpha-2 adrenergic receptors, with the likelihood of tissue-dependent heterogeneity as well. As outlined earlier in this volume (see chapter by Bylund), a subdivision into alpha-2A and alpha-2B subtypes has been proposed. The human cerebral cortex has only alpha-2A receptors, whereas the caudate nucleus contains approximately equal amounts of both subtypes (Petrash and Bylund, 1986). Based on pharmacologic and biochemical parameters, Nahorski et al. (1985) have tentatively proposed the existence of two distinct alpha-2 adrenergic receptor binding sites, one of which resembles in many respects the alpha-1 adrenergic receptor. They have speculated further on the existence of a single primitive alpha-adrenergic receptor that, during evolution, has resulted in a slight difference to accommodate, in a more effective manner, the effector system it subserves in different tissues and species. In this respect, it is interesting to note that Hieble and Woodward (1984) reported that the canine splenic vasculature appears to possess a single postjunctional alpha-adrenergic receptor with high affinity for both alpha-1 and alpha-2 adrenergic receptor antagonists, a situation also observed in feline mesenteric arteries (Skarby et al., 1983).

2.3. Postsynaptic vs Presynaptic alpha-2 Adrenergic Receptors

A good correlation exists between the affinities of a series of antagonists at presynaptic alpha-2 adrenergic receptors in guinea pig atrium and postsynaptic alpha-2 adrenergic receptors in canine saphenous vein, as shown in Fig. 2. This correlation holds true for members of several chemical classes, including yohimbine alkaloids (yohimbine and rauwolscine), imidazolines (phentolamine), tetralones (BE-2254), 3-benzazepines (SK&F 86466), benzodioxans (piperoxan and idazoxan), and tetrahydroisoquinolines (SK&F 72223). Similarly, in vivo studies have provided no evidence to suggest significant differential antagonism of either pre- or postjunctional alpha-2 adrenergic receptors by any of the currently available antagonists (Paciorek et al., 1984; Docherty and Hyland, 1985). These results are consistent with the notion that there exists no pre/postsynaptic alpha-2 adrenergic receptor heterogeneity. Using a pharmacologic approach, however, several groups have proposed that pre- and postjunctional alpha-2 adrenergic receptors can be differentiated. de Jonge et al. (1981a) found that several 2,5-disubstituted imidazolines were selective agonists at prejunctional alpha-2 adrenergic receptors relative to postjunctional alpha-2 adrenergic receptors. The interpretation of these data is open to question, however, since studies with agonists are complicated by potential differences in agonist intrinsic efficacy, tissue receptor density, and therefore differences in receptor reserve for the particular agonists studied (i.e., differences in pre- and postjunctional alpha-2 adrenergic receptor density and/or reserve), and this will alter the sensitivity of the agonists and possibly result in inappropriate receptor selectivities. Alabaster and Peters (1984) proposed that the benzoquinolizine derivative, Wy 26703, was a selective antagonist of prejunctional alpha-2 adrenergic receptors in the rat atrium and rat vas deferens compared to postjunctional alpha-2 adrenergic receptors in the rabbit saphenous vein. These data may instead be explained by species differences in alpha-2 adrenergic receptors, however, as discussed above. More recent studies with a series of alpha-2 adrenergic receptor antagonists of the 3-benzazepine class have provided more convincing evidence for the pharmacologic subclassification of pre- and postjunctional alpha-2 adrenergic receptors. The alpha-2 adrenergic receptor antagonist, SK&F 104078 (6-chloro-9- [(3-methyl-2-butenyl)oxy]-3-methyl-1H-2,3,4,5-tetrahydroh -benzazepine), has been shown to have at least 100-fold higher affinity for postjunctional alpha-2 adrenergic receptors in canine saphenous vein relative to prejunctional alpha-2 adrenergic receptors in guinea pig atrium (Fig. 2) (Ruffolo et al., 1987; Hieble et al., 1986b). The fact that SK&F 104078 can clearly discriminate between pre- and postjunctional alpha-2 adrenergic receptors, both in vitro and in vivo, establishes that these receptors do not represent one homogeneous class. The postjunctional selectivity of SK&F 104078 does not result from species differences, since postjunctional alpha-2 adrenergic receptor blockade with SK&F 104078 has been shown in both the canine and rabbit saphenous veins, the canine saphenous artery, and in the human platelet, and the lack of prejunctional alpha-2 adrenergic receptor blockade has been shown in atria from dog, rat, guinea pig, and

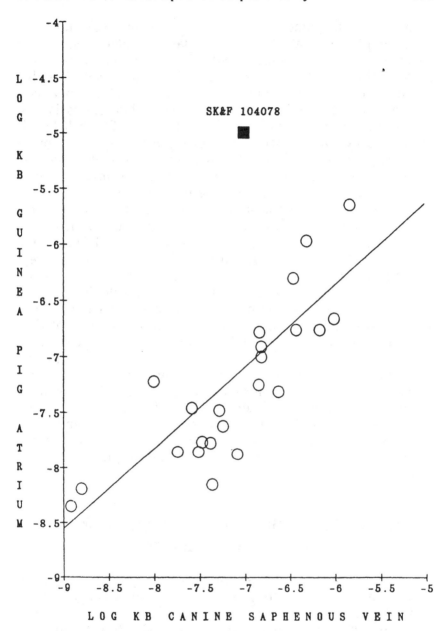

Fig. 2. Correlation between the affinities of a series of alpha-2 adrenergic receptor antagonists for prejunctional alpha-2 adrenergic receptors in field-stimulated guinea pig atrium and postjunctional alpha-2 adrenergic receptors in canine saphenous vein. SK&F 10,4078 is an alpha-2 adrenergic receptor antagonist with selectively high affinity for postjunctional alpha-2 adrenergic receptors.

rabbit and in the guinea pig ileum (Ruffolo et al., 1987). It is critical to emphasize that SK&F 104078 has equivalent dissociation
constants for postjunctional vascular alpha-2 adrenergic receptors
in the canine saphenous artery and vein, strongly suggesting that
the ability of SK&F 104078 to block selectively postjunctional vascular alpha-2 adrenergic receptors is not limited to venous tissues.

In vivo studies in the pithed rat show that SK&F 104078 is
more effective in blocking the postsynaptic vascular alpha-2
adrenoceptor-mediated pressor effects of B-HT 933 than the presynaptic effects of B-HT 933 on neurogenic tachycardia (Ruffolo et
al., 1987). These results confirm the postjunctional selectivity of
SK&F 104078 seen in the in vitro models. The postjunctional
alpha-2 adrenergic receptor selectivity seen in the pithed rat is
remarkable, since several groups (Docherty and Hyland, 1985;
Berridge et al., 1984) have shown many alpha-2 adrenergic receptor antagonists, including phentolamine, rauwolscine, yohimbine, and idazoxan, to produce a significantly greater inhibition of
the prejunctional neuroinhibitory response to alpha-2 adrenergic
receptor stimulation compared to pressor responses mediated by
postjunctional vascular alpha-2 adrenergic receptors. Yet, under
these conditions using the pithed rat, which has an obvious prejunctional bias, SK&F 104078 is still selective in blocking postjunctional vascular alpha-2 adrenergic receptors relative to prejunctional alpha-2 adrenergic receptors. Thus, it appears that there is
sufficient experimental evidence to propose that pre- and postjunctional alpha-2 adrenergic receptors in the periphery are heterogeneous and may require further pharmacologic subclassification.

3. Structure–Activity Relationships of alpha-2 Adrenergic Receptors

Since discovery of the alpha-2 adrenergic receptor, an intense effort has been mounted to develop specific agonists and antagonists as potential therapeutic agents (Timmermans and van
Zwieten, 1982; Langer et al., 1985), and a significant amount of
structure–activity data is beginning to emerge in pursuit of these
goals. Compounds that are either agonists or antagonists have
been discovered that represent a broad spectrum of chemical
classes. Within these, a great range of selectivities has been demonstrated by physiologic and radioligand binding experiments.
Representative examples of the most important structural types of
alpha-2 adrenergic receptor-selective agonists and antagonists are

shown in Figs. 3 and 4, respectively. Selectivity for a particular subtype of receptor is a function of many variables, especially for agonists. Yet agonists, such as B-HT 920, have been shown to possess alpha-2/alpha-1 adrenergic receptor selectivities of at least 100-fold (Kobinger and Pichler, 1981; van Meel et al., 1981; Starke et al., 1975), while alpha-2 adrenergic receptor antagonists, such as rauwolscine, are equally selective (Timmermans and van Zwieten, 1982). As such, these agents have become extraordinarily valuable tools for dissecting and understanding the extremely complex and interrelated structure–activity relationships that exist for the alpha-2 adrenergic receptors.

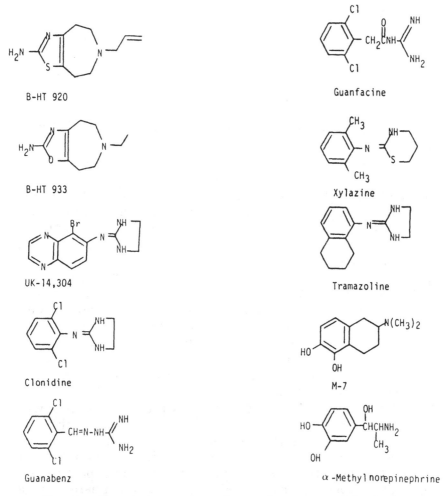

Fig. 3. Chemical structures of several important alpha-2 adrenergic receptor agonists.

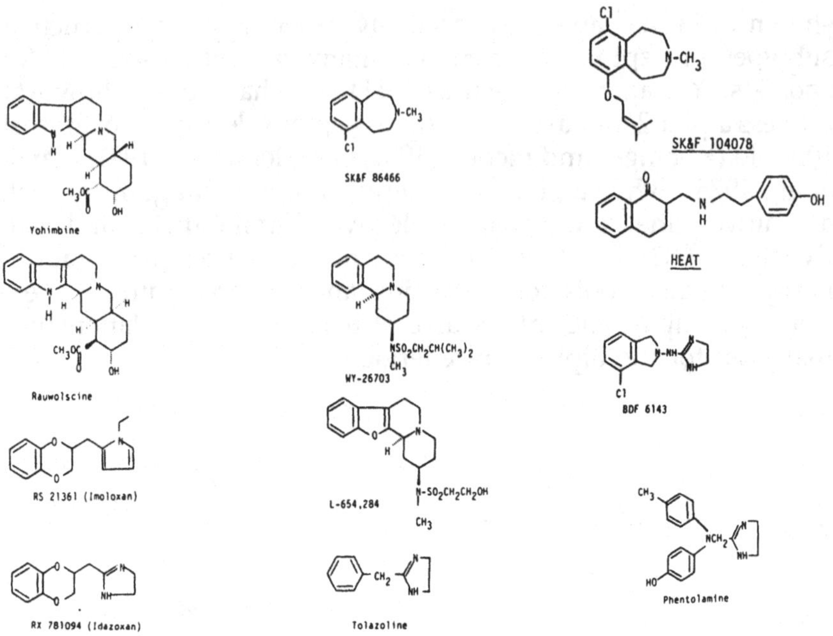

Fig. 4. Chemical structures of several important alpha-2 adrenergic receptor antagonists.

Structure–activity relationships have not been as thoroughly investigated for the alpha-2 adrenergic receptor as for the alpha-1 adrenergic receptor because of several factors. The paucity of data relating to the structural requirements of alpha-2 adrenergic receptors is undoubtedly attributed to the fact that the alpha-2 adrenergic receptor was initially believed to be present only prejunctionally on nerve terminals of postganglionic sympathetic neurons in the periphery (Starke et al., 1974,1975,1977; Starke, 1977) and in prejunctional synapses in the central nervous system (Anden et al., 1976; Drew, 1976; Starke and Montel, 1973; Doxey et al., 1977). The presynaptic location of the alpha-2 adrenergic receptor adds an additional complexity to the determination of efficacy and affinity measurements for agents that interact with this receptor. Determination of these parameters at a presynaptic receptor is an inherently more complex process compared with their determination in postsynaptic test systems in which an effector organ response is measured, such as in vascular smooth muscle. Hence, many of the structure–activity relationships obtained for agonists on presynaptic alpha-2 adrenergic receptors are not complete (Kenakin, 1984). In addition, pharmacologic analysis of agonist affinity constants is complicated by the presence of endogenous norepinephrine-mediating auto-inhibition of neurotransmitter release, which may compete with exogenous agonists for the same presynaptic alpha-2 adrenergic receptor population. Thus, two agonists of different affinity and efficacy competing for the same alpha-2 adrenergic receptor may complicate the results; in physiologic studies and clear determinations of affinity constants for agonists at the presynaptic alpha-2 adrenergic receptor can only be done when autoinhibition by endogenous norepinephrine does not occur. Hence few early studies have dealt effectively with structure–activity relationships for presynaptic alpha-2 adrenergic receptors.

When we evaluate structure–activity relationships for alpha-2 adrenergic receptor agonists, it is important that one keep in mind the two separate parameters of drug action, *affinity* and *efficacy*. Affinity is a measure of how tightly a ligand binds to its receptor. This parameter does not in any way relate to whether this binding process produces a response as a result of the receptor–ligand interaction. Thus, one cannot discriminate between agonists and antagonists based upon binding constants alone. Efficacy, on the other hand, relates directly to the ability of an agonist to activate the receptor and elicit a response. It is a measure of how efficiently the agonist stimulates or activates that receptor subsequent to, and independent of, binding to the receptor. For any given series of agonists, the structure–activity relationships for efficacy may not be, and in general are not, the same as for affinity, and no relationship between affinity and efficacy can be proposed (Ruffolo et al., 1979a,b; Ruffolo, 1982). As such, it is necessary to consider the structure–activity relationships as applied to either affinity or efficacy as separate entities (Ruffolo, 1982,1983a).

Unfortunately, as stated earlier, there are few data available on the efficacy of agonists acting at presynaptic alpha-2 adrenergic receptors. Much of our discussion of structure–activity relationships is gleaned from physiologic measurements of EC_{50} data or affinity measurements obtained through radioligand binding experiments. Although inconsistencies do arise between data that have been generated using radioligand binding techniques in comparison to data obtained in functional studies of isolated tissues (see below), the results are, in general, in substantial agreement (Wood et al., 1979; U'Prichard and Snyder, 1979; Timmermans et al., 1984) and do provide a rational basis for the development of structure–activity relationships. Alternatively, a large amount of data relating to quantitative structure–activity relationships (QSAR) have been obtained for alpha-2 adrenergic re-

ceptor agonists of the imidazoline class in vivo using pressor/
depressor responses as indices of activity, and these data also
provide a rich source of structural correlations (Timmermans et
al., 1980a).

It is the intent of this section to summarize the key features of
the structure–activity relationships of alpha-2 adrenergic receptor
agonists and antagonists. In addition, for the antagonists, an area
in which there is much current work underway, we have reviewed the current state of the art relating to antagonists and
have presented a model of alpha-2 adrenergic receptor blockade
based upon a computer-aided molecular modeling-derived hypothesis from a structurally diverse selection of compounds that
are alpha-2 adrenergic receptor antagonists.

Finally, compounds that stimulate the alpha-2 adrenergic receptor fall into a broad range of structural classes (Fig. 3). Two of
the most widely studied classes are the phenethylamines, as represented by alpha-methyl-norepinephrine, and the imidazolines,
such as clonidine. Much synthetic work has focused on these two
structural classes in relation to their clinical utility as centrally acting antihypertensive agents, and as such, we have concentrated
on the structure–activity relationships of agonists of the phenethylamine and imidazoline classes and have attempted to emphasize those differences in structure–activity that arise between
these two classes of compounds. There are, however, other important chemical classes of alpha-2 adrenergic receptor agonists,
represented by compounds such as M-7 (Drew, 1980; Hicks and
Cannon, 1980), B-HT 920 and B-HT 933 (Kobinger and Pichler,
1980; Pichler et al., 1980; Timmermans and van Zwieten 1980b;
Kobinger et al., 1980), and UK 14,304 (van Meel et al., 1981) (Fig.
3). Much less is known about the structure–activity relationships
of these chemical classes of agonists. Because of their interesting
pharmacologic profiles, however, especially in regard to their
selectivities for alpha-2 vis-à-vis alpha-1 adrenergic receptors, they
now represent some of the most important pharmacologic tools
used for delineating the localization, structure, and function of
alpha-2 adrenergic receptors.

The vast majority of alpha-adrenergic receptor agonists can
be grouped into the two main classes: (1) the phenethylamines,
which are exemplified by compounds such as norepinephrine,
phenylephrine, SK&F(−)-89748, amidephrine, and methoxamine, and (2) the imidazolines, such as clondine, cirazoline, St
587, naphazoline, tetrahydrozoline, and oxymetazoline. As a general rule, compounds of the phenethylamine type are often selective for the alpha-1 adrenergic receptor. In contrast, many

imidazolines demonstrate selectivity for the alpha-2 adrenergic receptor. As can be expected for such generalizations, there are many notable exceptions to these rules of preference. For example, the selectivity of alpha-methylnorepinephrine and 6-fluoro-norepinephrine, both being phenethylamines, is for the alpha-2 adrenergic receptor (Ruffolo et al., 1982a; Shepperson et al., 1981; Kirk and Creveling, 1984), and the selectivity of cirazoline and St 587, both of which are imidazolines, is for the alpha-1 adrenergic receptor (van Meel et al., 1981; de Jonge et al., 1981b).

The structure–activity relationships for these two classes of agonists are similar in many respects, but there are also some critically significant differences between them. These differences may be the result of differences in the way in which the imidazolines and phenethylamines bind to the same site of the alpha-2 adrenergic receptor (Ruffolo, 1983a). Alternatively, it is possible that the differences reflect binding to related, but not identical, parts of the receptor active site or perhaps even to different "subtypes" of the alpha-2 adrenergic receptor (Ruffolo, 1983a). Although it is not possible at present to distinguish unequivocally between these possibilities, there is general agreement that the interaction of certain phenethylamines with the alpha-2 adrenergic receptor is distinctly different from the interactions of the imidazolines with the same receptor (Ruffolo, 1984c Ruffolo et al., 1979a, 1980; Hieble et al., 1986b; Miller et al., 1980), and as such, the structure–activity of these two classes of compounds interacting with the alpha-2 adrenergic receptor will be presented separately.

3.1. Stereochemical Requirements of alpha-2 Adrenergic Receptors

3.1.1. CONFIGURATIONAL DEMANDS MADE BY ALPHA-2 ADRENERGIC RECEPTORS FOR PHENETHYLAMINES

Norepinephrine possesses one point of asymmetry at the beta-carbon atom (i.e., position 1), and the pharmacologic activity at alpha-2 adrenergic receptors resides predominately in the R(−)-enantiomer (Ruffolo, 1983b; Ruffolo et al., 1982a). The S(+)-enantiomer of norepinephrine is also active at alpha-2 adrenergic receptors, but is nearly 500 times less potent than the R(−)-enantiomer. Studies of several phenethylamines with asymmetric beta-carbon atoms indicate that the R(−)-enantiomer is always a more potent agonist at alpha-2 adrenergic receptors than the S(+)-enantiomer.

Although the stereochemical demands made by alpha-1 and alpha-2 adrenergic receptors for phenethylamines are qualitatively similar, significant quantitative differences between these two alpha-adrenergic receptor subtypes exist (Ruffolo et al., 1982a). Based on the observation that larger isomeric activity ratios for the enantiomers of norepinephrine are obtained at alpha-2 adrenergic receptors relative to alpha-1 adrenergic receptors, it has been proposed that the stereochemical demands made by alpha-2 adrenergic receptors for optically active phenethylamines are, in general, more stringent than those made by alpha-1 adrenergic receptors (Ruffolo et al., 1982a).

The enantiomers of alpha-methyldopamine have been used to evaluate the stereochemical demands made by alpha-2 adrenergic receptors for phenethylamines with asymmetry at the alpha-carbon atom (i.e., 2 position). Although both enantiomers of alpha-methyldopamine are weak agonists at alpha-1 adrenergic receptors, showing little or no difference in activity between the enantiomers, a marked selectivity of the 2S(+)-enantiomer has been demonstrated for the alpha-2 adrenergic receptor (Ruffolo and Waddell, 1982). In addition to the high degree of stereoselectivity shown by the alpha-2 adrenergic receptor (but not by alpha-1) for the enantiomers of alpha-methyldopamine, significant differences have also been demonstrated between these enantiomers in their alpha-1/alpha-2 adrenergic receptor selectivities; 2R(−)-alpha-methyldopamine showed only a two-fold selectivity for alpha-2 adrenergic receptors, whereas 2S(+)-alpha-methyl-dopamine displayed a 23-fold selectivity for alpha-2 adrenergic receptors (Ruffolo and Waddell, 1982). Asymmetry at the alpha-carbon atom of phenethylamines, therefore, appears to be a major factor in determining the alpha-adrenergic receptor subtype selectivity for the phenethylamines, and this becomes especially important when one considers that this position also, regardless of chirality, represents the single most important position for substitution in phenethylamines for discriminating between alpha-1 and alpha-2 adrenergic receptors. It has been shown that the alpha-2 adrenergic receptor, in contrast to the alpha-1 adrenergic receptor, has the unique ability to interact with an alpha-methyl group substituted in the phenethylamine series when this group is present and in the optimal 2S stereochemical configuration. It has been proposed that an additional recognition site may exist on the alpha-2 adrenergic receptor that can interact with and/or accommodate the alpha-methyl group of phenethylamines so substituted (Ruffolo, 1983b).

The stereochemical requirements made by alpha-2 adrenergic receptors for phenethylamines with asymmetry at both the alpha and beta carbon atoms are more complex, since four stereoisomers (i.e., two enantiomeric pairs) exist. The four possible isomers of alpha-methylnorepinephrine are shown in Fig. 5. Both alpha-1 and alpha-2 adrenergic receptors display a high preference for the 1R,2S(−)-*erythro* isomer of alpha-methylnorepinephrine (Patil and Jacobowitz, 1968; Ruffolo et al., 1982a). By comparing the stereochemical demands made by alpha-1 and alpha-2 adrenergic receptors for phenethylamines with asymmetry at both the alpha and beta carbon atoms, one may observe an important difference between these two alpha-adrenergic receptor subtypes. The isomeric activity ratio for the enantiomeric, 1R,2S(−)-*erythro* and 1S,2R(+)-*erythro* stereoisomers of alpha-methylnorepinephrine is approximately 60-fold at alpha-1 adrenergic receptors, and 550-fold at alpha-2 adrenergic receptors (Ruffolo et al., 1982a). These results, which are consistent with those presented above for the enantiomers of norepinephrine and alpha-methyldopamine, indicate that the stereochemical demands made by alpha-2 adrenergic receptors for phenethylamines are significantly more stringent than those made by alpha-1 adrenergic receptors (Ruffolo, 1984c,d).

Based on studies with alpha-methyldopamine, it was hypothesized that the alpha-2 adrenergic receptor could recognize, bind, and/or accommodate alpha-methyl substituents of phenethylamines when oriented in the optimal 2S stereochemical configu-

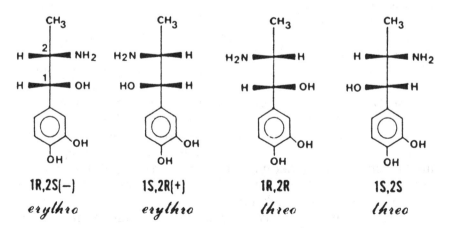

1R,2S(−)   1S,2R(+)   1R,2R   1S,2S
erythro   *erythro*   *threo*   *threo*

Fig. 5. Fischer projections of the four possible stereoisomers of alpha-methylnorepinephrine.

ration and that the alpha-1 adrenergic receptor did not have this ability. This hypothesis may be extended to include the interactions of alpha-methylnorepinephrine with alpha-1 and alpha-2 adrenergic receptors, where it has been proposed that the 1R,2S-(−)-*erythro*-isomer binds to the alpha-2 adrenergic receptor by a four-point interaction (i.e., catechol, beta-hydroxyl, amino, and alpha-methyl groups) as opposed to the alpha-1 adrenergic receptor, where only three functional groups (i.e., catechol, beta-hydroxyl, and amino) may be involved in the binding of this same compound (Ruffolo, 1983a). This hypothesis is consistent with the observation that 1R,2S(−)-*erythro*-alpha-methylnorepinephrine is highly selective for alpha-2 adrenergic receptors, and will also account for the fact that R(−)-norepinephrine is equipotent with, or slightly more potent than, 1R,2S(−)-*erythro*-alpha-methylnorepinephr ine at alpha-1 adrenergic receptors (Ruffolo et al., 1982a), whereas 1R,2S(−)-*erythro*-alpha-methylnorepinephrine is significantly more potent than R(−)-norepinephrine at alpha-2 adrenergic receptors (Ruffolo et al., 1982a; Ruffolo and Waddell, 1982).

3.1.2. Configurational Demands Made by alpha-2 Adrenergic Receptors for Imidazolines

Optically active centers in imidazolines are rare. Several examples, however, of optically active imidazolines are known and provide insight into how this class of agonists interacts with alpha-2 adrenergic receptors. An optically active catecholimidazoline (Fig. 6) has been synthesized and resolved into the individual enantiomers, and their effects were evaluated at the alpha-2 adrenergic receptor. The activity at alpha-2 adrenergic receptors is greatest for the R(−)-enantiomer (Ruffolo et al., 1983a,b), and in this regard, these imidazolines bear some resemblance to the phenethylamines. The isomeric activity ratio for the catecholimidazoline enantiomers is only approximately six-fold at alpha-2 adrenergic receptors, however, in contrast to isomeric activity ratios of between 100- and 1000-fold observed for the phenethylamines. Thus, quantitative differences exist between the stereochemical demands made by the alpha-2 adrenergic receptor for imidazolines and phenethylamines, such that the alpha-2 adrenergic receptor displays greater discriminatory properties for the optically active phenethylamines.

The optically active catecholamidine derivatives in Fig. 6, which are structurally and chemically related to the catecholimidazolines described above, were also synthesized and resolved into the R(−)- and S(+)-enantiomers in order to investigate the

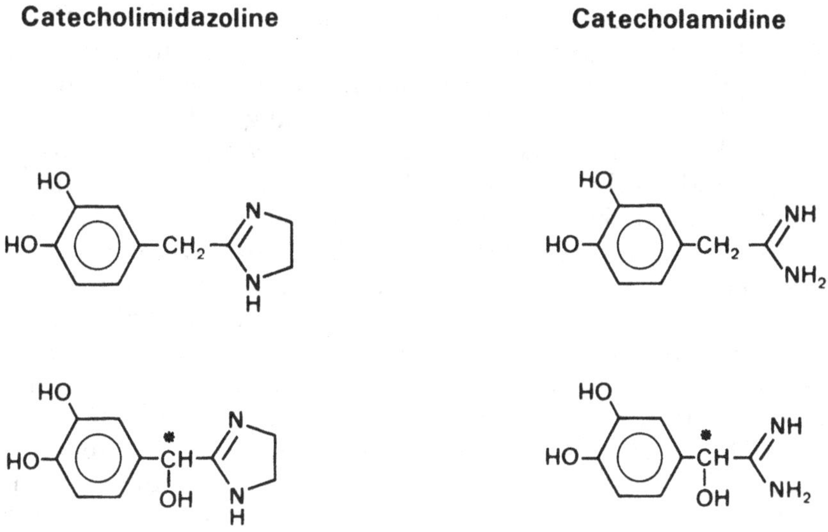

Catecholimidazoline   **Catecholamidine**

A   B

Fig. 6. Chemical structures of an optically active catecholimidazoline (A) and an optically active catecholamidine (B) used to investigate the Easson-Stedman hypothesis. The asterisk denotes the point of
asymmetry.

stereochemical demands made by alpha-2 adrenergic receptors
(Ruffolo et al., 1983c). Studies in vitro and in vivo clearly indicate
a significant preference for the R($-$)-enantiomer by the alpha-2
adrenergic receptor. As was the case for the optically active catecholimidazoline derivatives, however, the isometric activity difference between the R($-$)- and S($+$)-enantiomers of the catecholamidine derivative was small, on the order of only threefold
(Ruffolo et al., 1983c). Again, these findings support the notion
that the stereochemical demands made by alpha-2 adrenergic receptors for optically active imidazolines and imidazoline-like derivatives are weak relative to optically active phenethylamines
(Ruffolo, 1983b,1984c,d).

3.1.3. EASSON-STEDMAN HYPOTHESIS

The most important theory governing the activity of agonists possessing one point of asymmetry at the benzylic carbon atom (i.e.,
beta-carbon atom of phenethylamines) at alpha-2 adrenergic receptors is the Easson-Stedman hypothesis (Easson and Stedman,
1933). This hypothesis predicts a three-point attachment for the
binding of a sympathomimetic amine possessing an asymmetric

benzylic carbon atom to alpha-2 adrenergic receptors (Ruffolo, 1983,b,1984c). For phenethylamines, these groups are proposed to be (1) the basic nitrogen atom common to all sympathomimetic amines, (2) the phenyl group whose binding to the receptor is proposed to be enhanced by *meta* and/or *para* phenolic hydroxyl groups, and (3) the benzylic hydroxyl group of the beta-carbon atom (Beckett, 1959). According to the Easson-Stedman hypothesis, these three functional groups are in a most favorable stereochemical configuration for interaction with alpha-2 adrenergic receptors for only the R(−)-enantiomer of a phenethylamine (Ruffolo, 1983a,b). As far as the S(+)-enantiomer and corresponding beta-desoxy derivative are concerned, the beta-hydroxyl group is either incorrectly oriented or absent, and therefore not available for interaction with the alpha-2 adrenergic receptor. Thus, only a two-point attachment is considered possible for these isomers, and this would presumably explain the lower activities of the S(+)-enantiomer and corresponding desoxy derivative relative to the R(−)-enantiomer, and would also account for the fact that the S(+)-enantiomer and corresponding beta-desoxy derivative are equal in activity to each other at alpha-2 adrenergic receptors. Thus, the Easson-Stedman hypothesis predicts the following rank order of potency for optically active sympathomimetic amines interacting with the alpha-2 adrenergic receptor: R(−) > S(+) = desoxy. The Easson-Stedman hypothesis has been shown to apply to all phenethylamines interacting with alpha-2 adrenergic receptors that have been studied to date (Ruffolo et al., 1984c).

The applicability of the Easson-Stedman hypothesis to the alpha-2 adrenergic receptor-mediated effects of imidazolines has recently been challenged. The catecholimidazoline in Fig. 6, which possesses an asymmetric carbon atom at a position analogous to the beta-carbon atom of the phenethylamines, has been synthesized and resolved in order to test the Easson-Stedman hypothesis as it applies to the alpha-2 adrenergic receptor-mediated effects of the imidazolines. The rank order of potency for these compounds was found to be: desoxy > R(−) > S(+) (Ruffolo et al., 1983a,b), and additional studies with optically active clonidine derivatives are also consistent with these findings (Ruffolo et al., 1983d; Ruffolo and Waddell, 1983; Banning et al., 1984). Identical results were also obtained for the analogous optically active catecholamidine derivative in Fig. 6 (Ruffolo et al., 1983c). This order of potency is clearly different from that predicted by the Easson-Stedman hypothesis and indicates that this hypothesis does not apply to the alpha-2 adrenergic receptor-mediated effects of imidazolines or imidazoline-like compounds, in spite of the fact that it does accurately account for the alpha-2 adrenergic receptor-mediated effects of the phenethylamines. Such findings may indicate that imidazolines and phenethylamines interact differently with alpha-2 adrenergic receptors. (Ruffolo et al., 1983a,b,c,d; Ruffolo and Waddell, 1983).

3.1.4. CONFORMATIONAL REQUIREMENTS OF ALPHA-2 ADRENERGIC RECEPTORS FOR PHENETHYLAMINES AND IMIDAZOLINES

The relative positions in space of the three important functional groups (i.e., phenyl, beta-hydroxyl, and aliphatic nitrogen) of a sympathomimetic amine required for binding to, and activation of, the alpha-2 adrenergic receptor is determined from an analysis of the conformational demands made by these receptors. Initial attempts to define the conformational requirements of alpha-2 adrenergic receptors for phenethylamines were based upon considerations of energetically preferred conformations. Theoretical calculations and X-ray crystallographic studies indicate that the preferred conformation of R($-$)-norepinephrine in solution and in the solid state is the extended-*trans* conformation in which the amino and phenyl groups are at a dihedral angle of 180°, as shown in Fig. 7 (Kier, 1968, 1969; Kier and Truitt, 1970; Pullman et al., 1972). This conformation may also be stabilized by an intra-

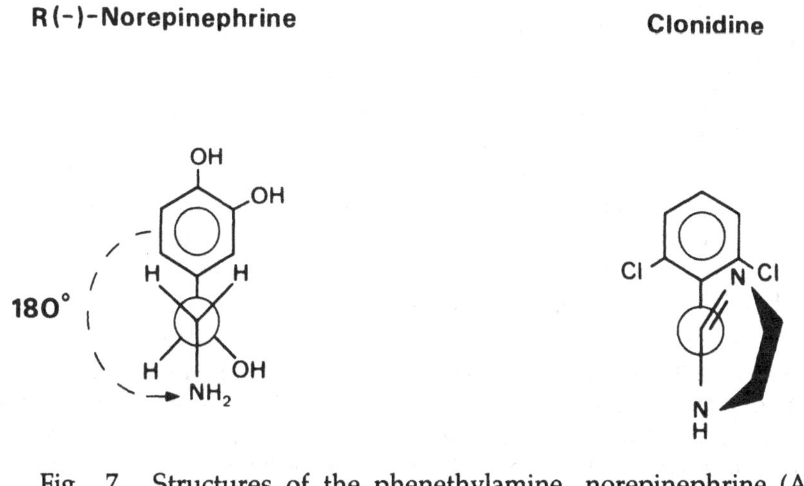

R(-)-Norepinephrine   Clonidine

Fig. 7. Structures of the phenethylamine, norepinephrine (A) and the imidazolidine, clonidine (B), showing the most energetically stable conformations, which also are believed to be the conformations required by the alpha-2 adrenergic receptor.

molecular electrostatic or hydrogen bonding interaction between the amino and beta-hydroxyl groups and represents an energy minimum and hence greater stability and a greater probability of existence at any point in time (Ison et al., 1973; Portoghese, 1967). The 1R,2S(–)-*erythro* enantiomer of the selective alpha-2 adrenergic receptor agonist, alpha-methylnorepinephrine, also exists predominantly in the extended-*trans* conformation (Kier, 1968).

The preferred conformation of phenethylamines in solution and in the solid state has led to speculation that this conformation is also required for binding to, and activation of, the alpha-2 adrenergic receptor. The preferred conformation of an agonist in solution or in the solid state is not necessarily the same conformation required for interaction with the receptor, however, since other less-preferred conformations may still exist for a sufficient period of time for activation of the receptor. By using conformationally restricted phenethylamine analogs, such as 2-(3,4-hydroxyphenyl)-cyclobutylamine, it has been established that the *trans*-extended conformation, which is the predominant conformation is, in fact, also that conformation required for binding to, and activation of, the alpha-2 adrenergic receptor (Ruffolo et al., 1982b; Ruffolo, 1983a,b,1984c). Figure 7 provides the structures of the phenethylamine norepinephrine and the imidazoline clonidine, and shows the most energetically stable conformation of these agents, which also is believed to be the conformation required by the alpha-2 adrenergic receptor.

Among the most selective alpha-2 adrenergic receptor agonists of the phenethylamine class, and indeed, one of the very few phenethylamines that is appreciably selective for the alpha-2 adrenergic receptor, is alpha-methylnorepinephrine (Starke et al., 1975). As indicated above, this high degree of selectivity for the alpha-2 adrenergic receptor is derived from the critical alpha-methyl substituent, which can impart marked alpha-2 adrenergic receptor selectivity when oriented in the optimal 2S stereochemical configuration (Ruffolo and Waddell, 1982). By drawing the most active configuration of alpha-methylnorepinephrine, which is the 1R,2S(–)-*erythro* enantiomer (Ruffolo et al., 1982a), in what we believe is the most active conformation, which is also the extended-*trans* form (Ruffolo et al., 1982b), as shown in Fig. 8, one can begin to appreciate the nature of the interaction of alpha-methylnorepinephrine with alpha-2 adrenergic receptors. The active enantiomer of alpha-methylnorepinephrine when in the optimum conformation presents a highly polarized molecule, both in terms of substitution pattern and physiochemical properties. alpha-Methylnorepinephrine presents itself to the alpha-2 adren-

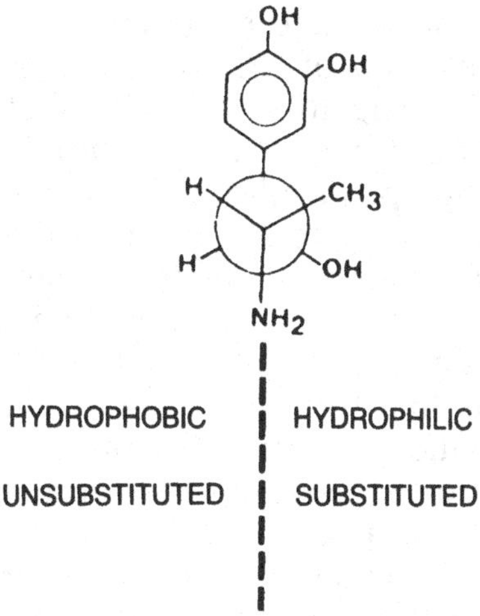

Fig. 8. Newman projection of the selective alpha-2 adrenergic receptor agonist of the phenethylamine class, alpha-methylnorepinephrine, for the most potent configuration (i.e., 1R,2S (−)-*erythro*) in the preferred conformation (i.e., extended-*trans*). The projection shows both a highly substituted, hydrophilic side, as well as a relatively unsubstituted, lipophilic side.

ergic receptor with both highly substituted and unsubstituted sides, which correspond to both hydrophilic and hydrophobic sides, respectively. Since the active site(s) of the alpha-adrenergic receptor is believed to be hydrophilic (Triggle, 1976), it would appear that the highly substituted side of alpha-methylnorepinephrine is involved in the process of binding to, and activation of, the alpha-2 adrenergic receptor. This interpretation is consistent with the observations of Timmermans et al. (1981) and Ruffolo and coworkers (Ruffolo et al., 1985; Ruffolo and Messick, 1985), who report that alpha-2 adrenergic receptor selectivity may be obtained in an agonist by substitution on only one side of the molecule, whereas selectivity for the alpha-1 adrenergic receptor will often occur when substitutions are made on both sides of the molecule. Again, consistent with this view is the observation that the *threo*-enantiomers of alpha-methylnorepinephrine, which are substituted on both sides of the molecule (Fig. 5), are not only less potent as alpha-2 adrenergic receptor agonists, but are also significantly less selective for the alpha-2 adrenergic receptor (Ruffolo et al., 1982a) than is the 1R,2S(−)-*erythro* enantiomer, which is substituted exclusively on one side of the molecule and is therefore highly selective for the alpha-2 adrenergic receptor.

One of the first attempts made to define the molecular conformation required for interaction of the imidazolines with alpha-adrenergic receptors was made by Pullman et al. (1972). In a quantum mechanical study of the conformational properties of naphazoline, these investigators concluded that the most stable conformation was one in which the naphthyl and imidazoline rings were mutually perpendicular with a dihedral angle of 90°. This conformation would place the aromatic ring and one of the imidazoline nitrogen atoms at a dihedral angle of approximately 180° (Fig. 7), similar to what has been observed for the phenethylamines in the solid state and in solution. Consistent with this observation are reports that the free base of clonidine in solution prefers the conformation in which the phenyl and imidazoline rings assume a mutually perpendicular arrangement (deJonge and van Dam, 1980). X-Ray crystallographic studies of clonidine hydrochloride in the solid phase also show a nearly perpendicular arrangement between the phenyl and imidazoline rings (Cody and DeTitta, 1979). Although the perpendicular arrangement of the phenyl and imidazoline rings of clonidine has been attributed to restriction of free rotation about the nitrogen bridge by steric hindrance resulting from the two relatively bulky *ortho* chlorine substituents (Timmermans et al., 1977), it has recently been suggested that even unsubstituted benzylimidazolines and phenyliminoimidazolines may also assume the same perpendicular arrangement of the phenyl and imidazoline rings in solution (deJong and van Dam, 1980).

Although the preferred conformations of the imidazolines in solution and in the solid state have been established and shown to be similar to those observed for the phenethylamines, conformationally restricted or rigid imidazoline derivatives have not been studied in a highly quantitative manner in order to determine the conformational preference of the alpha-2 adrenergic receptor for imidazolines.

3.2. Structure–Activity Relationships of the alpha-2 Adrenergic Receptor for the Phenethylamines

3.2.1. SUBSTITUTION AT THE AROMATIC RING

Much of the work that has been done to study the effects of substitution on the aromatic ring of phenethylamines in regard to affinity and efficacy of agonists at alpha-adrenergic receptors has addressed itself in large part toward alpha-1 adrenergic receptors (Ruffolo, 1983a). There are a few data in the literature, however, that pertain to the activation of alpha-2 adrenergic receptors by phenethylamines. The data in Table 1 (Wikberg, 1978,1979a,b)

Table 1
alpha-2 Adrenergic Receptor Activity of beta-Hydroxyl-Substituted
Phenethylamines[a]

Compound	Guinea pig ileum	
	ia	pD$_2$
Norepinephrine analogs		
Norepinephrine	1.0	6.7
Norfenfrine	0.2	4.5
Octopamine	0.3	4.8
Phenethanolamine	0.1	<4.0
Epinephrine analogs		
Epinephrine	1.0	7.2
Phenylephrine	0.3	4.7
Synephrine	0.4	4.5

(continued)

Table 1 (*continued*)

Compound	Guinea pig ileum	
	ia	pD_2

alpha-Methylnorepinephrine analogs

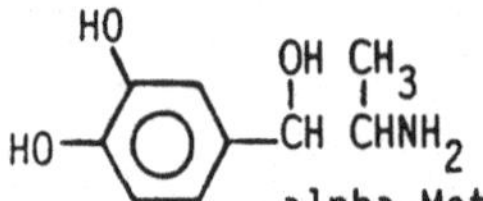

alpha-Methylnorepinephrine

| | 1.0 | 7.3 |

| 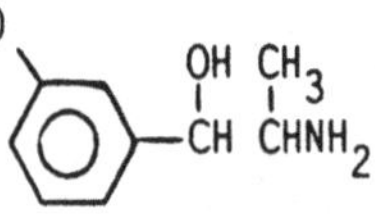 | 0.3 | 5.2 |

| 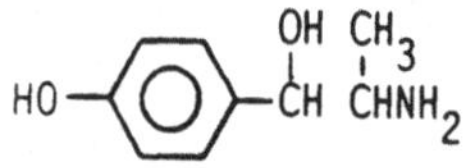 | 0.8 | 4.4 |

| 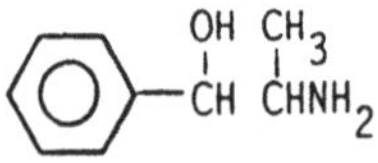 | 1.1 | 3.2 |

Methoxamine

| 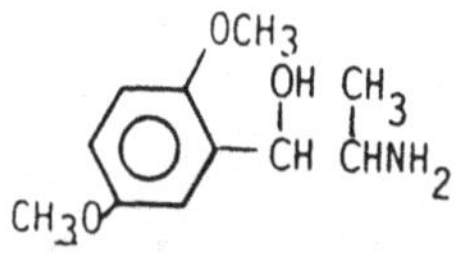 | 1.1 | 3.4 |

[a]Abbreviations: ia, intrinsic activity; pD_2, log EC_{50}.

illustrate some of the general features observed in the structure–activity relationships of phenethylamines for the alpha-2 adrenergic receptor. For activation of the alpha-2 adrenergic receptor, a catechol moiety on the aromatic ring of a phenethylamine is optimum, if not essential. Compounds that are substituted only with a *meta* hydroxyl are less active, whereas those with a *para* hydroxyl substitution are weaker still. Nonhydroxylated phenethylamines, such as methoxamine or phenethanolamine, are essentially inactive as alpha-2 adrenergic receptor agonists (Ruffolo and Waddell, 1983). In addition to the decreases in potency that are observed when the hydroxyl substitution pattern is changed from the catechol characteristic of the endogenous ligand, there are also dramatic changes in the intrinsic activity or efficacy of these compounds. Although all the catechol-containing phenethylamines are full agonists (i.e., intrinsic activity = 1), all of the monohydroxy-substituted phenethylamines are partial agonists with significantly lower intrinsic activities relative to the catechol congeners.

The substitution of fluorine for hydrogen on the aromatic ring of norepinephrine exerts marked changes in alpha-adrenergic receptor subtype specificity (Table 2). The 5-fluoro isomer is

Table 2
Adrenergic Receptor Selectivity of Fluoronorepinephrine Derivatives in Rat Brain[a]

	alpha-1	alpha-2	beta-1	beta-2
		IC_{50}, μM		
	3	0.8	0.8	11
	3	0.5	>100	>100
	10	1.0	3	10
	>100	10	0.7	10

[a]Data from Nimit et al. (1980).

equipotent to norepinephrine as an alpha-2 adrenergic receptor agonist and significantly more potent as a beta-adrenergic receptor agonist. On the other hand, the 2-fluoro isomer is a selective beta-adrenergic agonist, with little alpha-2 adrenergic receptor activity, whereas the 6-fluoro isomer is a pure alpha-adrenergic receptor agonist that is completely devoid of beta-adrenergic receptor agonist activity (Cantacuzene et al., 1979; Kirk and Creveling, 1984; Nimit et al., 1980). Shepperson et al. (1981) have demonstrated in a variety of physiologic test systems that 6-fluoronorepinephrine is a selective alpha-2 adrenergic receptor agonist.

Based upon the marked differences in alpha-adrenergic receptor subtype selectivity of 2- and 6-fluoronorepinephrine, DeBernardis et al. (1985) proposed that the selectivity of these agents was based on a conformational preference induced by the electrostatic repulsion between the aromatic fluorine and the side chain benzylic hydroxyl group. The high affinity for the alpha-2 adrenergic receptor of some of the rigid compounds that have no benzylic hydroxyl group suggests that the role of the beta-hydroxyl group of 6-fluoronorepinephrine in alpha-2 adrenergic receptor binding is to provide a favorable conformation for interaction with the receptor, rather than to produce a direct interaction with the receptor itself. Among both 2- and 6-electrostatic repulsion-based conformational prototypes (ERBCOPs) (Fig. 9), affinity for the alpha-2 adrenergic receptor is generally higher than that of the fluoronorepinephrine derivatives as determined by radioligand binding data (Tables 3 and 4). In every case reported, the data for the compounds based upon the conformer of 6-fluoronorepinephrine that was predicted upon electrostatic repulsion grounds has a higher affinity for the alpha-2 adrenergic receptor than that derived from the 2-fluoronorepinephrine. This suggests that electrostatic repulsion may predict preferred conformations of the aromatic ring for interactions of phenethylamines with the alpha-2 adrenergic receptor. It must be kept in mind, however, that these numbers reflect only affinity for the alpha-2 adrenergic receptor, and do not predict whether the interaction that is described will be a productive one in terms of intrinsic efficacy or agonist activity or whether these compounds are antagonists of the alpha-2 adrenergic receptor.

These conformational predictions are directly related to the effects that have been observed at the alpha-1 adrenergic receptor in regard to interaction with methoxamine and the corresponding 2-aminotetralins (DeMarinis et al., 1981, 1982). Here it was proposed that the hydrogen bonding of the benzylic hydroxyl to the 6-methoxy group of methoxamine resulted in a conformation of

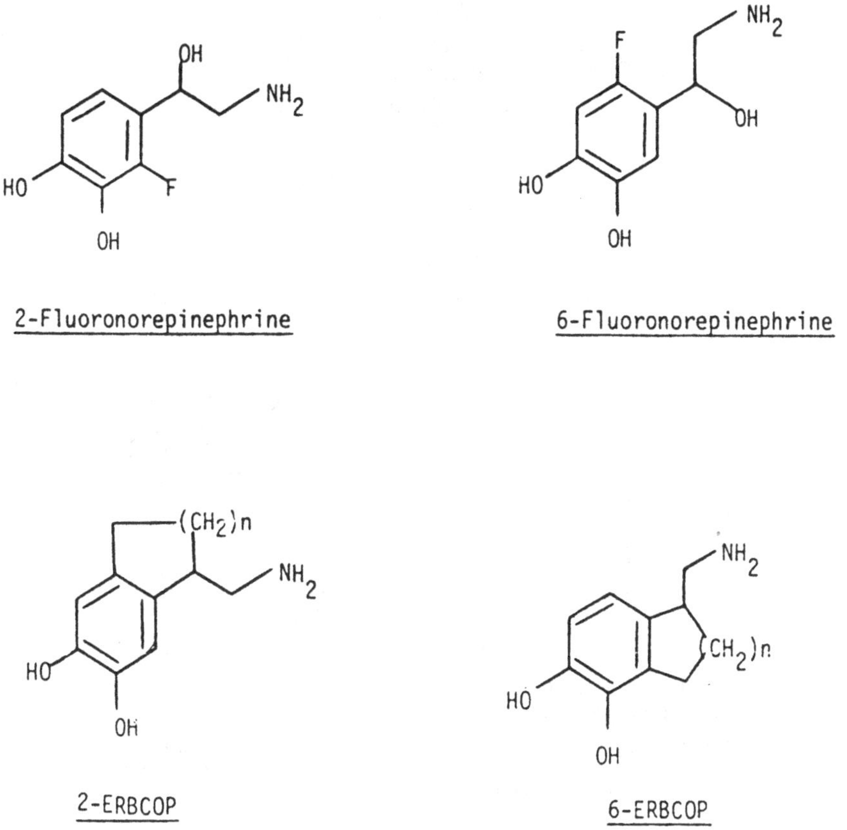

2-Fluoronorepinephrine   6-Fluoronorepinephrine

2-ERBCOP   6-ERBCOP

Fig. 9. Electrostatic repulsion based conformational prototypes (ERBCOPs) of 2- and 6-fluoronorepinephrine.

the amine that could be mimicked by the 2-aminotetralin structure to produce potent and selective agonists of the alpha-1 adrenergic receptor that did not have a benzylic hydroxyl substituent. Whether the specific conformational bias of the beta-hydroxyl group, either in terms of repulsion by fluorine or attraction through hydrogen bonding with methoxyl, is a key determinant in the alpha-1/alpha-2 adrenergic receptor selectivity of phenethylamines is a point that deserves further clarification.

3.2.2. SUBSTITUTION OF THE BETA-CARBON ATOM

The only significant substitution at the beta-carbon atom of phenethylamines with regard to activity at alpha-2 adrenergic receptors is the hydroxyl group. Substitution of a beta-hydroxyl group on the phenethylamine will result in a 100–1000-fold increase in affinity at the alpha-2 adrenergic receptor, but does not change

Table 3
alpha-Adrenergic Receptor Activity
of Conformationally Defined Prototypes
of 6-Fluoronorepinephrine[a]

n	R_1	R_2	K_i, nM	
			alpha-1	alpha-2
1	CH_3	H	8,600	15
2	H	H	13,000	4.7
2	CH_3	H	7,900	44
1	CH_3	H	2,000	7.0
1	CH_3	CH_3	6,900	25
2	H	CH_2CH_3	7,900	200
6-Fluoronorepinephrine			650	35
Norepinephrine			390	37
Clonidine			520	30

[a]Data from DeBernardis et al. (1985).

Table 4
alpha-Adrenergic Receptor Activity
of Conformationally Defined Prototypes
of 2-Fluoronorepinephrine[a]

n	R_1	R_2	K_i, nM	
			alpha-1	alpha-2
1	H	H	26,000	520
1	H	CH_3	13,000	860
2	H	H	15,000	5,100
2	H	CH_3	8,700	1,300
2-Fluoronorepinephrine			25,000	1,400
Norepinephrine			390	37
Clonidine			520	30

[a]Data from DeBernardis et al. (1985).

intrinsic efficacy (Ruffolo and Waddell, 1983; Ruffolo et al., 1982a). As indicated previously, the beta-hydroxyl substitution plays a pivotal role in the Easson-Stedman hypothesis, which has been shown to apply to all phenethylamines tested that interact with the alpha-2 adrenergic receptor. The optimum stereochemical configuration for the beta-hydroxyl group of a phenethylamine is R.

3.2.3. SUBSTITUTION OF THE ALPHA-CARBON ATOM

Substitution at the alpha-carbon atom represents the most important substitution in governing the alpha-1 vs. alpha-2 adrenergic receptor selectivity of the phenethylamines. In general, small alkyl substitutions are most important, and they must be in the 2S absolute configuration for maximum potency and alpha-2 adrenergic receptor selectivity. The marked increase in activity at alpha-1 adrenergic receptors produced by alpha-methyl substitution is usually accompanied by a decrease in activity at alpha-1 adrenergic receptors, further contributing to the increase in alpha-2 adrenergic receptor selectivity observed in phenethylamines so substituted (Ruffolo and Waddell, 1982). The primary effect of the alpha-methyl substituent on a phenethylamine is to increase affinity for alpha-2 adrenergic receptors, with little or no effect on intrinsic efficacy. The reduction in activity at alpha-1 adrenergic receptors caused by alpha-methyl substitution results both from decreases in affinity and efficacy (Ruffolo, 1983a). As indicated previously, substitution at the alpha-carbon atom is the single most important position with regard to increasing the potency and selectivity of phenethylamines for the alpha-2 adrenergic receptor (Ruffolo et al., 1984b). Similar results are also obtained with alpha-ethyl substitution of the phenethylamines (Wikberg, 1978).

3.2.4. SUBSTITUTION AT THE ALIPHATIC NITROGEN ATOM

Little is known regarding the effects of N-substitution of phenethylamines on alpha-2 adrenergic receptor agonist activity. N-Methyl substitution of norepinephrine (to form epinephrine) produces an increase in alpha-2 adrenergic receptor agonist activity similar to that observed for the alpha-1 adrenergic receptor. Furthermore, N-ethyl substitution produces a very slight decrease in alpha-2 adrenergic receptor agonist activity, consistent with the observations made for the alpha-1 adrenergic receptor that increasing the size of the N-substituent beyond methyl reduces agonist activity. Substitutions larger than ethyl have not been studied systematically in great detail, and it is therefore difficult, but not impossible, to make generalizations. Important information with regard to larger N-methyl substituents may be derived from studies with the inotropic agent, dobutamine, whose interaction with alpha-1 and alpha-2 adrenergic receptors has been studied in detail. Dobutamine, which possesses the general catecholamine nucleus with a large, bulky N-substituent, is devoid of agonist activity at alpha-2 adrenergic receptors (Ruffolo and Yaden, 1983). Although the decrease in activity at alpha-2 adrenergic receptors results primarily from a loss of intrinsic efficacy, radioligand binding studies to alpha-2 adrenergic receptors in rat cerebral cortex shows also a marked reduction in affinity (Ruffolo et al., 1984a). Likewise, the alpha-1 and beta-adrenergic receptor antagonist, labetolol, is a close structural analog of dobutamine that also possesses a large N-substituent, and like dobutamine, labetolol has extremely low affinity for alpha-2 adrenergic receptors (Brittain et al., 1982). It appears, therefore, that N-substituents larger than methyl reduce alpha-2 adrenergic receptor agonist activity of phenethylamines by decreasing both affinity and intrinsic efficacy at alpha-2 adrenergic receptors.

Although large N-substituents also decrease both affinity and intrinsic efficacy at the alpha-1 adrenergic receptor (Ruffolo, 1983a), the selective affinity of dobutamine and labetolol for alpha-1 versus alpha-2 adrenergic receptors demonstrates that large N-substituents may preferentially interfere with binding to the alpha-2 adrenergic receptor.

3.3. Structure–Activity Relationships of the alpha-2 Adrenergic Receptor for the Imidazolines

In the last two decades, many alpha-2 adrenergic receptor studies have been carried out on the imidazolines and related structural types. Beginning with the prototype alpha-2 adrenergic receptor agonist clonidine, an intense effort was made both chemically and biologically to characterize and optimize the desirable characteristics of this molecule by the synthesis and testing of hundreds of diverse structures related to clonidine. The structure–activity relationships of much of this early work were developed through measurements of pressor/depressor responses in whole animals with little understanding of the mechanistic basis for the activity. The discovery that alpha-adrenergic receptors were a nonhomogenous population that could be subdivided into the alpha-1 and alpha-2 adrenergic receptor subtypes, based on their interaction with a series of selective agonists and antagonists, allowed these early results to serve as the foundation upon which to build a separate structure–activity relationship for the imidazolines and related structures that demonstrated a distinct preference for the alpha-2 adrenergic receptor in both in vitro (Table 5) and in vivo (Table 6) models.

Chemical requirements, in terms of molecular connectivity, for activation of alpha-2 adrenergic receptors appear to be relatively loose, in marked contrast to the phenethylamines, in which the structural and stereochemical requirements are strict. There are numerous examples of chemical systems, many of which only in a broad sense can be viewed as analogs of the prototypic clonidine, that have agonist activity at alpha-2 adrenergic receptors (Timmermans et al. 1980a; Kobinger and Pichler, 1977; Armah, 1985; Hicks and Cannon, 1980; Laubie et al., 1985; Ruffolo, 1983a).

Although many types of heterocyclic and open chain structures possess alpha-2 adrenergic receptor agonist activity, the most detailed structure–activity relationship has been carried out on analogs closely related to clonidine. In these studies, both aromatic and imidazoline rings have been systematically substituted. In most cases, data on efficacy are still lacking, and results reflect affinities obtained through radioligand binding studies and in some cases on EC_{50} determinations from physiologic measurements. Although it may be difficult to extrapolate this structure–activity relationship to all other systems that possess alpha-2 adrenergic receptors, these data do provide a basic framework around which to study alpha-2 adrenergic receptor activation.

3.3.1. SUBSTITUTION AT THE AROMATIC RING

Many derivatives of phenylaminoimidazoline, the basic nucleus of clonidine-like imidazolines, have been studied at alpha-2

Table 5
Potencies of alpha-1 and alpha-2
Adrenergic Receptor Agonists In Vitro[a]

	EC_{20} alpha-1, nM	EC_{20} alpha-2, nM	$\dfrac{EC_{20}\ \text{alpha-2}}{EC_{20}\ \text{alpha-1}}$
Methoxamine	750	24,000	32.4
Phenylephrine	54	1,700	31.5
Noradrenaline	7.3	12	1.6
Adrenaline	3.2	1.9	0.59
Naphazoline	38	16	0.42
Oxymetazoline	18	3.1	0.17
Clonidine	65	10	0.15
alpha-Methylnor-adrenaline	61	8.1	0.13
Tramazoline	56	3.7	0.07

[a]Data from Starke et al. (1974, 1975).

Table 6
Potencies of alpha-1 and alpha-2 Adrenergic Receptor Agonists In Vivo[a]

	Decr. heart rate of 50 beats/min, μg/kg	Incr. in diast. blood press of 50 mm Hg, μg/kg	BP/HR
Xylazine	153.9	553.2	3.64
Clonidine	7.5	5.5	0.73
Oxymetazoline	5.4	0.7	0.13
Naphazoline	94.9	3.3	0.03
Methoxamine	1061.7	46.0	0.05
Phenylephrine	>100	4.8	<0.1

[a]Data from Drew (1976).

adrenergic receptors (Table 7). The presence of an aromatic ring is generally necessary for activity, since 2-methylimidazoline does not bind to the alpha-2 adrenergic receptor. Several compounds that do not have aromatic rings have also, however, been shown to be alpha-2 adrenergic receptor agonists (Laubie et al., 1985; Guicheney et al., 1981). Activity of phenylaminoimidazoline at alpha-2 adrenergic receptors is increased dramatically by addition of a chlorine atom at either the 2, 3, or 4 position, with the greatest activity observed with the 2-chloro derivative (Table 7). It is also clear that a methyl group may substitute for chlorine with no loss in affinity. Agonist activity at alpha-2 adrenergic receptors is further enhanced by the addition of a second chlorine atom with the greatest activity observed for the 2,6-dichloro derivative (clonidine). The 2,3-, 2,4-, and 2,5-dichloro analogs are also still extremely potent, however, indicating that, as long as one chlorine atom is at the 2 position, the second chlorine atom may be at any point on the phenyl ring with significant alpha-2 adrenergic receptor agonist activity being retained. The same holds true if a chlorine atom is at the 2 position and a methyl group is placed at any other point on the phenyl ring. When chlorine atoms are placed at the 3,4 positions, activity nearly equal to that of clonidine is obtained, but when the chlorine atoms are placed at the 3,5 positions, reductions in alpha-2 adrenergic receptor affinity result.

A significant decrease in alpha-2 adrenergic receptor affinity has been observed when both *ortho* chlorine atoms of clonidine are replaced by fluorine or bromine, although these compounds must still be considered to be relatively potent (Table 7). 2,6-Diethyl and 2-methyl,6-ethyl-phenylamino-imidazoline possess similar affinities as clonidine, indicating again that small alkyl

Table 7
Affinities for alpha-2 Adrenergic Receptors of Various
Imidazolines Substituted at the Aromatic Ring

		Rat brain		Rabbit ear artery
		$-\log K_i^a$	$-\log K_i^b$	pD_2^c
R			Inactive	
H		6.9	Inactive	>5.8
2-Cl		7.9	—	7.1
3-Cl		7.3	—	—
4-Cl		7.5	—	7.3
2,6-diCl (Clonidine)		8.6	8.8	7.7
2,3-diCl		—	8.6	—
2,4-diCl		—	8.7	—
2,5-diCl		8.0	8.3	—
3,4-diCl		—	—	7.3
3,5-diCl		7.0	—	—
2,6-diF		7.7	—	—
2-Me		8.0	—	—
2,6-diMe		8.3	—	7.8
2,6-diEt (St-91)		7.6	8.0	—
2-Me, 5-F		8.3	—	—
2-F, 6-Br		—	8.6	—
3,4-di (OH)		—	—	8.4
UK 14,304		9.1	8.9	—

[a]Data from deJonge and Soudijn (1981).
[b]Data from Jarrott et al. (1980).
[c]Data from Hieble and Pendleton (1979).

groups may substitute for chlorine at the alpha-2 adrenergic receptor with no perceptible loss in affinity.

Thus, in general, agonist activity at alpha-2 adrenergic receptors is markedly enhanced when the basic phenylaminoimidazoline nucleus is alkyl- or halogen-substituted, and activity is greatest for di-substituted compounds. At most positions, methyl and ethyl groups may substitute for halogens with little or no loss in affinity for the alpha-2 adrenergic receptor. For the di-substituted derivatives, it is apparent that as long as one substituent (usually a halogen or small alkyl group) is at the 2 position, the other substituent, whether a halogen or methyl group, may be placed at any other position of the phenyl ring with the resulting activity being similar to that of clonidine (Malta et al., 1980), although there are some notable exceptions to this generalization (deJonge et al., 1982). As a result, most di-substituted phenylaminoimidazolines have nearly identical affinities for the alpha-2 adrenergic receptor regardless of the positions of substitution. In addition, polycyclic aromatic aminoimidazolines, such as UK 14,304, are also potent and highly selective alpha-2 adrenergic receptor agonists. This suggests that the structural requirements of alpha-2 adrenergic receptors are, in general, not stringent for imidazolines that are multisubstituted at the phenyl ring. This is in marked contrast to the halogenated derivatives of norepinephrine, which vary dramatically in alpha-2 adrenergic receptor agonist activity and where the positions of substitution on the aromatic ring appear to be critical (Nimit et al., 1980; Cantacuzene et al., 1979).

It has been proposed that 2,5- and 2,3-di-substitution of the aromatic ring are important factors in distinguishing between imidazoline compounds that are selective for the alpha-1 and alpha-2 adrenergic receptors, respectively (deJonge et al., 1982; Ruffolo et al., 1985). Thus, di-substitution at positions 2,5 predisposes toward alpha-1 adrenergic receptor selectivity, whereas 2,3-di-substitution results in selectivity for alpha-2 adrenergic receptors (deJonge et al., 1982; Ruffolo and Messick, 1985; Ruffolo et al., 1985). This differential selectivity based on the position of di-substitution results from an apparent requirement of alpha-1 adrenergic receptors for substituents to be on both sides of the aromatic ring of imidazolines, as is the case with 2,5- and 3,5-disubstituted compounds, whereas alpha-2 adrenergic receptors prefer substituents to be on one side of the aromatic ring, as in the case of 2,3-di-substituted compounds (Ruffolo and Messick, 1985; Ruffolo et al., 1985). The 2,5-, 3,5- and 2,3-dimethoxy tolazoline derivatives adhere to this hypothesis (Table 8). As predicted, 2,5- and 3,5-dimethoxytolazoline are potent alpha-1 adrenergic receptor agonists, whereas 2,3-dimethoxytolazoline is a potent and selective alpha-2 adrenergic receptor agonist.

Recently, Fuder et al. (1986) carried out some of the first quantitative studies of aromatic hydroxy substitution on the affinity and efficacy of several imidazolines interacting with presynaptic alpha-2 adrenergic receptors. It is apparent from an inspection of the data in Table 9 that in all cases the efficacy of the imidazolines is generally greatly reduced relative to the pheneth-

Table 8
Pharmacologic Selectivities of Dimethoxy-Substituted Tolazoline
Derivatives at alpha-1 and alpha-2 Adrenergic Receptors

Dimethoxy-Substituted Tolazolines

Substitution	Adrenergic activity
2,3-Dimethoxy	alpha-2 Adrenergic receptor agonist
3,4-Dimethoxy	alpha-2 Adrenergic receptor antagonist
2,5-Dimethoxy	alpha-1 Adrenergic receptor agonist
3,5-Dimethoxy	alpha-1 Adrenergic receptor agonist

Table 9
Affinities and Efficacies of Various Imidazoline Derivatives
at Presynaptic alpha-2 Adrenergic Receptors

Compound	$-\log EC_{50}^{a}$	$-\log K_A^{a}$	Efficacy^a
Norepinephrine	7.8	5.9	100
alpha-Methylepinephrine	6.9	4.8	98
Phenylaminoimidazoline	<5	<5	<1

(continued)

Table 9 (*continued*)

Compound	$-\log EC_{50}{}^a$	$-\log K_A{}^a$	Efficacy[a]
3,4-Dihydroxyphenyl-Amino-2-imidazoline	8.3	8.2	2.0
Tolazoline	Antagonist	6.5[b]	0
3,4-Dihydroxytolazoline	7.3	6.9	3.5
Xylometazoline	7.9	7.3	4.4
Oxymetazoline	7.9	7.3	4.4

[a]Data from Fuder et al. (1986).
[b]Data from Wikberg (1979a).

ylamines. This is parallel to what has been observed for the alpha-1 adrenergic receptor, in which phenethylamines generally behave as full agonists, whereas the imidazolines have markedly reduced efficacies and are therefore partial agonists (Ruffolo and Waddell, 1983; Ruffolo et al., 1979a,b; Besse and Furchgott, 1976). For all the compounds studied by Fuder et al. (1986), with the exception of phenylaminoimidazoline, which was totally inactive, the affinities of the imidazolines were markedly higher than those of the phenethylamines. Thus, the combination of high affinity and lower efficacy for the imidazolines and high efficacy and low affinity for the phenethylamines resulted in a series of compounds that had comparable EC_{50} values at presynaptic alpha-2 adrenergic receptors.

3.3.2. SUBSTITUTION AT CARBON OR NITROGEN BRIDGE

Available data suggest that optimum agonist activity at alpha-2 adrenergic receptors requires a nitrogen atom separating the phenyl and imidazoline rings. Replacement of the nitrogen atom with either a carbon or sulfur atom, however, produces only a very minor decrease in activity. Replacing the nitrogen bridge with an oxygen atom abolishes alpha-2 adrenergic receptor agonist activity (Timmermans and van Zwieten, 1980c).

Substitution on the nitrogen atom separating the phenyl and imidazoline rings of clonidine with either an allyl or cyclopropylmethyl group abolishes alpha-2 and alpha-1 adrenergic receptor agonist activity (Kobinger et al., 1979; Stahle et al., 1980). These compounds, however, are still pharmacologically active as specific bradycardiac agents that work through a mechanism other than stimulation or blockade of alpha-adrenergic receptors.

The optimum carbon bridge length separating the phenyl and imidazoline rings of tolazoline-like imidazolines is one methylene unit. Increasing the bridge length to two carbon atoms decreases alpha-2 adrenergic receptor agonist activity (Ruffolo, 1983a). These results are consistent with the observation that replacing the nitrogen atom of clonidine with -CH--N-NH- reduces alpha-2 adrenergic receptor agonist activity (Timmermans and van Zwieten, 1980c). However, the reason for the high affinity of lofexidine [which has a bridge of $-O-CH(CH_3)$] at alpha-2 adrenergic receptors is not clear. The possibility exists that these expanded bridges allow the imidazoline ring to fold back and position one of the imidazoline nitrogens in the same relative position as the aliphatic nitrogen of norepinephrine or the imidazoline nitrogen of clonidine.

For the alpha-2 adrenergic receptor, a methylene bridge may substitute for the nitrogen bridge with no significant loss of agonist activity. When tolazoline-like imidazolines (i.e., methylene bridge) are hydroxy-substituted at the methylene bridge, a position analogous to the crucial beta-carbon atom of the phenethylamines, a significant reduction in affinity for the alpha-2 adrenergic receptor, is observed (Ruffolo et al., 1983a,b). No change in relative efficacy or intrinsic activity accompanies this decrease in affinity. This is analogous to the marked reduction in alpha-2 adrenergic receptor-mediated effects of clonidine-like imidazolines that are allyl- or cyclopropylmethyl-substituted at the nitrogen bridge. More striking, however, is the contrast with the phenethylamines, which, when likewise hydroxy-substituted at the analogous carbon atom (i.e., beta-carbon atom), show a marked increase in alpha-2 adrenergic receptor agonist activity of approximately two orders of magnitude (Ruffolo et al., 1982a). This increase in activity observed for the phenethylamines represents an increase predominantly in affinity and not efficacy. In addition, the isomeric activity ratios between the enantiomers of the hydroxyl-substituted methylene bridged imidazolines are small (Ruffolo et al., 1982a) relative to these observed for phenethylamines (Ruffolo et al., 1982a). These results have led Ruffolo et al. (1983a,b) to propose that the imidazolines and phenethylamines may interact differently with the alpha-2 adrenergic receptor.

3.3.3. Substitution at the Imidazoline Ring

The effects of substitution at the imidazoline ring on alpha-2 adrenergic agonist activity are presented in Table 10. It must be emphasized that these results are taken from radioligand binding

Table 10
Effects of Substitution at the Imidazoline Ring on Affinity
for the alpha-2 Adrenergic Receptor

Compound		Inhibition of ^{3}H-clonidine binding in rat brain		
		$-\log K_i^a$	$-\log K_i$ (calc)[b]	$-\log K_i^c$
	Clonidine	8.57	8.77	8.66
		8.25		
			7.97	8.13
		Inactive		
		Inactive		

(continued)

Table 10 (*continued*)

| Compound | Inhibition of ^{3}H-clonidine binding in rat brain | | |
	$-\log K_i{}^a$	$-\log K_i$ (calc)[b]	$-\log K_i{}^c$
(structure)		Inactive	
(structure)	7.68	8.05	
(structure)	Inactive		
(structure)	6.91	7.07	
(structure)		7.07	6.95
(structure)		Inactive	
(structure)			<5.00
(structure)	8.25		
(structure)	7.68		
(structure) Xylazine	6.87		

[a]Data from deJonge and Soudijn (1981).
[b]Data from Jarrott et al. (1980).
[c]Data from Summers et al. (1981).

studies using ^{3}H-clonidine and, as such, only pertain to affinity for the alpha-2 adrenergic receptor. As Jarrott et al. (1979) point out, these studies do not give an overall picture of alpha-2 adrenergic receptor agonist activity since efficacy cannot be obtained from these types of investigations. Therefore, as far as the imidazoline ring is concerned, only the structural requirements for binding (affinity) to the alpha-2 adrenergic receptor may be addressed.

For optimum affinity at alpha-2 adrenergic receptors, the imidazoline ring should be unsubstituted. Furthermore, additional unsaturation at the 4,5 positions to generate the imidazole results in a significant reduction in affinity. Substitution on one of the imidazoline nitrogen atoms destroys activity at alpha-2 adrenergic receptors. Table 10 shows that methyl or acyl N-substitution of the imidazoline ring markedly decreases affinity for the alpha-2 adrenergic receptor, analogous to the decrease in alpha-2 adrenergic receptor activity produced by N-substitution of the phenethylamines. N,N-dimethyl substitution at the imidazoline ring likewise abolishes alpha-2 adrenergic receptor activity.

Opening the imidazoline ring is associated with a significant loss of activity, since the guanidine and N,N-dimethyl guanidine analogs exhibit a marked decrease in affinity for the alpha-2 adrenergic receptor (Table 10).

Replacing one of the imidazoline nitrogens with a carbon atom (pyrrolidine) is associated with only a slight reduction in affinity. This suggests that only one of the nitrogen atoms of the imidazoline ring is required for interaction with the alpha-2 adrenergic receptor. As was the case with the intact imidazoline ring, N-methyl substitution of the pyrrolidine ring abolishes activity. Replacing one imidazoline nitrogen with an oxygen (oxazolidine) or sulfur (thiazolidine) reduces affinity for the alpha-2 adrenergic receptor, with the greatest reduction resulting from the oxygen substitution. When the thiazolidine ring is expanded to the corresponding six-membered ring (dihydrothiazine), as in xylazine, a further reduction in affinity for the alpha-2 adrenergic receptor is observed. A dramatic loss of affinity for the alpha-2 adrenergic receptor is associated with 4-keto substitution of the imidazoline ring.

It is clear from the foregoing discussion that for the alpha-2 adrenergic receptor, optimum affinity is achieved with the intact imidazoline ring. No change in this ring has been observed to enhance affinity for alpha-2 adrenergic receptors.

3.4. Structure–Activity Relationships of the alpha-2 Adrenergic Receptor for Antagonists

There has recently been an intense research effort to search for more potent and selective alpha-2 adrenergic receptor antagonists. Although the therapeutic effectiveness of selective alpha-2 adrenergic receptor antagonists is yet to be established, their role in the treatment of variety of disease states is under investigation (Timmermans and van Zwieten, 1982; Langer et al., 1985; DeMarinis et al., 1985; Roesler et al., 1986; Roach et al., 1985; Chapleo, 1986; Nakadate et al., 1980; Ruffolo et al., 1987; Hieble et al., 1987).

There are many types of structures that have the capacity to block alpha-2 adrenergic receptors, and representative examples of the more important classes are shown in Fig. 4. Vast differences in alpha-2 adrenergic receptor selectivity of antagonists have been demonstrated among these compounds, and selectivities between alpha-1 and alpha-2 adrenergic receptors of over 20,000-fold have been reported (Timmermans et al., 1980b; Kobinger and Pichler, 1981; Starke, 1981). Some antagonists, such as phentolamine, are nonselective and block with equally high affinity both alpha-1 and alpha-2 adrenergic receptors. Others, such as rauwolscine and idazoxan, are relatively selective antagonists of the alpha-2 adrenergic receptor. As discussed previously, some representatives of the benzazepine class, such as SK&F 104078, show marked ability to discriminate not only between alpha-1 and alpha-2 adrenergic receptors, but also between pre- and postjunctional alpha-2 adrenergic receptors (DeMarinis et al., 1986; Sulpizio et al., 1986; Ruffolo et al., 1987; Hieble et al., 1986b).

Many classes of alpha-2 adrenergic receptor antagonists are known, but data describing the structure–activity relationships have been published for only a few of these classes. In other cases, we are aware of perhaps only a single representative member that has been adequately characterized, and we have used these isolated examples in an attempt to build a generalized model of the alpha-2 adrenergic receptor.

3.4.1. YOHIMBINE ALKALOIDS

The yohimbine alkaloids are some of the best known alpha-1 and alpha-2 adrenergic receptor antagonists. Baldwin et al. (1985), in a study of structure–affinity relationships for a series of natural and semisynthetic yohimbine analogs, demonstrated the importance of the C-16 carbomethoxy substituent for alpha-1/alpha-2 adrenergic receptor selectivity. These radioligand binding experiments confirmed and extended the work of Weitzell et al. (1979), who, in a series of physiologic studies, demonstrated the alpha-1 adrenergic receptor selectivity for the C-16 axially substituted corynanthine vs. the alpha-2 adrenergic receptor selectivity of the equatorially substituted isomers, yohimbine and rauwolscine (Ta-

ble 11). Based on their data, a positive electronic interaction between the C-16 substituent and the alpha-2 adrenergic receptor
was postulated.

In the development of a photoaffinity probe for the alpha-2
adrenergic receptor, Lanier et al. (1986) prepared a number of amides of both rauwolscine and yohimbine. These amides possessed
an alkyl spacer arm attached to a large photoactive phenyl ring.
These derivatives of both yohimbine and rauwolscine possessed
high affinity for the alpha-2 adrenergic receptor of rat kidney,
with the derivatives of rauwolscine demonstrating 15–100-fold
higher affinity than the yohimbine analogs. What is particularly
noteworthy concerning these molecules is the fact that incorpora-
tion of an extremely large substituent into the alkaloid framework
did not significantly reduce affinity of the rauwolscine derivatives
for the alpha-2 adrenergic receptor (Table 12).

3.4.2. BENZOQUINOLIZINES

A comparison of the pA_2 values for antagonism of clonidine-
induced inhibition of the electrically stimulated contraction of the
rat vas deferens (alpha-2 adrenergic receptor-mediated) and
antagonism of contractions induced by methoxamine in the rat
anococcygeus muscle (alpha-1 adrenergic receptor-mediated)
showed a series of substituted benzoquinolizines to be selective
alpha-2 adrenergic receptor antagonists (Lattimer et al., 1984),

Table 11
Affinities of Various Yohimbine Analogs
for alpha-1 and alpha-2
Adrenergic Receptors[a]

Compound		alpha-1[b]	alpha-2[c]
		K_i, nM	
Yohimbine		216	49
Yohimbine (C_{16}-R; C_7-R')			
R	R'		
CH_2OH	OH	230	122
CH_3	OH	170	149
CO_2Me	H	54	42
Rauwolscine		800	19
Corynanthine		19	810
Apoyohimbine		20	14

[a]Data from Baldwin et al. (1985).
[b]Inhibition of ^{3}H-prazosin binding.
[c]Inhibition of ^{3}H-clonidine binding.

Table 12
Affinities of Rauwolscine Carboxamides
for alpha-2 Adrenergic Receptors[a]

C-16 Substituent	K_i, nM[b]
CO_2CH_3 (Rauwolscine)	4.7

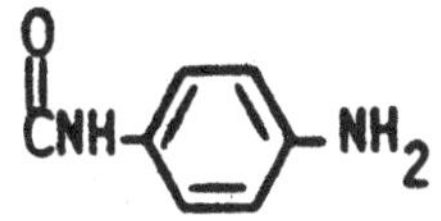

1.9

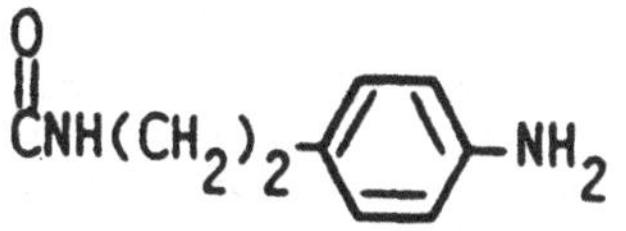

13.0

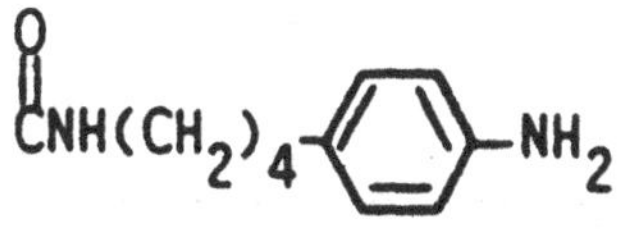

16.0

[a]Data from Lanier et al. (1986).
[b]K_i for inhibition of ^{3}H-rauwolscine binding from rat kidney.

which were more potent than yohimbine. These physiologic measurements were confirmed through radioligand binding studies (Table 13), where alpha-2/alpha-1 adrenergic receptor selectivity was also demonstrated (Cheung et al., 1984). The benzoquinolizines represent a structural fragment of the yohimbine skeleton in which the sulfonamide is positioned such that it can mimic the carbomethoxy group of the indole alkaloids, yohimbine, and rauwolscine. In a related study, Alabaster and Peters (1984) showed the benzoquinolizine derivative, WY 26703, to be nearly 100-fold weaker as antagonist of UK 14,304-induced contraction in rabbit saphenous vein than as an antagonist of UK 14,304-induced inhibition of neurotransmission in rat atrium or rat vas deferens, whereas yohimbine and phentolamine were equipotent in both models. Initially, these data were believed to demonstrate that the benzoquinolizines, represented by WY 26703, could discriminate between the pre- and postsynaptic alpha-2 adrenergic receptors. Subsequently, it was demonstrated that these apparent differences represented species differences in the alpha-2 adrenergic receptors of rats vs. rabbits, rather than between pre- and postjunctional alpha-2 adrenergic receptors (Lattimer and Rhodes, 1985; Waterfall et al., 1985). In related studies, Lattimer

Table 13
Affinities and Selectivities of Various Benzoquinolizines
for alpha-1 and alpha-2 Adrenergic Receptors

| | R | K_B, nM | | Selectivity |
		alpha-2	alpha-1	alpha-2/alpha-1
Wy 25309	CH_3	16	912	59
Wy 26392	$CH_2CH_2CH_3$	8.3	457	55
Wy 26703	$CH_2CH(CH_3)_2$	7.1	324	47
Yohimbine		26	263	10

	K_i, nM[b]		
	H^3-Rauwolscine alpha-2	H^3-Prazosin alpha-1	alpha-2/alpha-1
Wy 25309	7.4	3495	474
Wy 26392	5.5	1751	320
Wy 26703	9.2	1951	212
Yohimbine	11.1	1015	92

[a]Data from Lattimer et al. (1984).
[b]Data from Cheung et al. (1984).

and Rhodes (1985) compared the potencies of idazoxan, imiloxan, yohimbine, rauwolscine, and three benzoquinolizines at presynaptic alpha-2 adrenergic receptors in the vas deferens of rats and rabbits. Both yohimbine and rauwolscine were equally potent in blocking the alpha-2 adrenergic receptor-mediated agonist effects of clonidine or UK 14,304. In contrast, the benzoquinolizines, as well as the benzodioxans, were found to be weaker antagonists of either clonidine or UK 14,304 at the presynaptic alpha-2 adrenergic receptors of the rabbit vas deferens relative to the presynaptic alpha-2 adrenergic receptors in the rat vas deferens (Table 13). These in vitro results, as well as some in vivo studies (Paciorek et al., 1984), imply that this class of compound may discriminate alpha-2 adrenergic receptor differences between species rather than differences between pre- and postsynaptic alpha-2 adrenergic receptors.

The alpha-adrenergic receptor specificity of some hexahydro-benzofuroquinolizines has recently been reported (Huff et al., 1986a; Vacca et al., 1986). These structures are related to both the yohimbine alkaloids and the quinolizines (Table 14). In this series, the indole aromatic portion of the yohimbine alkaloids has been replaced by a benzofuran ring, whereas the carbomethoxy group has been replaced with a sulfonamide group similar to the structures of the benzoquinolizines. In both the benzoquinoli-zines and the hexahydrobenzofuroquinolizines, the methine proton of the ring fusion and the carbon bearing the sulfonamide have a *cis* relationship to each other. The alpha-2 adrenergic receptor antagonist properties of the hexahydrobenzofuroquinoli-zine derivatives reside primarily in one enantiomer having the 2R,12bS absolute configuration, which is analogous to that of the yohimbine alkaloids (Ambady and Kartha, 1973). The most active enantiomer is an extremely potent antagonist of alpha-2 adrener-gic receptors (Table 15). Vacca et al. (1986) proposed that the ele-ments of the sulfonamide were copolanar and were perpendicular to the six-membered ring. This suggested that spirocyclic com-pounds of the type in Table 16 would also have high affinity for alpha-2 adrenergic receptors, and preliminary published results support this hypothesis.

3.4.3. BENZODIOXANS

Based upon the structure of the relatively nonselective alpha-adrenergic receptor antagonist piperoxan, Caroon et al. (1981) prepared a series of 2-[(1,4-benzodioxan-2-yl)-alkyl] imidazoles,

Table 14

Species Differences in Prejunctional alpha-2 Adrenergic Receptors in Rabbit and Rat Vas Deferens

	$-\log K_B{}^a$			
	Rabbit vas deferens		Rat vas deferens	
	vs. clonidine	vs. UK 14,304	vs. clonidine	vs. UK 14,304
Wy 26392	6.45	—	8.08[b]	—
Wy 26703	6.34	6.23	8.16[b]	8.11
Wy 25309	6.12	—	7.81[b]	—
RX781094	5.90	6.45	8.04	7.97
RS21361	5.08	—	6.38	—
Yohimbine	7.73	7.51	7.58[b]	7.20
Rauwolscine	8.10	—	7.65	—

[a]Data from Lattimer and Rhodes (1985).
[b]Data from Lattimer et al. (1984).

Table 15
Affinities of Hexahydrobenzofuroquinolines for Central and
Peripheral alpha-1 and alpha-2 Adrenergic Receptors

		K_i, nM^{a,b}		K_B, nM^{a,c}	
		^{3}H-Clonidine	^{3}H-Prazosin	vs. Clonidine	vs. Methoxamine
Stereochemistry		alpha-2	alpha-1	alpha-2	alpha-1
2 RS	12b RS	1.36	210	1.7	102
2 R	12b S	0.77	110	0.79	10
2 S	12b R	100	8000	40	2510
Yohimbine		49	220	22	302

[a]Data from Huff et al. (1986b).
[b]For displacement from calf cerebral cortex.
[c]Determined in rat vas deferens.

which were subsequently evaluated for their activity as antagonists of the presynaptic (alpha-2) and postsynaptic (alpha-1) receptors of the isolated rat vas deferens (Table 17). Although all
the compounds were markedly less potent than the reference
standards, yohimbine and rauwolscine, one analog in particular
showed significantly greater selectivity for the alpha-2 adrenergic
receptor than either standard, since it was devoid of alpha-1
adrenergic receptor antagonist activity. This compound (imiloxan) was selected for further clinical evaluation.

In studies on a related class of alpha-2 adrenergic receptor antagonists, Chapleo et al. (1983) described a series of benzodioxan
analogs from which one compound, idazoxan (RX781094), was
selected for clinical trials as an antidepressant. Many of these
analogs (Table 18) showed significantly greater potency as alpha-2
adrenergic receptor antagonists than that of the related imidazoles (Table 17). Further substitution at the 2-position (Stillings et
al., 1985) in many cases produced compounds with potencies and
selectivities significantly greater than those of the parent compound, idazoxan (Table 18). Substitution at the position does not
appear to be favorable in analogous benzodioxans, such as

Table 16
Affinities of Various Spirocyclic Antagonists
for alpha-1 and alpha-2 Adrenergic Receptors

| | | K_i, nM[a] | |
| | | ^{3}H-Clonidine | ^{3}H-Prazosin |
R_1	R_2	alpha-2	alpha-1
H	$CH_3NSO_2CH_2CH_2OH$	0.77	110
		9.1	490
		2.9	900
		1.3	314
Yohimbine		49	220

[a]Data from Vacca et al. (1986).
[b]Displacement from rat cerebral cortex.

piperoxan (Stillings et al., 1986). Modification of the 1,4-benzodi-
oxan ring system by replacing one of the oxygen atoms with carbon or increasing or decreasing the ring size, markedly, reduced
affinity for the alpha-2 adrenergic receptor (Chapleo et al., 1983).
2-Substituted dihydrobenzofurans, however, are selective antag-

Table 17
Affinities and Selectivities of Various Benzodioxanyl Imidazoles
for alpha-1 and alpha-2 Adrenergic Receptors

Structure		$-\log K_B{}^{a,b}$		Selectivity
		alpha-1	alpha-2	alpha-2/alpha-1
n	R_1			
0	H	<4.00	5.60	>40
1	H	4.22	6.10	76
2	H	4.20	6.21	102
1	CH_3	4.33	6.20	74
1	CH_2CH_3	<4.00	6.71	>500
1	$CH_2CH_2CH_3$	4.20	6.57	234
Yohimbine		6.21	7.71	32
Rauwolscine		6.27	7.92	45
Piperoxan		6.40	7.40	10

[a]Data from Caroon et al. (1981).
[b]Determined in rat vas deferens.

onists of the alpha-2 adrenergic receptor, although they show
greatly reduced affinity relative to the benzodioxans (Chapleo et
al., 1984). Similar to the structure–activity relationships seen in
the benzodioxan series, substitution at the 2-position is also a crucial factor in determining the alpha-2 adrenergic receptor-
blocking potency and selectivity of these compounds (Doxey et
al., 1983a,b,1984; Gadie et al., 1983).

3.5. Model of the alpha-2 Adrenergic Receptor Based on the Structural Requirements for alpha-2 Adrenergic Receptor Antagonists

Competitive antagonists are believed to exert their activity by
binding, in a noncovalent manner, to the agonist site, thereby
blocking access by the agonist. Most of the alpha-2 adrenergic receptor antagonists discussed above show competitive kinetics, indicating that they all act at the same or similar site, although not
necessarily in the same orientation. Noncovalent binding of an
antagonist to a receptor active site results if the antagonist is

Table 18

Comparison of the Affinities and Selectivities of Various 2-Alkyl Derivatives of the Benzodioxan, Idazoxan (RX781094), for alpha-1 and alpha-2 Adrenergic Receptors

Compound	R	K_B, nM[a]		Selectivity	K_i, nM[b]		Selectivity
		alpha-2[c]	alpha-1[d]	alpha-2/alpha-1	alpha-2[c]	alpha-1[c]	alpha-2/alpha-1
RX781094	H (Idazoxan)	4.7	1000	214	3.1	91	29
RX801079	CH_3	7.6	2820	371	4.4	449	102
RX811033	CH_2CH_3	2.0	871	447	1.0	174	172
RX811054	$(CH_2)_2CH_3$	1.2	380	331	1.0	107	108
RX811005	$C=CH_2$ CH_3	10	15100	1514	16.3	1974	110

[a]Data from Doxey et al. (1983b).
[b]Data from Gadie et al. (1983).
[c]Determined in rat vas deferens.
[d]Determined in rat anococcygeus.
[e]Determined in rat cerebral cortex.

chemically complementary to the site, such that, for example, hydrogen bonding interactions can be formed, lipophilic regions of ligand and receptor are in close proximity, and local dipoles in the two molecules are favorably situated. Although the alpha-2 adrenergic receptor antagonists represent a variety of structural classes, they do exhibit chemically similar profiles that suggest that they may act not only at the same site, but in a similar orientation to one another.

The affinity of a molecule for a receptor is a function not of its internal connectivity, but of the overall chemical "face" that it presents to the surrounding active site. The important structural properties for affinity are the three-dimensional shape and the electronic character in each region. In spite of their varying atomic connectivity, the competitive alpha-2 adrenergic receptor antagonists in Fig. 4 contain regions of similar shape and electronic character, such as a planar aromatic region not far removed from a basic nitrogen atom, suggesting that they may bind to the receptor in a similar orientation to one another, despite the apparent diversity in their chemical constitution. Definition of a common rule for orienting (superpositioning) multiple antagonist classes can provide a more complete active site description and allow conclusions to be drawn from structure–activity relationships that ultimately allow for structural similarities to be extended from one diverse structural class to another.

Three-dimensional orientation rules can be investigated using computer-aided molecular modeling, in particular molecular graphics and conformational energy calculations. The process of developing a three-dimensional orientation rule is complicated by the fact that most molecules can adopt many shapes caused by conformational flexibility. In order to simplify this problem, it is logical to begin with rigid molecules, and then to extend the analysis in a stepwise fashion to chemical classes having progressively greater flexibility, in order to restrict the number of possible conformations that might be considered, and to develop a self-consistent model.

Yohimbine (Fig. 10) illustrates the three major regions found in alpha-2 adrenergic receptor antagonists; an aromatic region, a basic nitrogen with at least one protonation site available, and an optional region of modest polarity (semipolar region) sometimes adjacent to a bulky lipophilic area. Yohimbine itself is a nearly flat, angular molecular whose N-H bond forms an angle of 72° with the aromatic plane. The basic hydrogen atom (pK_a = 7.13) (Lambert et al., 1978) lies at a distance of 5.35Å from the center of the electron-rich aromatic region. The ester group, both in the

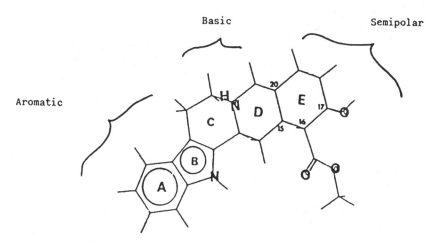

Fig. 10.   Chemical structure of yohimbine indicating the aromatic, basic, and semipolar regions.

crystal structure and from molecular mechanics calculations [SYBYL SIMPLEX method (Tripos Associates, 1984)], lies roughly perpendicular to the plane of the DE rings, placing one of the oxygens in the ester below the major plane of the molecule (Fig. 11A).

alpha-2 Adrenergic receptor antagonists that are isomers or derivatives of yohimbine show very similar structural character. The stereoisomeric rauwolscine (Fig. 4) (Weitzell et al., 1979; Tanaka and Starke, 1980), which is more potent and more selective (vs. alpha-1) than yohimbine as an alpha-2 adrenergic receptor antagonist, is derived from yohimbine by two inversions; one at the DE ring fusion (C20) and one at the site of the ester substituent (C16). The net effect of the inversions, as illustrated in the calculated low-energy conformer of Fig. 11B, is to shift the position of most of the E ring, while leaving the position of the ester unchanged. Retention of the position of the ester appears to be critical, since corynanthine, a yohimbine stereoisomer formed by a single inversion at C16, places the ester group in an axial position (Fig. 12) and has radically reduced alpha-2 adrenergic receptor antagonist activity, but is nonetheless a potent alpha-1 adrenergic receptor antagonist. Apoyohimbine (Fig. 11C), in which the ester group occupies an intermediate position, is potent at both the alpha-1 and alpha-2 adrenergic receptors (Baldwin et al., 1985). It has been proposed that the enhancement of alpha-2 adrenergic receptor antagonist activity that results from an appropriately positioned ester is caused by an energetically fa-

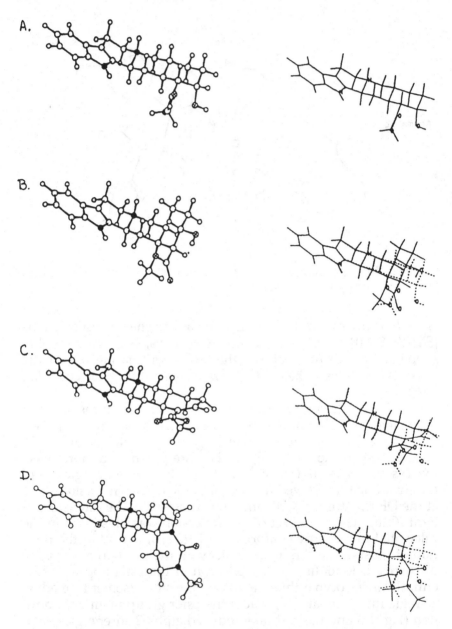

Fig. 11. Ball and stick models of the proposed active conformations of alpha-2 adrenergic receptor antagonists and superposition of the active conformers on yohimbine: (A) yohimbine; (B) rauwolscine; (C) apoyohimbine; (D) spiro derivative; (E) L654284; (F) WY 26392; (G) BE-2254; (H) BDF 6143; (I) idazoxan; (J)fluorinated pyridylpiperidine; (K) SK&F 86466; (L) tolazoline; (M) imiloxan; (N) phentolamine; (P) mianserin.

Fig. 11 (*continued*)

E.

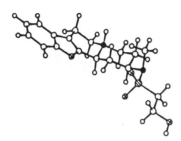

F.

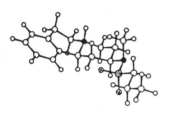

G.

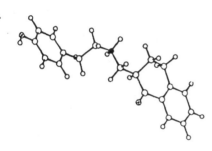
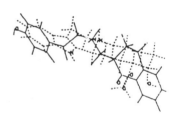

H.

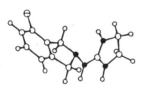

I.

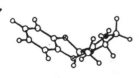

Fig. 11 (*continued*)

J.

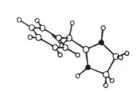

K.

L.

M.

N.

P.

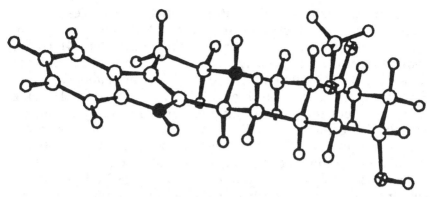

Fig. 12. Ball and stick model of corynanthine, illustrating the differences in location of the ester group relative to yohimbine (11A) and rauwolscine (11B).

vorable receptor interaction, such as hydrogen bonding, in this region.

The benzofuran derivatives of yohimbine (Table 16) are superimposed on yohimbine in Fig. 11D and E, illustrating the overlay of the aromatic regions, N-H bonds, and the polar regions. At least in the isolated system, benzofuran is electronically similar to indole in being electron rich, although the small local dipole in benzofuran points toward the oxygen, whereas the larger one of indole points into the ring. Both the spiro compound (Fig. 11D) and the sulfonamide derivative, L654284 (Fig. 11E), occupy a volume in the semipolar region beyond that which is required by yohimbine, suggesting a steric tolerance in this region. This same region is occupied by substituents of several other classes, including the groups incorporated as yohimbine affinity labels (Table 12). The size and polar character of groups in this region suggest the region may correspond to the receptor mouth through which the molecule accesses the binding site.

The benzoquinolizine, WY 26392 (Table 13) (Alabaster and Peters, 1984), superimposes well on the N-H and semipolar regions of yohimbine (Fig. 11F), although the aromatic region in the benzoquinolizines is smaller. As with the alpha-2 adrenergic receptor agonists of the imidazoline class, a variety of structures are tolerated in the aromatic region of alpha-2 adrenergic receptor antagonists, and several other classes of alpha-2 adrenergic receptor antagonists (below) also have modest aromatic regions. In the case of the benzoquinolizines, the presence of two strong polar interacting groups (the N-H and the sulfonamide) probably provide for high affinity in spite of only modest interaction in the aromatic region. As with other classes, bulky groups, such as isobutyl, are well tolerated in the semipolar region.

Previous analysis of conformational preferences among alpha-1 adrenergic receptor antagonists (DeMarinis et al., 1986) led to a postulated active conformation of BE-2254 (Fig. 4) (Hicks, 1981) that situated the phenyl group, N-H bond and naphthone regions of BE 2254-similarly to the respective aromatic, N-H, and E ring regions in corynanthine, the stereoisomer of yohimbine that is a potent and selective alpha-1 adrenergic receptor antagonist. In this conformation (Fig. 11G), the naphthone region lies somewhat below the E ring, and hence does not interfere with the ester site in yohimbine. Rather, the naphthone overlaps the region of steric tolerance required by other alpha-2 adrenergic receptor antagonists, in particular the spirobenzofuran derivative (Fig. 11D). We hypothesize that similar conformers of BE-2254 act at both alpha-1 and alpha-2 adrenergic receptors, and that, in fact, several different conformers in the naphthone half of the molecule may be permitted in the considerable steric tolerance that characterizes the E ring region.

BDF 6143 (Fig. 4) provides an example of a potent and selective alpha-2 adrenergic receptor antagonist that includes an aromatic region and a basic N-H group, but lacks the semipolar region. Although the electronic character of this molecule is quite complex, molecular mechanics calculations indicate that the two two-membered rings are sterically precluded from the coplanarity that might be preferred for extended conjugation. Thus the imidazoline ring and bridging NH must lie in a plane at an angle to the major plane of the isoindole system. With the plane between the two ring systems at 63° (Fig. 11H), the N-H bond and aromatic regions of BDF 6143 superimpose well on those regions of yohimbine. The major difference is some added bulk required for BDF 6143 around the outer periphery of yohimbine in the region of the nitrogen. This bulk may account for the alpha-2 adrenergic receptor selectivity of the molecule, although the preference may result from electronic effects in this distinctive structure.

Calculations on the benzodioxan derivatives, RX781094 [idazoxan (Fig. 4)], indicate that the single bond joining the two ring systems rotates freely, enabling a variety of conformations to exist. Optimal superposition of the N-H bond and the aromatic region are achieved when the imidazole ring forms a torsion angle of 150° relative to the proton on the asymmetric carbon atom (Fig. 11I). Only the active S-enantiomer provides this excellent fit. The R-enantiomer, when similarly oriented, has the H considerably further (5.8 Å) from the center of the aromatic system than in the S-enantiomer (5.4 Å) or yohimbine (5.4 Å). Like BDF 6143, the imidazoline of idazoxan protrudes beyond yohimbine in the NH region, which may account for the alpha-2 adrenergic receptor selectivity of idazoxan (Chapleo et al., 1983).

The fluorinated compound shown in Fig. 4 is another electronically complex selective alpha-2 adrenergic receptor antagonist. The pyridine ring is conjugated with, and presumably prefers to lie coplanar with, the connecting nitrogen atom in the piperazine ring, but the methylated nitrogen atom in the piperazine should remain basic. Substitution at the 3-position on the pyridine ring disfavors a coplanar arrangement of the rings, and the preferred conformation becomes a balance between steric interaction with the *ortho* substituent favoring a perpendicular arrangement, and conjugation with the pyridine system favoring a coplanar system. Molecular mechanics indicated an optimum angle of about 35° (Fig. 11J), which provides excellent superposition of the aromatic and N-H regions within the steric volume already known to be occupied by other alpha-2 adrenergic receptor antagonists.

The 6-chloro-*N*-methyl-2,3,4,5-tetrahydro-3-benzazepine, SK&F 86466 (Fig. 4), is among the simpler structures showing considerable alpha-2 adrenergic receptor selectivity (DeMarinis et al., 1985; Hieble et al., 1986a; Roesler et al., 1986). Crystallographic and molecular mechanics analyses of numerous benzazepines have indicated a preference for the chair conformation (Fig. 13A), although solution NMR analysis indicates considerable flexibility. The twist conformation (Fig. 13D), which is seen in a number of monounsaturated seven-membered rings and is favored in some heterosystems, is a better fit to yohimbine (Fig. 11K), since the twist is a more extended conformer (N to aromatic center distance = 4.4 Å in twist, 4.2 Å in chair) in which the proton to which the receptor presumably hydrogen bonds is less hindered than in the chair conformer. We speculate that the twist conformer is preferred for binding, although at a cost of some energy needed to induce this less-favored conformer, and that the chair conformer may have some activity as well.

Tolazoline (Fig. 4) is a relatively simple structure with antagonist activity at both the alpha-1 and the alpha-2 adrenergic receptors (Timmermans and van Zwieten, 1982). The fact that 3,4-dihydroxyl substitution of tolazoline converts the molecule to an alpha-2 adrenergic receptor agonist (Ruffolo and Waddell, 1983) supports the premise that a variety of antagonists are acting in a similar orientation to each other and to the agonists. Like BDF 6143, the arrangement in which two rings are connected by a single atom with a bond angle of 120° or less precluded coplanar arrangement of the rings, and the imidazoline ring of tolazoline lies, according to calculations, nearly perpendicular to the phenyl

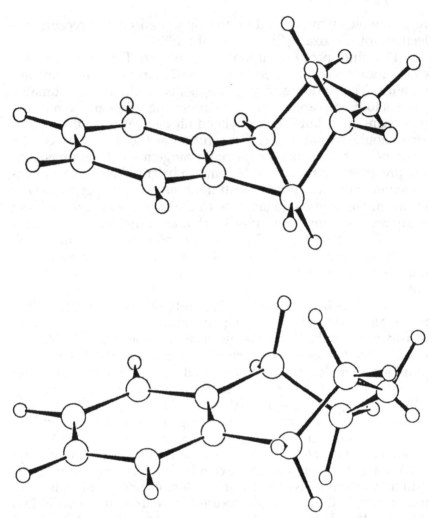

Fig. 13. Conformations of tetrahydro-3-benzazepine. Chair (above) and twist (below).

group, in agreement with the studies of deJong and van Dam (1980). The result is a very reasonable superposition of tolazoline on yohimbine (Fig. 11L). The compactness of tolazoline, which is evident in the superposition, probably accounts for its lack of alpha-1 vs. alpha-2 adrenergic receptor selectivity.

Imiloxan (RS21361, Fig. 4) also comprises two ring systems connected by a single atom, resulting in a noncoplanar arrangement of the rings (Caroon et al., 1981). Calculations indicate several low-energy structures, including structures in which the imidazole is adjacent to the 3-position and to the 1-position of the benzodioxan. Because of its shorter H-aromatic distance (5.4 Å vs. 6.7 Å with imidazole near the 3-position), the conformer with the imidazole adjacent to the 1-position gives a better fit to yohimbine (Fig. 11M). This arrangement leads to an angular molecule that, when properly situated, has the opposite curvature to yohimbine, and in effect, occupies the "bay" encompassed by yohimbine. The selectivity of imiloxan for the alpha-2 adrenergic receptor is certainly enhanced by the ethyl group, presumably for steric reasons, but the parent structure lacking the ethyl group still shows considerable preference for the alpha-2 adrenergic receptor.

Phentolamine (Fig. 4) (Besse and Furchgott, 1976; Ruffolo and Patil, 1979) is a nonselective alpha-1/alpha-2 adrenergic receptor antagonist. Both the crystal structure of the molecule and a similar computed low-energy conformer fit the model well. The major difference between the two conformers is a slight shift in the location of the *p*-tolyl group. Although either or both conformers may be active, the computed conformer (Fig. 11N) situates the tolyl group nearer to a region known to be sterically tolerated in alpha-2 adrenergic receptors and may be preferred at this site.

Mianserin (Fig. 4) is a rigid alpha-2 adrenergic receptor antagonist for which the crystal structure is superimposed on yohimbine (Fig. 11P). The C6 ring in mianserin, which is superimposed on the aromatic region of yohimbine, is the one in which the N-to-H vector points away from the aromatic plane as in the rigid yohimbine, rather than toward the plane. In this orientation, mianserin places considerable bulk in the "bay" region of yohimbine, above the yohimbine plane, in much of the same volume required by phentolamine and imiloxan, but does not violate any steric restrictions deduced for other alpha-2 adrenergic receptor antagonists.

3.5.1. SUMMARY OF THE MODEL

Analysis of the above alpha-2 adrenergic receptor antagonist structures suggests the following structural requirements made by alpha-2 adrenergic receptors for antagonists, which are summarized in Fig. 14.

> 1. A proton or a basic N and an aromatic region are required, oriented such that the H lies 5.3 ± 0.3 Å from the center of the aromatic system, and the N-H bond vector points away from the aromatic plane and forms an angle of 60 ± 16° with the plane.
> 2. The presence of an aromatic region with lipophilic substituents in all the alpha-2 adrenergic receptor antagonists examined implies that the active site is relatively lipophilic in that re-

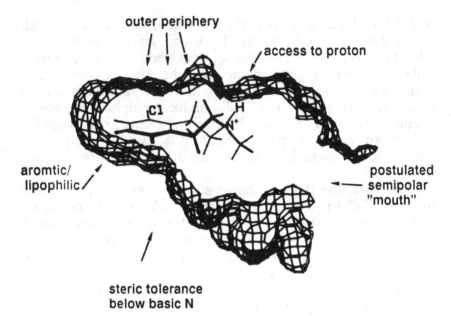

Fig. 14. Postulated stereochemical requirements of alpha-2 adrenergic receptors for antagonists based on molecular graphics analysis.

gion. Several factors implicate a polar region at the far end of the plane of the lipophilic region, however, based on the following observations: the acceptability of polar hydroxyls on the aromatic system in phentolamine and BE-2254, the ability to convert tolazoline to an agonist by addition of a catechol group to the phenyl ring, and the inactivity at alpha-2 adrenergic receptors of alpha-1 adrenergic receptor antagonists with more extended aromatic groups.

3. Considerable steric tolerance exists above the aromatic plane in the "bay" region of yohimbine, extending toward the semipolar region. It is proposed that this region corresponds to the mouth of the receptor.

4. The outer periphery of yohimbine corresponds to a sterically sensitive region that considerably influences selectivity for the alpha-2 adrenergic receptor.

5. The region below the N is more sterically tolerant in alpha-2 adrenergic receptors as opposed to alpha-1 adrenergic receptors.

4. Summary and Conclusions

During the past decade, great strides have been made in our understanding of the biochemistry and pharmacology of the alpha-2 adrenergic receptor. The alpha-2 adrenergic receptor plays a key role in biological function. This is evidenced by the fact that the alpha-2 adrenergic receptor plays a prominent functional role in most organs of the body and in the key systems responsible for survival of the organism and maintenance of optimum biological activity. This is most apparent in the cardiovascular system where alpha-2 adrenergic receptors appear to be involved in the maintenance and regulation of circulatory function.

Classically, most receptors have been characterized based on structure–activity relationships obtained for selective agonists and antagonists interacting with the receptor. Although there are many newer and more sophisticated approaches to receptor characterization, structure–activity relationships still provide important information regarding the chemical requirements made by the receptor for its occupation by ligands and its subsequent activation by those ligands possessing intrinsic efficacy, and therefore, agonist activity. It is also known for the alpha-2 adrenergic receptor that the structural demands that regulate the attachment of a drug to the receptor (i.e., affinity) differ dramatically from those chemical requirements that drugs must possess in order to stimulate the receptor (i.e., efficacy). Using newer techniques involving computer modeling and known structure–activity relationships for conformationally restrained alpha-2 adrenergic receptor antagonists, the topography and three-dimensional constitution of the active site of the alpha-2 adrenergic receptor can begin to be understood.

As described in the first and preceding chapter, it is now becoming apparent that alpha-2 adrenergic receptors do not represent one homogeneous population of receptors, but rather are composed of at least two subtypes of alpha-2 adrenergic receptors that may be differentiated by new and highly selective antagonists, which may distinguish between pre- and postsynaptic alpha-2 adrenergic receptors. In addition, several novel alpha-2 adrenergic receptor antagonists have the capacity to uncover species differences in alpha-2 adrenergic receptors. Clearly, systematic subclassification of alpha-2 adrenergic receptors will continue to become more sophisticated as additional pharmacologic tools become available.

It is attractive to realize that complementary experimental approaches, such as cloning of the alpha-2 adrenergic receptor, will eventually provide sufficient receptor protein to permit three-dimensional structural information regarding the ligand binding domain of the alpha-2 receptor(s). X-Ray crystallographic data obtained using free and ligand-bound receptor preparations will permit a rigorous test of the models described for the alpha-2 receptor binding site proposed based on structure–activity data.

REFERENCES

Alabaster, V. A. and Peters, C. J. (1984) Pre- and postjunctional α_2-adrenoceptors can be differentially antagonized. *Br. J. Pharmacol.* **81**, 163P.

Ambady, G. and Kartha, G. J. (1973) Crystal structure and absolute configuration of yohimbine hydrochloride. *J. Cryst. Mol. Struct.* **3**, 37–45.

Anden, N. E., Grabowska, M., and Strombom, K. (1976) Different α-adrenoceptors mediating biochemical and functional effects of clonidine and receptor blocking agents. *Naunyn Schmiedebergs Arch. Pharmacol.* **292**, 43–52.

Armah, I. B. (1985) Aganodine (BDF 7570): A novel peripherally active alpha$_2$-agonist with a preferential venous site of action. *Naunyn Schmiedbergs Arch. Pharmacol.* (suppl.) **329**, 317.

Baldwin, J. J., Huff, J. R., Randall, W. C., Vacca, J. P., and Zrada, M. M. (1985) Structure affinity relationships for yohimbine analogs at central alpha adrenoceptors. *Eur. J. Med. Chem.* **20**, 67–69.

Banning, J. W., Rice, P. J., Miller, D. D., Ruffolo, R. R., Hamada, A. H., and Patil, P. N. (1984) Differences in the Adrenoceptor Activation by Stereoisomeric Catecholimidazolines and Catecholamines, in *Neuronal and Extraneuronal Events in Autonomic Pharmacology* (Fleming, W. W., ed.) Raven, New York.

Berridge, T. L., Doxey, J. C., Roach, A. G., Strachen, D. A., and Virdee, N. K. (1984) Antagonist effects of Idazoxan and Yohimbine against several α_2-adrenoceptor agonists in pithed rats. *Br. J. Pharmacol.* **81**, 860.

Berthelsen, S., and Pettinger, W. A. (1977) Minireview: A functional basis for classification of alpha-adrenergic receptors. *Life Sci.* **21**, 595–606.

Besse, J. C. and Furchgott, R. F. (1976) Dissociation constants and relative efficacies of agonists acting on alpha adrenergic receptors in rabbit aorta. *J. Pharmacol. Exp. Ther.* **197**, 66–78.

Brittain, R. T., Drew, G. M., and Levy, G. (1982) The α- and β-adrenoceptor blocking potencies of labetolol and its individual stereoisomers in anesthetized dogs and in isolated tissues. *Br. J. Pharmacol.* **77**, 105–114.

Brown, G. L. and Gillepsie, J. S. (1957) The output of sympathetic transmitter from the spleen of the cat. *J. Physiol.* **138**, 81–102.

Bylund, D. B. (1985) Heterogeneity of alpha-2 adrenergic receptors. *Pharmacol. Biochem. Behav.* **22**, 835–843.

Bylund, D. B. (1987) Subtypes of alpha-2 Adrenergic Receptors in Human and Rat Brain, in *Epinephrine in the Central Nervous System* (U'Prichard, D., Stolk, J., and Fuxe, K., eds.) Oxford University Press, in press.

Cantacuzene, D., Kirk, K. L., McCulloh, D. H., and Creveling, C. R. (1979) Effect of fluorine substitution on the agonist specificity of norepinephrine. *Science* **204**, 1217–1219.

Caroon, J. M., Clark, R. D., Kluge, A. F., Olah, R., Repke, D. B., Unger, S. H., Michel, A. D., and Whiting, R. L. (1981) Structure–activity relationships for 2-substituted imidazoles as alpha-2-adrenoceptor antagonists. *J. Med. Chem.* **25**, 666–670.

Chapleo, C. B. (1986) The discovery of idazoxan. *Chem. Brit.* **22**, 313–314.

Chapleo, C. B., Myers, P. L., Butler, R. C. M., Doxey, J. C., Roach, A. G., and Smith, C. F. C. (1983) Alpha-adrenoceptor reagents. 1. Synthesis of some 1,4-benzodioxans as selective presynaptic alpha-2 adrenoceptor antagonists and potential antidepressants. *J. Med. Chem.* **26**, 823–831.

Chapleo, C. B., Myers, P. L., Butler, R. C. M., Davis, J. A., Doxey, J. C., Higgins, S. D., Myers, M., Roach, A. G., Smith, C. F. C., Stillings, M. R., and Welbourn, A. P. (1984) Alpha adrenoceptor reagents 2. Effects of modification of the 1,4-benzodioxan ring system on alpha-adrenoceptor activity. *J. Med. Chem.* **27**, 570–576.

Cheung, Y. D., Barnett, D. B., and Nahorski, S. R. (1982) [^{3}H]Rauwolscine and [^{3}H] yohimbine binding to rat cerebral and human platelet membranes; possible heterogeneity of α-adrenoceptors. *Eur. J. Pharmacol.* **84**, 79–85.

Cheung, L. D., Nahorski, S. R., Rhodes, K. F., and Waterfall, J. F. (1984) Studies of the alpha-2-adrenoceptor affinity and the alpha-2 to alpha-1 adrenoceptor selectivity of some substituted benzoquinolizines using receptor binding techniques. *Biochem. Pharmacol.* **33**, 1566–1565.

Cody, V. and DeTitta, G. T. (1979) The molecular conformation of clonidine hydrochloride, an α-adrenergic agonist. *J. Cryst. Mol. Struct.* **9**, 33–43.

DeBernardis, J. F., Kerkman, D. J., Winn, M., Bush, E. N., Arendsen, D. L., McClellan, W. J., Kyncl, J. J., and Basha, F. Z. (1985) Conformationally defined adrenergic agents. 1. Design and synthesis of novel alpha-2 selective adrenergic agents: Electrostatic repulsion based conformational prototypes. *J. Med. Chem.* **28**, 1398–1404.

deJonge, A. and Soudijn, W. (1981) Relationships between structure and alpha-adrenergic receptor affinity of clonidine and some related cyclic amidines. *Eur. J. Pharmacol.* **69**, 175–188.

deJonge, A. P. and van Dam, H. (1980) Ultraviolet photoelectron spectroscopy of cyclic amidines. 2. Electronic structure of clonidine and some related 2-(phenylimino) imidazolines with α-adrenergic activity. *J. Med. Chem.* **23**, 889–895.

deJonge, A., Timmermans, P. B. M. W. M., and van Zwieten, P. A. (1981a) Participation of cardiac presynaptic α_2-adrenoceptors in the bradycardic effects of clonidine and analogues. *Naunyn Schmiedebergs Arch. Pharmacol.* **317**, 8–12.

deJonge, A., van Meel, J. C. A., Timmermans, P. B. M. W. M., and van Zwieten, P. A. (1981b) A lipophillic, selective alpha-1 adrenoceptor agonist: 2-(2-Chloro-5-trifluoromethylphenylamino)-imidazoline (St 587). *Life Sci.* **28**, 2009–2016.

deJonge, A., Timmermans, P. B. M. W. M., and van Zwieten, P. A. (1982) Quantitative aspects of alpha-adrenergic effects induced by clonidine-like imidazolidines. I. Central hypotensive and peripheral hypertensive activities. *J. Pharmacol. Exp. Ther.* **222**, 705–711.

DeMarinis, R. M., Bryan, W. M., Shah, D. H., Hieble, J. P., and Pendleton, R. G. (1981) Alpha-adrenergic agents 1. Direct-acting alpha-1 agonists related to methoxamine. *J. Med. Chem.* **24**, 1432–1437.

DeMarinis, R. M., Shah, D. H., Hall, R. F., Hieble, J. P., and Pendleton, R. G. (1982) Alpha-adrenergic agents. 2. Synthesis and alpha-1 agonist activity of 2-aminotetralins. *J. Med. Chem.* **25**, 136–141.

DeMarinis, R. M., Hieble, J. P., and Matthews, W. D. (1985). Blockade of the postjunctional alpha-2 adrenoceptor in the treatment of hypertension. Abstracts of the American Chemical Society 189th Annual Meeting. Division of Medicinal Chemistry Symposium on Special Topics in Drug Discovery.

DeMarinis, R. M., Pfeiffer, F. R., Lavanchy, P., Hieble, J. P., Nichols, A., Gruber, F., Matthews, W. D., and Ruffolo, R. R. (1986) Evidence for the heterogeneity of pre- and postjunctional alpha-2 adrenoceptors: SK&F

104078, a selective postjunctional alpha-2 antagonist. Abstracts of the American Chemical Society 192nd Annual Meeting. Division of Medicinal Chemistry Abstract #15.

Docherty, J. R. and Hyland, L. (1985) No evidence for differences between pre- and postjunctional alpha-2 adrenoceptors. *Br. J. Pharmacol.* **86**, 335–339.

Docherty, J. R. and McGrath, J. C. (1980) A comparison of pre- and postjunctional potencies of several α-adrenoceptor agonists in the cardiovascular system and anococcygeus muscle of the rat. Evidence for two types of postjunctional α-adrenoceptor. *Naunyn Schmiedebergs Arch. Pharmacol.* **312**, 107–116.

Doxey, J. C., Roach, A. G., and Smith, C. F. C. (1983a) Studies on RX781094: A selective, potent and specific antagonist of alpha-2 adrenoceptors. *Br. J. Pharmacol.* **78**, 489–505.

Doxey, J. C., Roach, A. G., Strachan, D. A., and Virdee, N. (1983b) 2-Alkyl analogues of RX781094: Potent, selective antagonists at peripheral alpha-2 adrenoceptors. *Br. J. Pharmacol.* **79**, 311P.

Doxey, J. C., Roach, A. G., Strachan, D. A., and Virdee, N. K. (1984) Selectivity and potency of 2 alkyl analogues of the alpha-2 adrenoceptor antagonist idazoxan (RX781094) in peripheral systems. *Br. J. Pharmacol.* **83**, 713–722.

Drew, G. M. (1976) Effects of alpha-adrenoceptor agonists and antagonists on pre- and postsynaptically located alpha-adrenoceptors. *Eur. J. Pharmacol.* **36**, 313–320.

Drew, G. M. (1980) Postsynaptic alpha-2 adrenoceptors mediate pressor responses to 2-N,N-dimethylamino-5,6-dihydroxy-1,2,3,4-tetrahydronap hthnaph (M-7). *Eur. J. Pharm.* **65**, 85–87.

Drew, G. M. (1985) What do antagonists tell us about α-adrenoceptors? *Clin. Sci.* **68** (suppl. 10), 15s–19s.

Drew, G. M. and Whiting, S. B. (1979) Evidence for two distinct types of postsynaptic α-adrenoceptor in vascular smooth muscle in vivo. *Br. J. Pharmacol.* **67**, 207–215.

Dubocovich, M. L. and Langer, S. Z. (1974) Negative feed-back regulation of noradrenaline release by nerve stimulation in the perfused cat's spleen: Differences in potency of phenoxybenzamine in blocking pre- and postsynaptic adrenergic receptors. *J. Physiol.* **237**, 505–519.

Easson, L. H. and Stedman, E. (1933) Studies on the relationship between chemical constitution and physiological action. V. Molecular dissymmetry and physiological activity. *Biochem. J.* **27**, 1257–1266.

Feller, D. J. and Bylund, D. B. (1984) Comparison of α-adrenergic receptors and their regulation in rodent and porcine species. *J. Pharmacol. Exp. Ther.* **228**, 275–282.

Fuder, H., Braun, H-J., and Schimkus, R. (1986) Presynaptic alpha-2 adrenoceptor activation and coupling of the receptor-presynaptic effector system in the perfused rat heart: Affinity and efficacy of phenethylamines and imidazoline derivatives. *J. Pharmacol. Exp. Ther.* **237**, 237–245.

Gadie, B., Lane, A. C., McCarthy, P. S., Tulloch, I. F., and Walter, D. S. (1983) 2-Alkyl analogs of RX781094: Potent, selective antagonists at central alpha-2 adrenoceptors. *Br. J. Pharmacol.* **79**, 312P.

Guicheney, P., Dausse, J. P., and Meyer, P. (1981) Affinities respective du S

3341 et de la Clonidine pour les receptors adrenergic alpha-1 et alpha-2 du cerveau du rat. *J. Pharmacol.* **12**, 3, 255–262.

Hicks, P. E. (1981) Antagonism of pre- and postsynaptic alpha-adrenoceptors by BE 2254 (HEAT) and prazosin. *J. Auton. Pharmacol.* **1**, 391–397.

Hicks, P. E. and Cannon, J. G. (1980) Cardiovascular effects of 2-(N,-N-dimethyl)-amino-6,7-dihydroxy-1,2,3,4-tetrahydronaphthalene in pithed rats: Differential antagonism by yohimbine and prazosin. *J. Pharm. Pharmacol.* **32**, 786–787.

Hieble, J. P. and Pendleton, R. G. (1979) Effects of ring substitution on the pre- and postjunctional alpha-adrenergic activity of aryliminoimidazolidines. *Naunyn Schmiedebergs Arch. Pharmacol.* **309**, 217–224.

Hieble, J. P. and Woodward, D. F. (1984) Different characteristics of postjunctional α-adrenoceptor on arterial and venous smooth muscle. *Naunyn Schmiedebergs Arch. Pharmacol.* **328**, 44–50.

Hieble, J. P., DeMarinis, R. M., Matthews, W. D., and Fowler, P. J. (1986a) Selective alpha-2 adrenoceptor blockade by SK&F 86466: In vitro characterization of receptor selectivity. *J. Pharmacol. Exp. Ther.* **236**, 90–96.

Hieble, J. P., Ruffolo, R. R., Nichols, A., Gruber, F., Matthews, W., DeMarinis, R. M., Pfeiffer, F., and Lavanchy, P. (1986b) Pharmacological differentiation of pre- and postjunctional alpha-2 adrenoceptors. *J. Hypertens.* **4** (suppl. 6), S189–S192.

Hieble, J. P., Sulpizio, A. C., DeMarinis, R. M., and Ruffolo, R. R. (1987) Alpha-2 adrenoceptor antagonists: A new approach to cardiovascular therapy. *Blood Vess.*, in press.

Huff, J. R., Anderson, P. S., Baldwin, J. J., Clineschmidt, B. V., deSolms, S. J., Guare, J. P., Hunt, C. A., Lotti, V. J., Pettibone, D. J., Randall, W. C., Sakurai, Y., Sanders, W. M., Vacca, J. P., and Young, S. D. (1986a) A new class of spirocyclic alpha-2 adrenoceptor antagonists. Abstracts of the American Chemical Society 191st National Meeting. Division of Medicinal Chemistry Abstract #56.

Huff, J. R., Anderson, P. S., Baldwin, J. J., Clineschmidt, B. V., Guare, J. P., Lotti, V. J., Pettibone, D. J., Randall, W. C., and Vacca, J. P. (1986b) N-(1,3,4,6,7-12b-Hexahydro-2H-benzo[b]furo[2,3-a]quinoliN-(1,3,4 6, methyl-2-hydroxyethanesulfonamide: A potent and selective alpha-2 adrenoceptor antagonist. *J. Med. Chem.* **28**: 1756–1759.

Ison, R. R., Partington, P., and Roberts, G. C. K. (1973) The conformation of catecholamines and related compounds in solution. *Mol. Pharmacol.* **9**, 756–765.

Jarrott, B., Louis, W. J., and Summers, R. J. (1979) The characteristics of ^{3}H-clonidine binding to an α-adrenoceptor in membranes from guinea pig kidney. *Br. J. Pharmacol.* **65**, 663–670.

Jarrott, B., Summers, R. J., Culvenor, A. J., and Louis, W. J. (1980) Characterization of alpha-adrenoceptors in rat and guinea pig tissues using radiolabeled agonists and antagonists. *Circ. Res.* **46**, 115–120.

Kenakin, T. P. (1984) The classification of drugs and drug receptors in isolated tissues. *Pharmacol. Rev.* **36**, 165–222.

Kier, L. B. (1968) The preferred conformations of ephedrine isomers and the nature of the α-adrenergic receptor. *J. Pharmacol. Exp. Ther.* **164**, 75–81.

Kier, L. B. (1969) The preferred conformations of noradrenaline and a consideration of the alpha adrenergic receptor. *J. Pharm. Pharmacol.* **21**, 93–96.

180   Ruffolo et al.

Kier, L. B. and Truitt, E. B., Jr. (1970) The preferred conformation of dopamine from molecular orbital theory. *J. Pharmacol. Exp. Ther.* **174,** 94–98.

Kirk, K. L. and Creveling, C. R. (1984) The chemistry and biology of ring fluorinated biogenic amines. *Med. Res. Rev.* **4,** 189–220.

Kobinger, W., and Pichler, W. (1977) Pharmacological characterization of B-HT 933 (2-amino-6-ethyl-4,5,7,8-tetrahydro-6H-oxazolo-[5,4-d]-azep p dihydrochloride) as a hypotensive agent of the "clonidine-type." *Naunyn Schmiedebergs Arch. Pharmacol.* **300,** 39–46.

Kobinger, W. and Pichler, L. (1981) Alpha-1 and alpha-2 adrenoceptor subtypes: Selectivity of various agonists and relative distribution of receptors as determined in rats. *Eur. J. Pharmacol.* **73,** 313–321.

Kobinger, W. and Pichler, L. (1980) Investigations into different types of post- and presynaptic sites in rats. *Eur. J. Pharmacol.* **65,** 393–402.

Kobinger, W., Lillie, C., and Pichler, L. (1980) Central cardiovascular alpha-adrenoceptors. Relation to peripheral receptors. *Circ. Res.* **46,** 121–25.

Kobinger, W., Lillie, C., and Pichler, L. (1979) Cardiovascular actions of N-allyl-clonidine (St 567), a substance with specific bradycardic action. *Eur. J. Pharmacol.* **58,** 141–150.

Lambert, G. A., Lang, W. J., Friedman, E., Meller, E., and Gershon, S. (1978) Pharmacological and biochemical properties of isomeric yohimbine alkaloids. *Eur. J. Pharmacol.* **49,** 39–48.

Langer, S. Z. (1970) The metabolism of [^{3}H] noradrenaline released by electrical stimulation from the isolated nictitating membrane of the cat and from the vas deferens of the rat. *J. Physiol.* **208,** 515–546.

Langer, S. Z. (1974) Presynaptic regulation of catecholamine release. *Biochem. Pharmacol.* **23,** 1783–1800.

Langer, S. Z., Duval, N., and Massingham, R. (1985) Pharmacological and therapeutic significance of alpha adrenoceptor subtypes. *J. Cardiovasc. Pharmacol.* **7,** (suppl. 8), S1–S8.

Langer, S. Z., Adler, E., Enero, M. A., and Stefano, F. J. E. (1971) The role of the α-receptor in regulating noradrenaline overflow by nerve stimulation. *Proc. Int. Union Physiol. Sci.* **9,** 335.

Lanier, S. M., Hess, H. J., Grodsky, A., Homcy, C. J., and Graham, R. M. (1986) Binding properties of aminophenyl carboxamide derivatives of the alpha-2 adrenergic receptor antagonists, rauwolscine and yohimbine: Spatial and stereochemical considerations. *Fed. Proc.* **45,** 1570.

Lattimer, N. and Rhodes, K. F. (1985) A difference in the affinity of some selective alpha-2 adrenoceptor antagonists when compared on isolated vasa deferentia of rat and rabbit. *Naunyn Schmiedebergs Arch. Pharmacol.* **329,** 278–281.

Lattimer, N., McAdams, R. P., Rhodes, K. F., Sharma, S., Turner, S. J., and Waterfall, J. F. (1984) Alpha-2 adrenoceptor antagonism and other pharmacological antagonist properties of some benzoquinolizines and yohimbine in vitro. *Naunyn Schmiedebergs Arch. Pharmacol.* **327,** 312–318.

Laubie, M., Poignant, J. C., Scuvee-Moreau, J., Dabire, H., Dresse, A., and Schmitt, H. (1985) Pharmacological propeties of (N-dicyclopropyl-methyl) amino-2-oxazoline (S 3341), an alpha-2 adrenoceptor agonist. *J. Pharmacol.* **16,** 259–278.

Malta, E., Ong, J. S. B., Raper, C., Tawa, P. E., and Vaughan, G. N. (1980)

Structure–activity relationships of clonidine and tolazoline-like compounds at histamine and alpha-adrenoceptor sites. *Br. J. Pharmacol.* **69**, 679–688.

McGrath, J. C. (1982) Evidence for more than one type of postjunctional α-adrenoceptor. *Biochem. Pharmacol.* **31**, 467–484.

Miller, D. D., Hamada, A., Rice, P. J., and Patil, P. N. (1980) Optically active imidazolines and their interactions with alpha adrenergic receptors. *Abstracts of the American Chemical Society Division of Medicinal Chemistry* Abstract ;ns38 (March, 1980).

Nahorski, S. R., Barnett, D. B., and Cheung, Y. D. (1985) α-Adrenoceptor-effector coupling; Affinity states or heterogeneity of the α-adrenoceptor? *Clin. Sci.* **68** (suppl. 10), 39s–42s.

Nakadate, T., Nakaki, T., Muraki, T., and Kato, R. (1980) Adrenergic regulation of blood glucose levels: Possible involvement of postsynaptic alpha-2 type adrenergic receptors regulating insulin release. *J. Pharmacol. Exp. Ther.* **215**, 226–230.

Nimit, Y., Cantacuzene, D., Kirk, K. L., Creveling, C. R., and Daly, J. W. (1980) The binding of fluorocatecholamines to adrenergic and dopaminergic receptors in rat brain membranes. *Life Sci.* **27**, 1577–1585.

Paciorek, P. M., Pierce, V., Shepperson, N. B., and Waterfall, J. F. (1984) An investigation into the selectivity of a novel series of benzoquinolizines for alpha-2 adrenoceptors in vivo. *Br. J. Pharmacol.* **82**, 127–134.

Patil, P. N. and Jacobowitz, D. (1968) Steric aspects of adrenergic drugs. IX. Pharmacologic and histochemical studies on isomers of cobefrin (α-methyl norepinephrine). *J. Pharmacol. Exp. Ther.* **161**, 279–295.

Petrash, A. C. and Bylund, D. B. (1986) Alpha-2 adrenergic receptor subtypes indicated by [^{3}H] yohimbine binding in human brain. *Life Sci.* **38**, 2129–2137.

Pettinger, W. A. (1977) Unusual α-adrenergic receptor potency of methyldopa metabolites on melanocyte function. *J. Pharmacol. Exp. Ther.* **201**, 622–626.

Pichler, L., Plachita, P., and Kobinger, W. (1980) Effect of azepexole (B-HT 933) on pre- and postsynaptic alpha-adrenoceptors at peripheral and central nervous sites. *Eur. J. Pharmacol.* **65**, 233–241.

Portoghese, P. S. (1967) Stereochemical studies on medicinal agents. IV. Conformational analysis of ephedrine isomers and related compounds. *J. Med. Chem.* **10**, 1057–1063.

Pullman, B., Coubelis, J.-L., Courriere, Ph. D., and Gervois, J. P. (1972) Quantum mechanical study of the conformational properties of phenethylamines of biochemical and medicinal interest. *J. Med. Chem.* **15**, 17–23.

Roach, A. G., Berridge, D. T., Dexter, D. T., and Doxey, J. C. (1985) Effects of selective α_1 and α_2-adrenoceptor antagonists on plasma glucose and insulin levels in rats. *Fed. Proc.* **44**, 1391.

Roesler, J. M., Hieble, J. P., McCafferty, J. P., DeMarinis, R. M., and Matthews, W. D. (1986) Characterization of the antihypertensive activity of SK&F 86466, a selective alpha-2 antagonist in the rat. *J. Pharmacol. Exp. Ther.* **236**, 1–7.

Ruffolo, R. R. (1982) Important concepts of receptor theory. *J. Auton. Pharmacol.* **2**, 277–295.

Ruffolo, R. R. (1983a) Structure–Activity Relationships of alpha-Adrenoceptor Agents, in Adrenoceptors and Catecholamine Action Part B (Kunos, G., ed.) John Wiley, New York.

Ruffolo, R. R., Jr. (1983b) Stereoselectivity in Adrenergic Agonists and Adrenergic Blocking Agents, in Stereochemistry and Biological Activity of Drugs (Ariens, E. J., Soudijn, W., and Timmermans, P. B. M. W. M., eds.) Blackwell Scientific, Oxford.

Ruffolo, R. R., Jr. (1984a) α-Adrenoceptors. Monogr. Neural Sci. 10, 224–253.

Ruffolo, R. R., Jr. (1984b) Interaction of agonists with peripheral α-adrenergic receptors. Fed. Proc. 43, 2910–2916.

Ruffolo, R. R., Jr. (1984c) Stereochemical requirements for activation and blockade of α_1 and α_2-adrenoceptors. Trends Pharmacol. Sci. 5, 160–164.

Ruffolo, R. R., Jr. (1984d) Stereochemical Requirements for Activation and Blockade of α_1 and α_2-Adrenoceptors, in Receptor Again (Lamble, J. W. and Abbott, A. C., eds.) Elsevier, Amsterdam.

Ruffolo, R. R. (1985a) Relative agonist potency as a means of differentiating α-adrenoceptors and α-adrenergic mechanisms. Clin. Sci. 68 (suppl. 10), 9s–14s.

Ruffolo, R. R., Jr. (1985b) Distribution and function of peripheral α-adrenoceptors in the cardiovascular system. Pharmacol. Biochem. Behav. 22, 827–833.

Ruffolo, R. R. and Messick, K. (1985) Evaluation of the alpha-1 and alpha-2 adrenoceptor-mediated effects of a series of dimethoxy-substituted tolazoline derivatives in the cardiovascular system of the pithed rat. J. Pharmacol. Exp. Ther. 232, 94–99.

Ruffolo, R. R. and Patil, P. N. (1979) Kinetics of alpha adrenoceptor blockade by phentolamine in the normal and denervated rabbit aorta and rat vas deferens, Blood Vess. 16, 135–143.

Ruffolo, R. R., Jr. and Waddell, J. E. (1982) Stereochemical requirement of α_2-adrenergic receptors for α-methy substituted phenethylamines. Life Sci. 31, 2999–3007.

Ruffolo, R. R. and Waddell, J. E. (1983) Aromatic and benzylic hydroxyl substitution of imidazolines and phenethylamines: Differences in activity at alpha-1 and alpha-2 adrenergic receptors. J. Pharmacol. Exp. Ther. 224, 559–566.

Ruffolo, R. R. and Yaden, E. L. (1983) Vascular effects of the stereoisomers of dobutamine. J. Pharmacol. Exp. Ther. 224, 46–50.

Ruffolo, R. R., Dillard, R. D., Waddell, J. E., and Yaden, E. L. (1979a) Receptor interactions of imidazolines. II. Affinities and efficacies of hydroxy-substituted tolazoline derivatives in rat aorta. J. Pharmacol. Exp. Ther. 211, 74–79.

Ruffolo, R. R., Rosing, E. L., and Waddell, J. E. (1979b) Receptor interactions of imidazolines. I. Affinity and efficacy for alpha adrenergic receptors in rat aorta. J. Pharmacol. Exp. Ther. 211, 733–738.

Ruffolo, R. R., Yaden, E. L., Waddell, J. E., and Dillard, R. D. (1980) Receptor interactions of imidazolines. VI. Significance of carbon bridge separating phenyl and imidazoline rings of tolazoline-like alpha adrenergic imidazolines. J. Pharmacol. Exp. Ther. 214, 535–540.

Ruffolo, R. R., Yaden, E. L., and Waddell, J. E. (1982a) Stereochemical requirements of alpha-2 adrenergic receptors. J. Pharmacol. Exp. Ther. 222, 645–651.

Ruffolo, R. R., Anderson, K. S., and Miller, D. D. (1982b) Conformational requirements of alpha-2 adrenergic receptors. *Mol. Pharmacol.* **21**, 259–265.

Ruffolo, R. R., Jr., Rice, P. J., Patil, P. N., Hamada, A., and Miller, D. P. (1983a) Differences in the applicability of the Easson-Stedman hypothesis to the α_1 and α_2-adrenergic effects of phenethylamines and imidazolines. *Eur. J. Pharmacol.* **86**, 471–475.

Ruffolo, R. R., Jr., Patil, P. N., and Miller, D. (1983b) Adrenergic effect of optically active catecholimidazoline derivatives in pithed rat. *Naunyn Schmiedebergs Arch. Pharmacol.* **323**, 221–227.

Ruffolo, R. R., Jr., Banning, J. W., Patil, P. N., Hamada, A., and Miller, D. D. (1983c) Evaluation of the adrenergic effects of a novel optically active catecholamine *in vitro* and *in vivo*: Differential application of the Easson-Stedman hypothesis to α-1 and α_2-adrenoceptors. *J. Auton. Pharmacol.* **3**, 185–193.

Ruffolo, R. R., Timmermans, P. B. M. W. M., and van Zwieten, P. A. (1983d) Interaction of clonidine, its methylene-bridged analog, SK-1913, and the benzylic hydroxyl-substituted derivative, St 1965, with α- and β-adrenoceptors. *J. Pharmacol. Exp. Ther.* **226**, 469–476.

Ruffolo, R. R., Jr., Goldberg, M. R., and Morgan, E. L. (1984a) Interactions of epinephrine, dopamine, and their corresponding α-methyl-substituted derivatives with α- and β-adrenoceptors in the pithed rat. *J. Pharmacol. Exp. Ther.* **230**, 595–600.

Ruffolo, R. R., Jr., Messick, K., and Horing, J. S. (1984b) Interaction of the selective inotropic agents, ASL-7022, dobutamine and dopamine, with α- and β-adrenoceptors *in vivo*. *Naunyn Schmiedebergs Arch. Pharmacol.* **326**, 317–326.

Ruffolo, R. R., Jr., Morgan, E. L., and Messick, K. (1984c) Possible relationship between receptor reserve and the differential antagonism of α_1 and α_2-adrenoceptor mediated pressor responses by calcium channel antagonists. *J. Pharmacol. Exp. Ther.* **230**, 587–594.

Ruffolo, R. R., Messick, K., and Horng, J. S. (1985) Interactions of dimethoxy-substituted tolazoline derivatives with alpha-1 and alpha-2 adrenoceptors *in vitro*. *J. Auton. Pharmacol.* **5**, 71–79.

Ruffolo, R. R., Jr., Sulpizio, A. C., Nichols, A. J., DeMarinis, R. M., and Hieble, J. P. (1987) Arterial α_2-adrenoceptor blockade: A potentially new approach to antihypertensive therapy. *J. Cardiovasc. Pharmacol.*, in press.

Schimmel, R. J. (1976) Roles of α- and β-adrenergic receptors in control of glucose oxidation in hamster epididymal adipocytes. *Biochim. Biophys. Acta* **428**, 379–387.

Shepperson, N. B. and Langer, S. Z. (1981) The effects of the 2-amino-tetrahydronaphthalene derivative, M7, a selective α_2-adrenoceptor agonist in vitro. *Naunyn Schmiedebergs Arch. Pharmacol.* **318**, 10–13.

Shepperson, N. B., Purcell, T., Massingham, R., and Langer, S. Z. (1981) In vitro studies on 6-fluoronoradrenaline at several peripheral sympathetic neuroeffector junctions. *Naunyn Schmiedebergs Arch. Pharmacol.* **317**, 1–4.

Skarby, T. V. C., Anderson, K. E., and Edvinsson, L. (1983) Pharmacological characterization of postjunctional α-adrenoceptors in isolated feline cerebral and peripheral arteries. *Acta. Physiol. Scand.* **117**, 63–73.

Stahle, H., Daniel, H., Kobinger, W., Lillie, C., and Pichler, L. (1980) Chemistry, pharmacology and structure–activity relationships with a new type of imidazolines exerting a specific bradycardic action at a cardiac site. *J. Med. Chem.* **23**, 1217–1222.

Starke, K. (1972a) A sympathomimetic inhibition of adrenergic and cholinergic transmission in the rabbit heart. *Naunyn Schmiedebergs Arch. Pharmacol.* **274**, 18–45.

Starke, K. (1972b) Influence of extracellular noradrenaline on the stimulation-evoked secretion of noradrenaline from sympathetic nerves: Evidence for an α-receptor mediated feed-back inhibition of noradrenaline release. *Naunyn Schmiedebergs Arch. Pharmacol.* **275**, 11–23.

Starke, K. (1977) Regulation of noradrenaline release by presynaptic receptor systems. *Rev. Physiol. Biochem. Pharmacol.* **77**, 1–124.

Starke, K. (1981) Alpha-adrenoceptor subclassification. *Rev. Physiol. Biochem. Pharmacol.* **88**, 199–236.

Starke, K. and Montel, H. (1973) Involvement of alpha-receptors in clonidine-induced inhibition of transmitter release from central monoamine neurons. *Neuropharmacology* **12**, 1073–1080.

Starke, K., Endo, T., and Taube, H. D. (1975) Relative pre- and postsynaptic potencies of alpha-adrenoceptor agonists in the rabbit pulmonary artery. *Naunyn Schmiedebergs Arch. Pharmacol.* **291**, 55–78.

Starke, K., Montel, H., and Schumann, H. J. (1971a) Influence of cocaine and phenoxybenzamine on noradrenaline uptake and release. *Naunyn Schmiedebergs Arch. Pharmacol.* **270**, 210–214.

Starke, K., Montel, H., and Wagner, J. (1971b) Effect of phentolamine on noradrenaline uptake and release. *Naunyn Schmiedebergs Arch. Pharmacol.* **271**, 181–192.

Starke, K., Montel, H., Gayk, W., and Merker, R. (1974) Comparison of the effects of clonidine on pre- and postsynaptic adrenoceptors in the rabbit pulmonary artery. Alpha-sympathomimetic inhibition of neurogenic vasoconstriction. *Naunyn Schmiedebergs Arch. Pharmacol.* **285**, 133–150.

Starke, K., Taube, H. D., and Borowski, E. (1977) Commentary, presynaptic-receptor systems in catecholaminergic transmission. *Biochem. Pharmacol.* **26**, 259–268.

Stillings, M. R., Chapleo, C. B., Butler, R. C. M., Davis, J. A., England, C. D., Myers, M., Myers, P. L., Tweddle, N., Welbourn, A. P., Doxey, J. C., and Smith, C. F. C. (1985) Alpha adrenoceptor reagents. 3. Synthesis of some 2-substituted 1,4-benzodioxans as selective presynaptic alpha-2 adrenoceptor antagonists. *J. Med. Chem.* **28**, 1054–1062.

Stillings, M. R., England, C. D., Welbourn, A. P., and Smith, C. F. C. (1986) Effect of methoxy substitution on the adrenergic activity of three structurally related α2-adrenoceptor antagonists. *J. Med. Chem.* **29**, 1780–1783.

Sulpizio, A. C., Hieble, J. P., Nichols, A. J., DeMarinis, R. M., Pfeiffer, F. R., Lavanchy, P. G., and Ruffolo, R. R. (1986) Pharmacological evidence for the existence of discrete subtypes of alpha-2 adrenoceptors. *Pharmacologist* **28**, 134.

Summers, R. J., Barnett, D. B., and Nahorski, S. R. (1983) The characteristics of adrenoceptors in homogenates of human cerebral cortex labelled by (3H)-rauwolscine. *Life Sci.* **33**, 1105–1112.

Summers, R. J., Jarrott, B., and Louis, W. J. (1981) Displacement of
    ^{3}H-clonidine binding by clonidine analogs in membranes from rat cere-
    bral cortex. *Eur. J. Pharmacol.* **66**, 223–241.
Tanaka, T. and Starke, K. (1980) Antagonist/agonist preferring alpha-
    adrenoceptors or alpha-1/alpha-2 adrenoceptors. *Eur. J. Pharmacol.* **63**,
    191–194.
Thoenen, H., Hurliman, A., and Haefely, W. (1964) Dual site of action of
    phenoxybenzamine in the cat's spleen: Blockade of α-adrenergic recep-
    tors and inhibition of reuptake of neurally released norepinephrine.
    Experientia **20**, 272–273.
Timmermans, P. B. M. W. M. and van Zwieten, P. A. (1980a) Vasoconstric-
    tion mediated by postsynaptic $α_2$-adrenoceptor stimulation. *Naunyn
    Schmiedebergs Arch. Pharmacol.* **313**, 17–20.
Timmermans, P. B. M. W. M. and van Zwieten, P. A. (1980b) Postsynaptic
    $α_1$- and $α_2$-adrenoceptors in the circulatory system of the pithed rat; se-
    lective stimulation of the $α_2$-type by B-HT 933. *Eur. J. Pharmacol.* **63**,
    199–202.
Timmermans, P. B. M. W. M. and van Zwieten, P. A. (1980c) Clonidine and
    some bridge analogues, cardiovascular effects and nuclear magnetic
    resonance data. *Eur. J. Med. Chem.* **15**, 323–329.
Timmermans, P. B. M. W. M. and van Zwieten, P. A. (1982) Alpha-2
    adrenoceptor: Classification, localization, mechanisms, and targets for
    drugs. *J. Med. Chem.* **25**, 1389–1401.
Timmermans, P. B. M. W. M., deJonge, A., van Meel, J. C. A., Slothorst-
    Grisdijk, F. P., Lam, E., and van Zwieten, P. A. (1981) Characterization
    of α-adrenoceptor populations. Quantitative relationships between car-
    diovascular effects initiated at central and peripheral α-adrenoceptors.
    J. Med. Chem. **24**, 502–507.
Timmermans, P. B. M. W. M., Brands, A., and van Zwieten, P. A. (1977)
    Lipophilicity and brain disposition of clonidine and structurally related
    imidazolines. *Naunyn Schmiedebergs Arch. Pharmacol.* **300**, 217–226.
Timmermans, P. B. M. W. M., Kwa, H. Y., and van Zwieten, P. A. (1979)
    Possible subdivision of postsynaptic α-adrenoceptors mediating pres-
    sor responses in the pithed rat. *Naunyn Schmiedebergs Arch. Pharmacol.*
    310, 189–193.
Timmermans, P. B. M. W. M., de Jonge, A., Thoolen, M. J. M. C., Wilffert,
    B., Batnik, H., and van Zwieten, P. A. (1984) Quantitative relationships
    between alpha adrenergic activity and binding affinity of alpha
    adrenoceptor agonists and antagonists. *J. Med. Chem.* **27**, 495–503.
Timmermans, P. B. M. W. M., Hoefke, W., Stahle, H., and van Zwieten, P.
    A. (1980a) Structure activity relationships in clonidine-like imidazoline
    and related compounds. *Prog. Pharmacol.* **3**, 1–104.
Timmermans, P. B. M. W. M., van Meel, J. C. A., and van Zwieten, P. A.
    (1980b) Evaluation of the selectivity of alpha-adrenoceptor blocking-
    drugs for postsynaptic alpha-1 adrenoceptors and alpha-2 adrenocep-
    tors in a simple animal model. *J. Auton. Pharmacol.* **1**, 53–60.
Triggle, D. J. (1976) Chemical pharmacology of the synapse. (Triggle, D. J.
    and Triggle, C. R., eds.) Academic, New York.
Tripos Associates (1984) The SYBYL Molecular Modeling System V3.2. St.
    Louis, TRIPOS Associates.

U'Prichard, D. C. and Snyder, S. H. (1979) Distinct alpha-noradrenergic receptors differentiated by binding and physiological relationships. *Life Sci.* **24**, 79–88.

Vacca, J. P., Anderson, P. S., Baldwin, J. J., deSolms, S. J., Gould, N. P., Grace, J. P., Huff, J. R., Hunt, C. A., Randall, W. C., Sakurai, Y., Sanders, W. M., Smith, S. J., Wiggins, J. M., Young, S. D., and Zrada, M. M. (1986) Derivatives of L-654,284, A Potent and Selective alpha-2 Adrenoceptor Antagonist. Abstracts of the American Chemical Society 191st National Meeting. Division of Medicinal Chemistry Abstract # 55.

van Meel, J. C. A., deJonge, A., Timmermans, P. B. M. W. M., and van Zwieten, P. A. (1981) Selectivity of some alpha adrenoceptor agonists for peripheral alpha-1 and alpha-2 adrenoceptors in the normotensive pithed rat. *J. Pharmacol. Exp. Ther.* **219**, 760–767.

Waterfall, J. F., Rhodes, K. F., and Lattimer, N. (1985) Studies of alpha-2 adrenoceptor antagonist potency in vitro, comparisons in tissues from rats, rabbits, dogs and humans. *Clin. Sci.* **68**, (suppl. 10), 21S–24S.

Weitzell, R., Tanaka, T., and Starke, K. (1979) Pre- and postsynaptic effects of yohimbine stereoisomers on noradrenergic transmission in the pulmonary artery of the rabbit. *Naunyn Schmiedebergs Arch. Pharmacol.* **308**, 127–136.

Wikberg, J. E. S. (1979a) The pharmacological classification of adrenergic alpha-1 and alpha-2 receptors and their mechanisms of action. *Acta Physiol. Scand.* (suppl.) **468**, 1–89.

Wikberg, J. E. S. (1979b) Pre- and Postjunctional α-Receptors, in *Presynaptic Receptors* (Langer, S. Z., Starke, K., and Dubocovich, M. L., eds.) Pergamon, Oxford.

Wikberg, J. E. S. (1978) Differentiation between pre- and postjunctional α-receptors in guinea pig ileum and rabbit aorta. *Acta Physiol. Scand.* **103**, 225–239.

Wood, C. L., Annett, C. D., Clarke, W. R., Tsai, B. S., and Lefkowitz, R. J. (1979) Subclassification of alpha-adrenergic receptors by direct binding studies. *Biochem. Pharmacol.* **28**, 1277–1282.

Chapter 5

Functions Mediated by alpha-2 Adrenergic Receptors

Robert R. Ruffolo, Jr., Andrew J. Nichols, and J. Paul Hieble

1. Introduction

This chapter summarizes the diverse physiologic functions that are influenced by alpha-2 adrenergic receptor activation. A great deal of attention has been given to the cardiovascular system, although the physiologic consequences of alpha-2 receptor occupancy are described for other target organs as well. The diverse cellular consequences of alpha-2 adrenergic receptor activation are paralled by the demonstration of several possible biochemical signaling systems that are activated by these receptors, including inhibition of adenylate cyclase, alterations in Ca^{2+} translocation, and acceleration of Na^+/H^+ exchange. A similar diversity is known to exist for electrophysiologic consequences of alpha-2 adrenergic receptor occupancy (also explored in Chapter 6). What remains for future experimental data to determine is whether alpha-2 adrenergic receptors all share certain early biochemical and/or electrophysiological events that lead to differing distal effects depending on the particular target cell, or whether distinct signaling mechanisms prevail resulting in the distinct biochemical and physiological changes characteristic of varying target organs.

2. alpha-2 Adrenergic Receptors in the Cardiovascular System

alpha-2 Adrenergic receptors exist in many organs of the body, and the functions they mediate are only now beginning to be understood. Because of the prominent role that alpha-2 adrenergic receptors play in the cardiovascular system, their distribution and function in this important system will be discussed separately. Figure 1 is a highly schematic representation of the distribution of alpha-2 adrenergic receptors in the cardiovascular system. In some instances, alpha-1 adrenergic receptors are also discussed in order to emphasize several major differences in the distribution and function of the two alpha-adrenergic receptor subtypes.

2.1. Cardiovascular Effects Mediated by Central alpha-2 Adrenergic Receptors

Stimulation of central alpha-2 adrenergic receptors in the ventrolateral medulla (Fig. 1) induces a reduction in sympathetic outflow to the periphery, manifested as a reduction in arterial blood pressure accompanied by bradycardia. This response has been studied extensively over the past two decades, and several comprehensive reviews are available (Schmitt, 1971; Kobinger, 1978; van Zwieten et al., 1983; Ruffolo, 1984a). Quantitative structure–activity studies have shown excellent correlation between the alpha-2 adrenergic receptor agonist potency of a series of clonidine analogs and blood pressure reduction, provided a lipophilicity term is included to correct for penetration through the blood–brain barrier, which is required in order to gain access to the site of action within the central nervous system (Timmermans et al., 1980, 1981a; Ruffolo et al., 1982b,c).

The characteristic response to intravenous administration of an alpha-2 adrenergic receptor agonist in a normotensive or hypertensive animal is an immediate pressor response, caused by stimulation of peripheral arterial postjunctional alpha-1 and alpha-2 adrenergic receptors (Ruffolo et al., 1982c). This response is also seen in human subjects following intravenous administration of clonidine (Onesti et al., 1971). This pressor response is relatively short-lived and is followed by a slow decline in arterial blood pressure to levels lower than those observed prior to drug administration. This long-lasting depressor/antihypertensive response is a result of central alpha-2 adrenergic receptor stimulation. Heart rate declines immediately following administration and continues to be reduced for the duration of drug action. If the

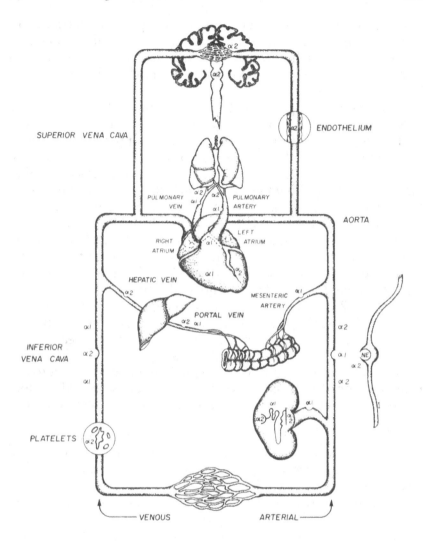

Fig. 1 Schematic representation of several important vascular beds in the cardiovascular system and the distribution of alpha-1 and alpha-2 adrenergic receptors. The indentations shown in the various arterial and venous beds represent junctional neuroeffector distributions of alpha-adrenergic receptors.

alpha-2 adrenergic receptor agonist is administered directly into the central nervous system or via the vertebral artery, which allows for easy access to the central nervous system, the initial pressor response is not observed (Ruffolo et al., 1982b,c; Timmermans et al., 1980). High oral doses of centrally acting alpha-2 adrenergic receptor agonists, such as clonidine or guanfacine, can also increase blood pressure via peripheral arterial alpha-adrenergic receptor stimulation (Davis et al., 1977) and provide an explanation for the "therapeutic window" seen with clonidine in antihypertensive therapy (Frisk-Holmberg et al., 1984).

Although the peripheral prejunctional action of alpha-2 adrenergic receptor agonists does not appear to make a major contribution to the antihypertensive activity of these compounds, a peripheral neuroinhibitory action of clonidine can be demonstrated in the cat on the cardiovascular response to electrical stimulation of sensory fibers in vivo (Walland, 1978). This peripheral action has been postulated to contribute to the antihypertensive effects of clonidine under conditions in which blood pressure escapes baroreceptor-mediated homeostatic control. A similar peripheral presynaptic contribution to blood pressure effects in human subjects has been proposed by Gunnar Wallin and Frisk-Holmberg (1981). Furthermore, the bradycardia associated with alpha-2 adrenergic receptor agonists may result, in part, from a peripheral presynaptic action at prejunctional alpha-2 adrenergic receptors on sympathetic nerves in the heart, since heart rate can be reduced in pithed rats (Drew, 1976; Misu and Kubo, 1982), and, in contrast to hypotension, alpha-2 adrenergic receptor agonist-induced bradycardia in the anesthetized rat does not require penetration into the central nervous system (deJonge et al., 1981).

The antihypertensive action of alpha-2 adrenergic receptor agonists is likely to result from stimulation of postsynaptic alpha-2 adrenergic receptors in the brainstem. Animal experiments have shown that catecholamine depletion with reserpine or destruction of sympathetic neurons by treatment with 6-hydroxydopamine does not generally attenuate the ability of alpha-2 adrenergic receptor stimulation to decrease sympathetic outflow (Haeusler, 1974; Kobinger and Pichler, 1976), although Dollery and Reid (1973) showed a slight attenuation by 6-hydroxydopamine. This would indicate that the central alpha-2 adrenergic receptor involved in this response is not located prejunctionally on a catecholaminergic neuron. A brainstem site is indicated, based on the inability of transection at the intercollicular level or at the pontomedullary junction to attenuate the antihypertensive activity of clonidine (Schmitt and Schmitt, 1969). Many experiments have been performed in an attempt to locate more precisely the site of action of alpha-2 adrenergic receptor agonists within the brainstem. Although the nucleus tractus solitarius has often been considered as the principal site of action of central alpha-2 adrenergic receptor agonists (Schmitt, 1971), recent studies using microinjections of clonidine suggest the lateral reticular nucleus in the ventrolateral medulla as a more likely candidate (Gillis et al., 1985). This nucleus is readily accessible from the ventral surface of the medulla, where alpha-2 adrenergic receptor agonists have been shown to be effective following local application (Bousquet and Guertzenstein, 1973; Scholtysik et al., 1975; Srimal et al., 1977; Gillis et al., 1985). Although guanfacine has been reported to be ineffective via this route (Scholtysik et al., 1975), the body of experimental evidence would suggest that all alpha-2 adrenergic receptor agonists are acting at this same central locus.

In addition to a reduction in sympathetic outflow, central alpha-2 adrenergic receptor stimulation can enhance parasympathetic outflow. This has usually been demonstrated as a potentiation of the reflex bradycardia induced by intravenous injection of a pressor agent such as angiotensin II (Kobinger, 1978; Connor et al., 1982). This action requires penetration of the alpha-2 adrenergic receptor agonist into the central nervous system (Kobinger, 1978), but the precise site of action has not yet been determined (Gillis et al., 1985).

Central alpha-2 adrenergic receptor stimulation has been utilized clinically as antihypertensive therapy. In addition to the directly acting central alpha-2 adrenergic receptor agonists discussed above, alpha-methyldopa, which has been extensively employed for over a decade, is now known to stimulate central alpha-2 adrenergic receptors following metabolic conversion to alpha-methylnorepinephrine (van Zwieten, 1980), which has much greater alpha-2 adrenergic receptor selectivity than norepinephrine (Ruffolo et al., 1982a; Ruffolo and Waddell, 1982; Jarrott et al., 1984). Following chronic treatment with alpha-methyldopa in rats, medullary norepinephrine stores are almost completely replaced by alpha-methylnorepinephrine (Conway et al., 1979), which is available for interaction with medullary alpha-2 adrenergic receptors to inhibit sympathetic outflow. The therapeutic and side-effect profile of alpha-methyldopa is similar to that observed with the directly acting central alpha-2 adrenergic receptor agonists (van Zwieten et al., 1983).

In addition to clonidine and alpha-methyldopa, guanfacine (BS 100-141) is now in general use as an antihypertensive drug. This compound has an in vitro pharmacologic profile similar to that of clonidine (Scholtysik, 1980), but appears to have a longer duration of action (Jain et al., 1985; Farsang et al., 1984a; Reid et al., 1983a). Clinical trials have been conducted with several other alpha-2 adrenergic receptor agonists, including St 600 (Kho et al.,

1975), tiamenidine (Campbell et al., 1980; Clifton et al., 1981), monoxidine (Planitz, 1984), lofexidine (Schultz et al., 1981; Wilkins et al., 1981; Lopez and Mehta, 1984), and B-HT 933 (azepexole) (Reid et al., 1983a). The latter compound is more selective than clonidine for alpha-2 vis-a-vis alpha-1 adrenergic receptors (van Meel et al., 1981a), and its antihypertensive activity confirms an alpha-2 adrenergic receptor-mediated mechanism. As in animal studies, the clinical cardiovascular profiles of the various alpha-2 adrenergic receptor agonists is relatively similar (Reid, 1985). Besides their antihypertensive indication, the sympatholytic action of the centrally active alpha-2 adrenergic receptor agonists may offer clinical benefit in congestive heart failure and angina pectoris, again through a centrally mediated reduction in sympathetic outflow. Although extensive evaluation for efficacy in these conditions has not yet been performed, preliminary trials in patients are encouraging (Giles et al., 1985).

Although the antihypertensive activity of clonidine and clonidine-like imidazolidines results from their pharmacologic selectivity for central alpha-2 adrenergic receptors, it is known also that the physicochemical properties of clonidine-like imidazolidines are critical in determining the antihypertensive efficacy of such compounds (for review, see Timmermans et al., 1980). Highly lipophilic imidazolidines, such as clonidine, which readily penetrate the the blood–brain barrier and gain access to their site(s) of action in the brainstem, are potent antihypertensive agents. Conversely, many imidazolidines with selectivities similar to those of clonidine for alpha-2 adrenergic receptors, but with low lipophilicity, do not readily penetrate the blood–brain barrier and are either weak antihypertensive agents or completely devoid of all antihypertensive activity. Such compounds are still effective in lowering blood pressure when injected beyond the blood–brain barrier into specific brain regions, such as the ventrolateral medulla, or when injected into the cerebral ventricles or cisterna magna (see Ruffolo, 1984a; Ruffolo et al., 1982b,c). Because these alpha-2 adrenergic receptor agonists with low lipophilicity are still active when the blood–brain barrier is bypassed, it has been concluded that one major factor affecting the antihypertensive activity of clonidine-like imidazolidines following systemic administration is their ability to penetrate the blood–brain barrier, and this, in turn, is highly dependent upon overall lipophilicity.

Many properties of a molecule will determine the overall lipophilicity, which, as indicated above, is critical for antihypertensive efficacy of clonidine-like imidazolidines. For these particular compounds, the most important determinant of lipophilicity is the extent of ionization occurring at physiological pH, and this property is governed by the ionization constant (K_a) (Timmermans and van Zwieten 1978; Timmermans et al., 1977, 1981a; Ruffolo et al., 1982b,c). Imidazolidines in the ionized species possess low lipophilicity and will penetrate the blood–brain barrier slowly, whereas the un-ionized form is highly lipophilic and will penetrate the blood–brain barrier rapidly. Thus, the ratio of the un-ionized:ionized species is a major determinant of the antihypertensive efficacy and potency of many clonidine-like imidazolidines. There exists an excellent correlation between the antihypertensive potencies of a series of clonidine-like imidazolidines and their pK_a (i.e., $-\log K_a$; Ruffolo et al., 1982b), such that those compounds with low pK_a values, and that are therefore significantly un-ionized at physiological pH, will penetrate the blood–brain barrier rapidly and be potent antihypertensive agents, whereas those imidazolidines with high pK_a values, and that are extensively ionized at physiological pH, will penetrate the blood–brain barrier to a lesser extent (or at a slower rate) and be weaker antihypertensive agents (Ruffolo et al., 1982b).

Clonidine is not metabolized to a great extent in humans (Lowenthal, 1980), and the limited metabolism that does occur does not take place in the brain. As a result, termination of the central antihypertensive effects of clonidine and clonidine-like imidazolidines is likely to be by diffusion out of the central nervous system. It has been demonstrated recently that the pK_a of clonidine-like imidazolidines, and therefore their ratios of un-ionized:ionized species, also governs the diffusion of these compounds out the brain (Ruffolo et al., 1982b). As expected, high proportions of the un-ionized species with high lipophilicity are associated with a more rapid exit from the central nervous system by diffusion through the blood–brain barrier in the reverse direction (Ruffolo et al., 1982b).

Based on the importance of the physicochemical properties of imidazolidines in governing their pharmacologic activity at central alpha-2 adrenergic receptors, the following scheme has been proposed to explain the antihypertensive effects of clonidine (Ruffolo, 1984c). Following systemic administration, clonidine will exist in the blood in an equilibrium between the ionized and un-ionized species, with the relative proportions of each species being determined by the pK_a and physiological pH. The ionized form will interact with postsynaptic vascular alpha-2 and to a lesser extent with alpha-1 adrenergic receptors (Ruffolo et al., 1982c) to mediate a transient pressor response, which is particularly apparent following intravenous administration. The un-ionized form (mainly) will penetrate the blood–brain barrier to gain access to the site(s) of action in the ventrolateral medulla. Again, a new equilibrium between the ionized and un-ionized form will be established within the central nervous system, the extent of which also being determined by the pK_a and the pH of cerebrospinal fluid. The ionized form (Ruffolo et al., 1982c) is believed to be the species responsible for activation of central alpha-2 adrenergic receptors to mediate the decrease in sympathetic outflow and the increase in parasympathetic outflow that ultimately produces the antihypertensive and bradycardic response. Although the ionized species appears to interact with the central alpha-2 adrenergic receptor, it is the un-ionized species that will penetrate the blood–brain barrier in the reverse direction and exit the central nervous system to terminate the antihypertensive response. Once in the periphery, the drug is subsequently removed from the blood by metabolism and/or excretion.

In contrast to the imidazolidines, the physicochemical properties of alpha-methyldopa play a lesser role in the antihypertensive activity of this compound. alpha-Methyldopa gains access to the central nervous system via the aromatic amino acid transport system. In the brain, alpha-methyldopa is sequentially decarboxylated and beta-hydroxylated to form 1R,2S(−)-erythro-alpha-methylnorepinephrine, which then activates medullary alpha-2 adrenergic receptors to inhibit sympathetic outflow and enhance parasympathetic outflow.

One of the concerns associated with antihypertensive therapy with centrally acting alpha-2 adrenergic receptor agonists is the "rebound hypertension" or "withdrawal" phenomenon that often occurs when treatment is abruptly terminated (Reid et al., 1977; Hansson et al., 1973). This phenomenon is characterized by tachycardia and abrupt rises in blood pressure, sometimes to levels greater than those observed before initiation of therapy (Hansson, 1983; Weber, 1980). Studies in animals have confirmed the presence of a hyperadrenergic state following abrupt termination of chronic clonidine therapy (Lewis et al., 1981; Engberg et al., 1982; Thoolen et al., 1983; Jarrott et al., 1984). Administration of an alpha-2 adrenergic receptor antagonist, such as yohimbine, can also precipitate this withdrawal phenomenon (Thoolen et al., 1983).

The withdrawal phenomenon observed following abrupt cessation of alpha-2 adrenergic receptor agonist therapy bears some similarity to opiate withdrawal (Hansson et al., 1973; Engberg et al., 1982; Jarrott et al., 1984), and appears to involve overactivity of locus ceruleus neurons (Engberg et al., 1982). This may represent a rebound phenomenon following chronic suppression of the firing rate of these neurons during chronic antihypertensive treatment. In view of the similarities and possible receptor interactions between alpha-2 adrenergic and opiate receptors (see below), it is not surprising that morphine can suppress, via a naloxone-sensitive mechanism, some of the cardiovascular rebound effects observed following termination of clonidine infusion in rats (Thoolen et al., 1983).

2.2. Peripheral Arterial alpha-2 Adrenergic Receptors

2.2.1. SYSTEMIC ARTERIAL CIRCULATION

Postjunctional alpha-2 adrenergic receptors are present on arterial smooth muscle, as demonstrated by the pressor response produced by alpha-2 adrenergic receptor agonists in the pithed rat. This response fulfills all the criteria for an alpha-2 adrenergic receptor-mediated response, in that it is produced by highly selective alpha-2 adrenergic receptor agonists, such as B-HT 920, B-HT 933, and UK 14,304, and is blocked competitively by selective alpha-2 adrenergic receptor antagonists, such as rauwolscine and idazoxan, and is almost completely insensitive to alpha-1 adrenergic receptor blockade with prazosin. Since the discovery of arterial alpha-2 adrenergic receptors in the pithed rat (Bentley et al., 1977; Docherty et al., 1979; Drew and Whiting, 1979; Timmermans et al., 1979), this preparation has been used extensively to characterize the postjunctional arterial alpha-2 adrenergic receptor. An added advantage to this preparation is the ability to compare pre- and postjunctional alpha-2 adrenergic receptor function in the same animal. Subsequent experiments have been performed to show the presence of postjunctional arterial alpha-2 adrenergic receptors in dogs (Constantine et al., 1980), cats, (Timmermans et al., 1983), rabbits (McGrath et al., 1982; van Meel et al., 1982), and humans (Elliott and Reid, 1983).

alpha-2 Adrenergic receptor-mediated pressor responses can be demonstrated in intact animals, although centrally mediated sympathoinhibition produced by those alpha-2 adrenergic receptor agonists that penetrate the blood–brain barrier may complicate the interpretation of results. The pressor response to intravenous administration of guanabenz, a selective alpha-2 adrenergic receptor agonist, to conscious rabbits is blocked by low doses of idazoxan (Hannah et al., 1983). In the conscious rabbit, rauwolscine or yohimbine are more effective in blocking the pressor response induced by norepinephrine than that elicited by phenylephrine, a selective alpha-1 adrenergic receptor agonist; conversely, prazosin is more potent in blocking the pressor response of phenylephrine (Hamilton and Reid, 1982). Similar experiments have been performed in normal human volunteers, using epinephrine as a nonselective alpha-adrenergic receptor agonist and phenlyephrine to activate only the alpha-1 adrenergic receptor. Prazosin and yohimbine were most effective in blocking the phenylephrine and epinephrine responses, respectively (Goldberg and Robertson, 1984).

In addition to pressor responses in intact or decentralized animals, arterial alpha-2 adrenergic receptors can be demonstrated in a variety of perfused vascular beds. Langer et al. (1981) have demonstrated that the increases in blood pressure induced by intraarterial administration of norepinephrine in the canine autoperfused hind limb are relatively resistant to prazosin blockade. Prazosin, at a dose that significantly attenuated the response to phenylephrine, has no effect on the response to guanabenz; in contrast, the response to guanabenz can be selectively inhibited by rauwolscine. Similar results have been obtained by Elsner et al., 1984.

Direct injection of clonidine into the femoral artery of an anesthetized dog produced a decrease in femoral arterial blood flow that was completely insensitive to blockade by prazosin, but sensitive to antagonism by yohimbine (Horn et al., 1982). Conversely, methoxamine and phenylephrine produced effects that were blocked by prazosin, but not by yohimbine. Similar experiments in the renal vascular bed showed less of an alpha-2 adrenergic receptor-mediated contribution (see below), suggesting that the density of arterial alpha-2 adrenergic receptors may vary in different vascular beds (Horn et al., 1982).

Experiments in the rabbit autoperfused hind limb have demonstrated the presence of arterial alpha-2 adrenergic receptors, based on a significant vasoconstrictor response produced by the selective alpha-2 adrenergic receptor agonist xylazine, which could be blocked by rauwolscine, but not by prazosin (Madjar et al., 1980).

In their initial studies with the isolated perfused rat hindquarters, Kobinger et al. (1980) did not detect an alpha-2 adrenergic receptor-mediated pressor response to B-HT920, although the preparation remained responsive to alpha-1 adrenergic receptor agonists. The principal difference between this preparation and the other perfused vascular beds described above is that a physiologic salt solution was used by Kobinger et al. (1980) to perfuse the tissue, as opposed to whole blood. Additional experiments from this group (Kobinger and Pichler, 1981) and others (van Meel et al., 1983) have shown that an alpha-2 adrenergic receptor-mediated vasoconstrictor response to B-HT 920 could be produced in hindquarters isolated from rats pretreated with reserpine. This observation is consistent with a report (Hicks and Waldron, 1982) showing supersensitivity to alpha-2 adrenergic receptor agonists in the pithed rat following reserpine pretreatment. This effect may reflect postreceptor changes as opposed to true denervation supersensitivity, since 6-hydroxydopamine combined with adrenal demedullectomy did not mimic the reserpine-induced supersensitivity changes in alpha-2 adrenergic receptor-mediated vascular response.

Despite the demonstration of arterial alpha-2 adrenergic receptors in a variety of in vivo models, and in autoperfused vascular beds, in vitro studies have not been able to provide conclusive evidence of alpha-2 adrenergic receptor-mediated arterial vasoconstriction. Although certain tissues, such as the rat aorta, are sensitive to clonidine-induced contraction, this effect has been shown to be mediated by alpha-1 adrenergic receptors, although these alpha-1 adrenergic receptors may differ somewhat from those found in other commonly studied arterial tissues (Ruffolo, 1985a). A norepinephrine-induced vasoconstrictor response that is insensitive to prazosin has been observed in certain human arteries, such as the palmar digital artery (Jauernig et al., 1978); further studies of this tissue, however, suggest the presence of an a typical alpha-1 adrenergic receptor, rather than an arterial alpha-2 adrenergic receptor (Stevens and Moulds, 1982). An in vitro study of isolated human arteries from several vascular beds has failed to demonstrate responsiveness to the highly selective alpha-2 adrenergic receptor agonist, UK 14,304 (Calvette et al., 1984).

Almost all other isolated arterial tissues studied respond only to higher concentrations of selective alpha-2 adrenergic receptor agonists, and this effect is generally sensitive to blockade by alpha-1 adrenergic receptor antagonists (see Langer and Hicks, 1984). One notable exception seems to be the canine basilar artery, in which clonidine produces contraction that is insensitive to blockade by corynanthine (Sakakibara et al., 1982), but is sensitive to inhibition by yohimbine. Similar conclusions in this tissue were obtained by Toda (1983), who found the response to norepinephrine to be insensitive to blockade by prazosin; the failure, however, of this selective alpha-1 adrenergic receptor antagonist to block the response to phenylephrine (Toda, 1983) suggests that the response seen in the canine basilar artery may not be a typical alpha-2 adrenergic receptor-mediated response. A recent report (Ito and Chiba, 1985) describes an alpha-2 adrenergic receptor-mediated contraction in the perfused ear artery of the dog. Xylazine, a moderately selective alpha-2 adrenergic receptor agonist, induced contraction that was preferentially antagonized by yohimbine. Conversely, phenylephrine-induced contraction was preferentially antagonized by prazosin. These experiments were not conducted under equilibrium conditions, however, since both the agonists and antagonists were administered as bolus injections into the perfusion flow; hence receptor dissociation constants for the antagonists could not be used to compare to other established systems in order to verify alpha-2 adrenergic receptor activity.

Despite the failure to demonstrate convincing contractile responses to alpha-2 adrenergic receptor agonists in most isolated arteries, studies with antagonists often suggest a contribution of alpha-2 adrenergic receptors in the vasoconstrictor response to nonselective agonists, such as norepinephrine (Skarby et al., 1983; Hieble and Woodward, 1984).

There are two possible explanations for the apparent paradox of arterial alpha-2 adrenergic receptors being clearly demonstrable in vivo, but not in vitro: (1) the alpha-2 adrenergic receptors are present only on very small resistance arteries or arterioles that are not routinely studied in vitro or (2) some factor present under in vivo, but not in vitro, conditions is required for expression of the response to arterial alpha-2 adrenergic receptors.

If alpha-2 adrenergic receptors are present only on smaller resistance arteries, alpha-2 adrenergic receptor-mediated increases in perfusion pressure should be evident in isolated perfused organs or vascular beds, since the entire arterial/arteriolar system is intact. As indicated above, the initial studies of Kobinger et al. (1980) could not demonstrate alpha-2 adrenergic receptor-mediated vasoconstriction in rat isolated hindquarters perfused with Tyrode's solution. The few other reported studies using saline-perfused organs report either only alpha-1 adrenergic receptor-mediated responses (perfused rat kidney) or a mixture of alpha-1 and alpha-2 adrenergic receptor-mediated responses (perfused cat spleen) (Langer and Hicks, 1984).

Changes in alpha-adrenergic receptor characteristics with decreasing arterial size have not been studied extensively, but there is evidence for abrupt changes in receptor characteristics and adrenergic receptor function in general, as lumen diameter decreases (Owen and Bevan, 1985; Oriowo et al., 1985). These changes may relate to anatomical differences in the walls of large and small blood vessels (Bevan et al., 1985). Further studies on alpha-2 adrenergic receptor responsiveness in both saline-perfused vascular beds and in isolated small arteries are warranted.

It is well known that alpha-2 adrenergic receptor-mediated contraction of the canine saphenous vein is dependent on extracellular calcium concentration and that removal of extracellular calcium will abolish the response (Jim and Matthews, 1985). In addition, it has also been reported that partial blockade of calcium channels with a relatively low concentration of nifedipine alters the B-HT 920 concentration–response curve in the canine saphenous vein by eliminating the abrupt increase in tension that occurs when a "threshold" concentration of the alpha-2 adrenergic receptor agonist is reached (Sulpizio and Hieble, 1985). Since alpha-2 adrenergic receptor-mediated contractile responses are highly dependent on extracellular calcium, the effects of activation of the calcium channel on the expression of an alpha-2 adrenergic receptor-mediated contraction in arterial tissue was studied. It has been observed recently that the plantar branch of canine saphenous *artery* contracted in a concentration-dependent manner to B-HT 920 in vitro only in the presence of a relatively low concentration of the calcium channel activator BAY K 8644, and the response was blocked selectively by rauwolscine, but not by prazosin (Sulpizio and Hieble, 1987). These findings indicate that the response to B-HT 920 in this arterial tissue is mediated by postjunctional vascular alpha-2 adrenergic receptors. These results may suggest that, in vivo, circulating factors that increase the probability of the calcium channel opening may be necessary for the expression of arterial responses to postjunctional alpha-2 adrenergic receptor stimulation, and that partial calcium channel activation may be necessary for the in vitro expression of postjunctional arterial alpha-2 adrenergic receptor-mediated responses.

It is now widely accepted that arterial vasoconstriction may be mediated by a mixed population of postsynaptic vascular alpha-1 and alpha-2 adrenergic receptors (Fig. 1). The physiologic function and/or distribution of these receptors is beginning to be understood. By using a variety of alpha-1-selective, alpha-2-selective, and nonselective alpha-adrenergic receptor antagonists, Yamaguchi and Kopin (1980) observed that the pressor responses to exogenously administered catecholamines were selectively antagonized by alpha-2 adrenergic receptor blockers. Conversely, the pressor response evoked by sympathetic nerve stimulation was selectively antagonized by alpha-1 adrenergic receptor blockers. These authors postulated that postsynaptic vascular alpha-adrenergic receptors located at the neuroeffector junction (i.e., junctional receptors) were of the alpha-1 subtype, whereas those located away from the neuroeffector junction (i.e., extrajunctional receptors) were of the alpha-2 subtype.

Support of the concept of junctional alpha-1 and extrajunctional alpha-2 adrenergic receptors in the arterial circulation has been obtained in perfused cat spleen in which increases in perfusion pressure elicited by nerve stimulation, norepinephrine, and phenylephrine were found to be differentially inhibited by selective alpha-1 and alpha-2 adrenergic receptor antagonists (Langer and Shepperson, 1982a). The responses to nerve stimulation and phenylephrine were abolished by the selective alpha-1 adrenergic receptor antagonist prazosin, with the response to norepinephrine being only partially inhibited. Based on the known alpha-1 adrenergic receptor selectivity of phenylephrine and the nonselective activity of norepinephrine, the results were compatible with the notion that neuronally released norepinephrine interacted with junctional alpha-1 adrenergic receptors, which could also be activated by exogenously administered phenylephrine and norepinephrine. Postsynaptic vascular alpha-2 adrenergic receptors in this model were proposed to reside extrajunctionally, since they were not activated by norepinephrine release from sympathetic nerves, but could be stimulated by exogenously administered norepinephrine. In further studies using neuronal-uptake inhibitors, Langer and Shepperson (1982a,b) have shown that postjunctional vascular alpha-1 adrenergic receptors are located in the vicinity of the neuronal uptake pump (uptake$_1$), and that postjunctional vascular alpha-2 adrenergic receptors are positioned away from this site. These results, and those obtained by Wilffert et al. (1982), strongly suggest the existence of junctional alpha-1 and extrajunctional alpha-2 adrenergic receptors located postsynaptically in the arterial circulation.

The physiologic role of the postsynaptic junctional alpha-1 adrenergic receptors appears to be in maintaining resting vascular tone. Presumably, these receptors, which are located in the vicinity of the neurovascular junction, would interact with endogenous norepinephrine liberated from sympathetic nerves. The physiologic role of the extrajunctional alpha-2 adrenergic receptors is not fully understood. It has been suggested that the extrajunctional alpha-2 adrenergic receptors would not normally interact with liberated norepinephrine since they are located at some distance away from the adrenergic nerve terminal, and the highly efficient neuronal uptake pump keeps synaptic levels of norepinephrine sufficiently low and thereby prevents diffusion of the neurotransmitter to the extrajunctional sites (Langer and Shepperson, 1982a). It has been proposed that the extrajunctional alpha-2 adrenergic receptors may respond to circulating epinephrine acting as a blood-borne hormone (Langer and Shepperson, 1982b). Although circulating catecholamines may be below the levels required to exert a physiologic effect, it has been suggested that in times of stress these levels may be elevated to threshold levels at which postsynaptic vascular alpha-2 adrenergic receptors are activated (Cutter et al., 1980). It also has been suggested that the contribution made by arterial extrajunctional alpha-2 adrenergic receptors to total peripheral vascular resistance may be greater in certain hypertensive states than in normotensive patients (Bolli et al., 1984; Jie et al., 1986), implying that postsynaptic vascular alpha-2 adrenergic receptors may play an important role in pathophysiological states such as hypertension and possibly congestive heart failure, in which circulating catecholamine levels are high (Levine et al., 1982). It is unclear at the present time whether epinephrine is, in fact, responsible for stimulating the extrajunctional alpha-2 adrenergic receptors in these states, since circulating levels of norepinephrine are high also and could account, at least in part, for their activation in disease states such as hypertension and congestive heart failure.

Although alpha-1 and alpha-2 adrenergic receptors coexist postjunctionally in arteries, and both subtypes mediate vasoconstriction, recent evidence suggests that alpha-1 and alpha-2 adrenergic receptors may be coupled differently to their respective vasoconstrictor processes (Ruffolo and Yaden, 1984). The nature of this difference between alpha-1 and alpha-2 adrenergic receptor coupling to arterial vasoconstriction may be reflected in quantitatively different "occupancy–vasoconstriction response" relationships, as well as qualitatively different mechanisms for the translation of stimuli into vasoconstrictor responses.

The relationship between alpha-1 adrenergic receptor occupancy by agonists and vasoconstrictor response has been studied in vitro for many years (Besse and Furchgott, 1976; Purdy and Stupecky, 1984; Ruffolo et al., 1979). In general, nonlinear occupancy–response relationships have been obtained in most arterial vessels studied, with approximately 6% alpha-1 adrenergic receptor occupancy by full agonists being required for a half-maximal vasoconstrictor response to be obtained. "Spare" alpha-1 adrenergic receptors are commonly found. A direct comparison of the occupancy–response relationships for alpha-1 and alpha-2 adrenergic receptor-mediated arterial vasoconstriction has not been performed in vitro because of the difficulty in demonstrating alpha-2 adrenergic receptor-mediated vasoconstriction in isolated arterial preparations (see above). From studies in vivo, however, in which alpha-2 adrenergic receptor-mediated arterial vasoconstriction is easily quantified, evidence for differences in alpha-1 and alpha-2 adrenergic receptor occupancy–response relationships has been accumulated. Irreversible alkylation of postsynaptic vascular alpha-1 adrenergic receptors by phenoxybenzamine in vivo produces marked rightward shifts in the dose–response curves of alpha-1 adrenergic receptor agonists with no depression of the maximum response, whereas alpha-2 adrenergic receptor alkylation by higher doses of phenoxybenzamine is associated with depressed maximum vasoconstrictor responses with only small rightward shifts in the dose–response curves to alpha-2 adrenergic receptor agonists (Hamilton et al., 1983; Reid et al., 1983b; Ruffolo and Yaden, 1984). It appears, therefore, that a more favorable relationship exists between alpha-1 adrenergic receptor occupancy and vasoconstrictor response than between alpha-2 adrenergic receptor occupancy and vasoconstrictor response, and that the degree of "receptor reserve" is greater for postjunctional vascular alpha-1 adrenergic receptors, at least for the particular agonists used in these studies (Hamilton et al., 1983; Reid et al., 1983b; Ruffolo and Yaden, 1984).

Recently, it has been shown that the pressor response mediated by postjunctional vascular alpha-1 adrenergic receptors is resistant to antagonism by calcium slow channel-blocking agents, suggesting that alpha-1 adrenergic receptors do not rely heavily upon extracellular calcium to produce arterial vasoconstriction. In contrast, vasoconstriction elicited by postjunctional arterial alpha-2 adrenergic receptors appears to be critically dependent upon extracellular calcium, as evidenced by the extreme sensitivity of this response to inhibition by the calcium channel-blocking agents (Cavero et al., 1983; van Meel et al., 1981b,c).

The proposal that alpha-1 and alpha-2 adrenergic receptor-mediated arterial vasoconstriction results from utilization of different calcium pools has met with some criticism (Ruffolo et al., 1984c; Hamilton et al., 1983) and is inconsistent with in vitro studies in which alpha-1 adrenergic receptor-mediated vasoconstriction is found to be highly sensitive to antagonism by calcium channel blockers (Awad et al., 1983; Beckeringh et al., 1984; van Breeman et al., 1982). As a possible explanation for the resistance of the alpha-1 adrenergic receptor-mediated vasoconstrictor process in vivo to antagonism by calcium channel blockers, it has been suggested (Reid et al., 1983b; Ruffolo et al., 1984c) that the large receptor reserve known to exist for postsynaptic vascular alpha-1 adrenergic receptors (Ruffolo and Yaden, 1984) would "buffer" this process from antagonism by any noncompetitive antagonist, including calcium channel blockers, in accord with receptor theory (Ariens and van Rossum, 1957). When the large alpha-1 adrenergic receptor reserve is removed by pretreatment with phenoxybenzamine (Ruffolo et al., 1984c) or benextramine (Nichols and Ruffolo, 1986) or when alpha-1 adrenergic receptor partial agonists are investigated for which no receptor reserve exists (Ruffolo et al., 1984c), alpha-1 adrenergic receptor-mediated arterial vasoconstriction appears to be equally sensitive as alpha-2 adrenergic receptor-mediated arterial vasoconstriction to inhibition by calcium channel antagonists. It appears, therefore, that both alpha-1 and alpha-2 adrenergic receptor-mediated arterial vasoconstrictor processes are highly dependent upon the mobilization of extracellular calcium (Ruffolo et al., 1984b). It has recently been proposed that alpha-1 adrenergic receptor-mediated arterial vasoconstriction produced by a full agonist may depend also, in part, on the mobilization of intracellular stores of calcium in addition to the translocation of extracellular calcium (Chiu et al., 1986), whereas alpha-2 adrenergic receptor-mediated vasoconstriction involves only the translocation of extracellular calcium (Timmermans and Thoolen, 1987).

The biochemical correlates of alpha-1 and alpha-2 adrenergic receptor-mediated arterial vasoconstriction are poorly understood. Although both alpha-adrenergic receptor subtypes may utilize predominantly extracellular calcium to evoke vasoconstrictor responses, it is unlikely that the steps leading to calcium translocation are the same for both the postsynaptic vascular alpha-1 and alpha-2 adrenergic receptors. In most alpha-1 adrenergic receptor-mediated systems (Michell, 1979), including the vasculature (Villalobos-Molina et al., 1982), calcium translocation is secondary to enhanced turnover of inositol phospholipids, the later induced as a direct consequence of agonist interaction with alpha-1 adrenergic receptors. In contrast, although vasoconstriction mediated by postjunctional vascular alpha-2 adrenergic receptors is equally dependent (or more so) upon mobilization of extracellular calcium, this process appears not to be secondary to increases in phoshatidylinositol turnover (Reese and Matthews, 1986). In many systems, alpha-2 adrenergic receptor-mediated responses are closely coupled to the inhibition of adenylate cyclase. It is not known at present, however, whether pressor responses elicited by postjunctional vascular alpha-2 adrenergic receptor agonists result from alpha-2 adrenergic receptor-mediated decreases in vascular cyclic-AMP levels.

2.2.2. CORONARY ARTERIAL CIRCULATION

Although the precise role of alpha-2 adrenergic receptor stimulation in the dynamic regulation of coronary blood flow is still unclear, it has been known for some time that following beta-adrenergic receptor blockade, alpha-adrenergic receptor agonists or cardiac sympathetic nerve stimulation can produce coronary artery vasoconstriction leading to an increase in coronary arterial resistance and a decrease in coronary artery blood flow. alpha-Adrenergic receptor agonists, such as phenylephrine (Pitt et al., 1967; Williams and Most, 1981), methoxamine (Hashimoto et al., 1960), and norepinephrine (Berne, 1958; Hashimoto et al., 1960; Gaal et al., 1966; Lioy, 1967; Malindzak et al., 1978) produce coronary artery vasoconstriction in the dog as well as in other species (Proctor, 1968; Glomstein et al., 1967; Broadley, 1970; Parratt, 1969). In animals pretreated with beta-adrenergic receptor-blocking agents, cardiac sympathetic nerve stimulation produces a fall in coronary artery blood flow that can be blocked by alpha-adrenergic receptor antagonists, demonstrating that alpha-adrenergic receptors can mediate vasoconstriction to endogenous as well as exogenous norepinephrine in the coronary circulation (Feigl, 1967, 1975; Hamilton and Feigl, 1976; Malindzak et al., 1978).

It has recently been suggested that alpha-2 adrenergic receptors may play a role in the alpha-adrenergic receptor-mediated regulation of coronary artery blood flow. In the presence of beta-adrenergic receptor blockade, intracoronary administration of the selective alpha-1 adrenergic receptor agonist phenylephrine and the selective alpha-2 adrenergic receptor agonist B-HT 933 produces a rapid decrease in coronary artery blood flow, and these effects are blocked by the alpha-1 and alpha-2 adrenergic receptor antagonists prazosin and rauwolscine, respectively (Holtz et al., 1982). These same investigators have demonstrated that the reduction in coronary artery blood flow elicited by exogenously administered norepinephrine is antagonized to a greater degree by rauwolscine than by prazosin, thus suggesting a more prominent role of alpha-2 adrenergic receptors in the regulation of coronary artery blood flow. The presence of alpha-1 adrenergic receptors on the large, epicardial coronary arteries has recently been demonstrated (Heusch et al., 1984), whereas alpha-2 adrenergic receptors appear to be located primarily on the smaller subendocardial resistance vessels of the coronary vascular bed (Kopia et al.,

1986). In addition, it has been found that the presence of a flow-limiting coronary artery stenosis can unmask a vasoconstrictor response mediated by sympathetic nerve stimulation, and that this response can be antagonized by the nonselective alpha-adrenergic receptor antagonist phentolamine, as well as by the selective alpha-2 adrenergic receptor antagonist rauwolscine, but not by the selective alpha-1 adrenergic receptor antagonist prazosin (Heusch and Deussen, 1983). These results suggest that alpha-adrenergic receptor-mediated coronary artery vasoconstriction may occur in the coronary circulation under pathologic circumstances (i.e., coronary artery disease, angina, or coronary artery vasospasm) and that alpha-2 adrenergic receptors might therefore represent a novel therapeutic target.

Recently, it has been shown (Kopia et al., 1986) that equieffective blocking doses of the alpha-1 adrenergic receptor antagonist prazosin and the alpha-2 adrenergic receptor antagonist idazoxan produce a rightward shift in the left circumflex coronary artery blood flow–frequency response curve, with idazoxan producing greater blockade than prazosin. A similar result is also obtained when a more prolonged period of stimulation is used. These results indicate that alpha-1 and alpha-2 adrenergic receptors coexist in the coronary circulation and that both alpha-adrenergic receptor subtypes mediate coronary artery vasoconstriction. Furthermore, the data suggest that postjunctional vascular alpha-2 adrenergic receptors may play a more important functional role than postjunctional vascular alpha-1 adrenergic receptors in the canine coronary circulation and that alpha-2 adrenergic receptors may be preferentially innervated (i.e., junctional), since they may be activated selectively by endogenous norepinephrine liberated from sympathetic nerves upon electrical stimulation. These results are consistent with the observations of Holtz et al. (1982), which show that the alpha-2 adrenergic receptor-blocking agent rauwolscine produced a greater degree of inhibition of in the coronary vasoconstrictor response to exogenously administered norepinephrine than did an equieffective blocking dose of prazosin.

In total, the results allow the following conclusions to be drawn regarding the coronary arterial circulation, which are illustrated schematically in Fig. 1: (1) postjunctional vascular alpha-1 and alpha-2 adrenergic receptors coexist in the coronary circulation, (2) both alpha-adrenergic receptor subtypes have the capacity to produce coronary artery vasoconstriction, (3) the postjunctional vascular alpha-2 adrenergic receptor may be more important than the postjunctional vascular alpha-1 adrenergic receptor in the regulation of coronary artery blood flow, (4) neuronally released norepinephrine may selectively activate postjunctional vascular alpha-2 adrenergic receptors relative to alpha-1 adrenergic receptors, (5) postjunctional vascular alpha-2 adrenergic receptors may have predominantly a junctional or synaptic distribution, with the postjunctional vascular alpha-1 adrenergic receptors residing extrajunctionally to a greater extent, and (6) alpha-2 adrenergic receptors may be located on subendocardial resistance vessels in the coronary arterial circulation, whereas the alpha-1 adrenergic receptors may be located predominantly in the larger epicardial conduit vessels.

2.2.3. PULMONARY ARTERIAL CIRCULATION

Postjunctional vascular alpha-1 and alpha-2 adrenergic receptors mediate vasoconstriction in the pulmonary circulation of the dog (Shebuski et al., 1986). This is evidenced by the dose-related increases in pulmonary perfusion pressure observed following the administration of the selective alpha-1 adrenergic receptor agonist methoxamine, which is highly sensitive to blockade with the alpha-1 adrenergic receptor antagonist prazosin and resistant to the alpha-2 adrenergic receptor antagonist rauwolscine. Similarly, the pulmonary pressor effects mediated by the selective alpha-2 adrenergic receptor agonist B-HT 933 were sensitive to blockade by rauwolscine and resistant to prazosin. The results indicate that postjunctional vascular alpha-1 and alpha-2 adrenergic receptors coexist in the pulmonary circulation of the dog and that both alpha-adrenergic receptor subtypes mediate vasoconstriction. Similar results have also been reported in the pulmonary vasculature of the cat (Hyman and Kadowitz, 1985). The maximal pulmonary pressor response attainable is greater with methoxamine than with B-HT 933, indicating that under conditions of normal pulmonary vascular tone alpha-1 adrenergic receptor-mediated responses may predominate over alpha-2 adrenergic receptor-mediated responses.

The greatest increase in pulmonary perfusion pressure is achieved with the nonselective alpha-adrenergic receptor agonist norepinephrine. Prazosin and rauwolscine both antagonize the increases in pulmonary perfusion pressure elicited by exogenously administered norepinephrine, indicating that norepinephrine has the capacity to stimulate both postjunctional alpha-1 and alpha-2 adrenergic receptors in the pulmonary vascular bed of the dog (Shebuski et al., 1986). Pulmonary pressor responses to endogenous norepinephrine released from sympathetic nerves by administration of the indirectly acting sympathomimetic amine tyramine are antagonized primarily by prazosin, with little or no effect of rauwolscine. It appears, therefore, that endogenous norepinephrine acts primarily on alpha-1 adrenergic receptors in the pulmonary vascular bed of the dog (Shebuski et al., 1986) and that endogenously released norepinephrine stimulates predominantly intrasynaptic alpha-1 adrenergic receptors, whereas exogenously administered norepinephrine stimulates both intrasynaptic alpha-1 as well as extrasynaptic alpha-2 adrenergic receptors in the canine pulmonary vascular bed. Similar conclusions have been made regarding the peripheral arterial circulation of the dog, in which preferential innervation of postjunctional vascular alpha-1 adrenergic receptors has been demonstrated (Langer et al., 1981).

The ability of selective alpha-2 adrenergic receptor agonists and exogenously administered norepinephrine to elicit increases in pulmonary perfusion pressure may indicate that, under some conditions, circulating catecholamines may play a role in maintaining or elevating pulmonary vascular tone by a mechanism involving, at least in part, postsynaptic vascular alpha-2 adrenergic receptors. Hyman et al. (1985) have infused epinephrine into the perfused pulmonary circulation of the cat (after propranolol treatment) to elicit a large rise in pulmonary perfusion pressure (10–20 mm Hg). Recently, Sawyer et al. (1985) have demonstrated that circulating catecholamines are responsible for alpha-2 adrenergic receptor-mediated pressor effects in the spontaneously hypertensive rat. Therefore, it may be postulated in certain disease states, such as congestive heart failure, in which pulmonary pressure is elevated and circulating catecholamine levels are high (Cohn et al., 1984), that alpha-2 adrenergic receptor-mediated increases in pulmonary vascular resistance may be secondary to the elevated circulating catecholamines.

It appears, therefore, that alpha-1 and alpha-2 adrenergic receptors can be identified in the pulmonary circulation of the dog and cat, and both alpha-adrenergic receptor subtypes mediate vasoconstriction. Norepinephrine released from sympathetic nerves activates primarily alpha-1 adrenergic receptors, whereas exogenously administered norepinephrine may activate both alpha-1 and alpha-2 adrenergic receptors under conditions of normal pulmonary vascular tone. Based on these observations, it has been suggested that postjunctional vascular alpha-1 adrenergic receptors in the pulmonary circulation are located at the vascular neuroeffector junction, whereas postjunctional vascular alpha-2 adrenergic receptors may be located extrajunctionally (Fig. 1) (Shebuski et al., 1986).

Under conditions of normal pulmonary vascular tone, post-junctional vascular alpha-1 and alpha-2 adrenergic receptors mediate vasoconstriction in the pulmonary circulation of the dog (Shebuski et al., 1986) and cat (Hyman and Kadowitz, 1985). Under these conditions (basal pressure = 10 ± 1 mm Hg), alpha-1 adrenergic receptor-mediated responses predominate over alpha-2 adrenergic receptor-mediated responses (Hyman and Kadowitz, 1985; Shebuski et al., 1986). When pulmonary vascular tone is elevated, even slightly, however, with a vasoconstrictor agent, responses to the selective alpha-2 adrenergic receptor agonist, B-HT 933, are markedly enhanced. (Shebuski et al., 1987; Hyman and Kadowitz, 1986). Furthermore, the enhanced responsiveness of alpha-2 adrenergic receptors is tone-dependent and highly selective for alpha-2 adrenergic receptors, since responses to the alpha-1 adrenergic receptor agonist methoxamine or to angiotensin II were not enhanced by elevating pulmonary vascular tone (Shebuski et al., 1987). The nature of the vasoconstrictor agent used to elevate pulmonary vascular tone does not influence the enhanced alpha-2 adrenergic receptor responsiveness, although the manner in which pulmonary vascular pressure is elevated is critically important. When pulmonary perfusion pressure is elevated by increased pulmonary blood flow as opposed to pulmonary vasoconstriction, responses to B-HT 933 are not enhanced as they are when vasoconstrictor agents were utilized to elevate pulmonary pressure (Shebuski et al., 1987). This observation indicates that pulmonary vascular smooth muscle tone, and not pulmonary pressure *per se*, is the major determinant of enhanced alpha-2 adrenergic receptor responsiveness in the pulmonary vasculature.

Thus, under conditions of elevated pulmonary vascular tone, alpha-2 adrenergic receptor-mediated responsiveness is enhanced markedly. Enhancement of pulmonary alpha-2 adrenergic receptor responsiveness is tone-dependent and selective for postjunctional vascular alpha-2 adrenergic receptors. Enhanced responsiveness of alpha-2 adrenergic receptors in the pulmonary circulation at high tone does not appear to be caused by generalized improvement in vascular excitation-contraction coupling, but may be caused by improved function of calcium channels linked to pulmonary vascular alpha-2 adrenergic receptors.

2.2.4. RENAL ARTERIAL CIRCULATION

The kidneys receive approximately 20% of the cardiac output and provide a significant contribution to total systemic vascular resistance. Their dense noradrenergic innervation extends to both the afferent and efferent arterioles (Barajas and Wang, 1979). Stimulation of the renal nerves and administration of alpha-adrenergic receptor agonists produce an increase in renal vascular resistance with redistribution of blood flow from the cortical to the medullary areas. This response is blocked by phenoxybenzamine or phentolamine (Cooke et al., 1972; Oswald and Greven, 1981), indicating the activation of alpha-adrenergic receptors. Initial in vivo studies of the alpha-adrenergic receptor subtype mediating renal vascular responses to exogenously administered agonists suggested an almost exclusive role of alpha-1 adrenergic receptors in the renal vasculature of the rat (Schmitz et al., 1981), cat (Drew and Whiting, 1979), and dog (Horn et al., 1982). More recent studies show that the alpha-2 adrenergic receptor agonist B-HT 933 produces renal vasoconstriction in the dog, and this response is blocked by the selective alpha-2 adrenergic receptor antagonist rauwolscine (A.J. Nichols and R.R. Ruffolo, unpublished observation), indicating the presence of postjunctional arterial alpha-2 adrenergic receptors in the renal vasculature of the dog (Fig. 1). It should be emphasized, however, that the renal vasoconstriction produced by B-HT 933 is much less than that observed in the hindlimb vasculature and is less than that produced by the alpha-1 adrenergic receptor agonist cirazoline in the renal arterial circulation. Significant renal vasoconstrictor responses are produced also by postjunctional alpha-2 adrenergic receptor activation in the rabbit in vivo (Hesse and Johns, 1984). In contrast, studies on isolated rabbit perfused afferent and efferent arterioles have identified vasoconstriction mediated by alpha-1 adrenergic receptors, but not by alpha-2 adrenergic receptors, with no significant difference in the sensitivity of afferent and efferent arterioles being observed (R.M. Edwards, personal communication). This lack of alpha-2 adrenergic receptor-mediated vasoconstriction in vitro may result from the relatively low density of postjunctional alpha-2 adrenergic receptors that exist in the renal arterial vasculature as suggested by the in vivo studies. Efferent renal nerve stimulation in the dog produces vasoconstriction via activation of alpha-1 adrenergic receptors (Osborn et al., 1983). It has been suggested, however, that, in the rabbit, alpha-2 adrenergic receptors may play a role in neurogenic vasoconstrictor responses (Hesse and Johns, 1984). Thus, despite evidence demonstrating the presence of postjunctional alpha-2 adrenergic receptors in the renal arterial vasculature and suggestions of slight species differences, postjunctional alpha-1 adrenergic receptors predominate.

2.2.5. MESENTERIC ARTERIAL CIRCULATION

The splanchnic circulation receives approximately 20–25% of the cardiac output and contains a similar proportion of the blood volume. The major part of the splanchnic blood supply is received by the mesenteric circulation, which supplies the small intestine and the upper two thirds of the large intestine via the superior mesenteric artery. Consequently, the mesenteric circulation has the potential to play a major role in the determination of total systemic vascular resistance. Sympathetic nerve stimulation and exogenous norepinephrine administration produce mesenteric arteriolar vasoconstriction via activation of alpha-adrenergic receptors (Granger et al., 1980). Studies using the *in situ* autoperfused superior mesenteric arterial bed of the rat suggest that only alpha-1 adrenergic receptors are present in the mesenteric vasculature, since vasoconstrictor responses to norepinephrine are blocked exclusively by low doses of prazosin and are relatively unaffected by yohimbine (Nichols and Hiley, 1985; Nichols, 1985). Similarly, the alpha-1 adrenergic receptor agonists phenylephrine, amidephrine, and cirazoline produce mesenteric arterial vasoconstriction, whereas the alpha-2 adrenergic receptor agonists xylazine, B-HT 920, and B-HT 933 do not (Nichols, 1985). Hiley and Thomas (1987), however, using the microsphere technique in the pithed rat, have shown that the mesenteric vasculature of the rat does indeed possess postjunctional vascular alpha-2 adrenergic receptors in addition to the previously identified alpha-1 subtype, with an apparent greater density of alpha-1 adrenergic receptors relative to alpha-2 adrenergic receptors (Fig. 1). Similarly, studies in the cat (Drew and Whiting, 1979) and the dog (Shepperson et al., 1982) have demonstrated a significant population of postjunctional vascular alpha-2 adrenergic receptors in the superior mesenteric arterial bed. In the cat, it appears that alpha-2 adrenergic receptors are less prominent in the mesenteric vasculature than in the hindlimb skeletal muscle vasculature (Drew and Whiting, 1979). No studies have been specifically designed to determine whether or not alpha-1 and alpha-2 adrenergic receptors are differentially distributed within the intestinal wall. Neuronally released norepinephrine, however, which presumably acts exclusively on alpha-1 adrenergic receptors in the resistance vessels, does not produce a significant redistribution of blood flow within the intestinal wall. Thus, it would appear that alpha-1 adrenergic receptors are relatively uniformly distributed throughout the arterial circulation in the gut wall. No information is available regarding the distribution of postjunctional vascular alpha-2 adrenergic receptors in the mesenteric arterial circulation.

2.2.6. CEREBRAL ARTERIAL CIRCULATION

The arteries supplying blood to the brain clearly have different pharmacologic characteristics compared to peripheral arteries. If the reactivity of the vertebral or carotid artery of the rabbit to norepinephrine is determined sequentially, a marked decrease in sensitivity is seen just prior to the entry of the vessel into the subarachnoid space (Bevan, 1979). The point of transition corresponds to the change in embryological origin of the proximal and distal portions of each of these blood vessels. Although most studies have been performed on the relatively large cerebral arteries, such as the basilar artery or the middle cerebral artery, these may be the most important sites for an alpha-2 adrenergic receptor-mediated effect, since, as in other vascular beds, such as the pulmonary circulation, the contribution of alpha-adrenergic receptors may decrease with decreasing vascular diameter (Bevan et al., 1985).

Although cerebral blood vessels have extensive and active sympathetic innervation (Duckles, 1980), the alpha-adrenergic receptor-mediated responses of these vessels to sympathetic nerve stimulation are small compared to those of peripheral vessels (McCalden, 1981). This may be related either to insensitivity of the alpha-adrenergic receptor or to a reduced alpha-adrenergic receptor number (Bevan, 1984). Nevertheless, there is evidence that the sympathetic nervous system can modulate cerebral blood flow in the conscious animal through an alpha-adrenergic receptor-mediated effect as measured by hypothalamic washout of radioactive xenon in the rabbit (Rosendorff et al., 1976).

In vitro characterization of alpha-adrenergic receptors on cerebral blood vessels has not yet yielded a uniform picture. Radioligand binding studies show the presence of both [³H]prazosin and [³H]yohimbine binding sites in membranes from human and monkey cerebral arteries. In contrast, only [³H]yohimbine sites could be detected in canine and bovine cerebral arteries. Furthermore, the B_{max} for [³H]yohimbine was higher in canine (450 fmol/mg protein) and bovine (670 fmol/mg) cerebral arteries compared to human (240 fmol/mg) and monkey (200 fmol/mg) cerebral arteries (Usui et al., 1985). This suggests an increased alpha-2 adrenergic receptor contribution in the canine and bovine cerebral vessels.

Physiologic support for this hypothesis has been provided by the observations that clonidine produces a yohimbine-sensitive contraction in the isolated canine basilar artery (Sakakibara et al., 1982). Yohimbine inhibits the contractile response to norepinephrine in canine, but not in monkey and human, cerebral arteries. Conversely, the contractile response to norepinephrine in monkey and human cerebral arteries is sensitive to prazosin (Sakakibara et al., 1982; Toda, 1983). The vascular alpha-adrenergic receptor of the canine cerebral artery may be atypical, however, since phenylephrine will produce a response in this tissue that is insensitive to blockade by prazosin (100 nM), but is antagonized by a moderate concentration of yohimbine (K_B <50 nM) (Toda, 1983). The magnitude of contraction in the canine cerebral artery induced by norepinephrine and other alpha-1 adrenergic receptor agonists is also lower than that seen in human and monkey cerebral arteries (Usui et al., 1985). This is also consistent with a relative lack in alpha-1 adrenergic receptor number observed in radioligand binding studies in the canine cerebral artery.

The cat cerebral artery also has predominantly alpha-2 adrenergic receptors located postjunctionally, since rauwolscine is approximately 100-fold more potent than prazosin in blocking vasoconstriction responses elicited by norepinephrine (Skarby, 1984).

In addition to responses via alpha-1 and alpha-2 adrenergic receptors, rabbit cerebral arteries also show a vasoconstrictor response to high concentrations (>100 μM) of norepinephrine, and this response is insensitive to blockade by phenoxybenzamine and other alpha-adrenergic receptor antagonists (Duckles et al., 1976). It has been proposed that this response may not be mediated by specific receptors, since the effect produced by norepinephrine at these concentrations is not stereospecific. Nevertheless, the response to high concentrations of norepinephrine has been postulated to play a role in cerebrovascular neurotransmission, because of the limited number of alpha-adrenergic receptors present on these vessels (Bevan et al., 1985). This type of response also has been reported in peripheral blood vessels following phenoxybenzamine treatment.

Much information regarding the role of the alpha-adrenergic receptor subtypes in mediating vasoconstriction of cerebral arteries still remains to be elucidated. Nevertheless, at least in certain species, the alpha-2 adrenergic receptor appears to be demonstrable in radioligand binding studies, by vasoconstriction induced by alpha-2 adrenergic receptor agonists and by blockade of the response to the physiologic neurotransmitter norepinephrine by selective alpha-2 adrenergic receptor antagonists (Fig. 1).

2.2.7. ENDOTHELIUM

It has been demonstrated recently that vascular endothelial cells mediate relaxation of arterial smooth muscle in response to certain vasodilators, such as acetylcholine, bradykinin, and substance P, by the release of the so-called endothelium-derived relaxing factor (EDRF) (Furchgott, 1983). It has been proposed that activation of alpha-2 adrenergic receptors on endothelial cells stimulates the release of EDRF (Egleme et al., 1984; Cocks and

Angus, 1983; Matsuda et al., 1985), an action that would tend to antagonize vasoconstriction produced by activation of postjunctional vascular alpha-adrenergic receptors. Thus, removal of endothelial cells from rat aorta produces an increase in responsiveness to alpha-adrenergic receptor agonists (Egleme et al., 1984). The mechanism by which endothelium removal enhances alpha-adrenergic receptor responsiveness in this tissue may not result from the removal of alpha-2 adrenergic receptor-stimulated release of EDRF, however, since enhancement of alpha-adrenergic receptor-mediated vasoconstriction by removal of the endothelium in rat aorta is not related to the alpha-adrenergic subtype specificity of the agonist used, inasmuch as both alpha-1 and alpha-2 adrenergic receptor agonists produce this effect (Malta et al., 1986; Martin et al., 1986). Furthermore, removal of the alpha-1 adrenergic receptor reserve for phenylephrine markedly enhances the potentiating effect of endothelium removal (Martin et al., 1986). Combined with the failure of clonidine to produce relaxation of precontracted endothelium-intact rat aorta (Martin et al., 1986), these data suggest that spontaneous release of EDRF from rat aortic endothelium depresses contractility of the vascular smooth muscle by functional antagonism of the response to alpha-1 adrenergic receptor activation, with the depression of responsiveness being inversely related to agonist efficacy (Malta et al., 1986; Martin et al., 1986). In other vascular preparations, however, there is convincing evidence that endothelial cells lining arteries do possess alpha-2 adrenergic receptors that mediate the release of EDRF. Cocks and Angus (1983) demonstrated that removal of endothelium enhanced the contractile response produced by norepinephrine in canine and porcine circumflex coronary artery, and that after blockade of alpha-1 adrenergic receptors, norepinephrine could produce yohimbine- and idazoxan-sensitive relaxation of precontracted arteries only in the presence of an intact endothelium. Additional studies have shown that alpha-2 adrenergic receptors mediate release of EDRF from carotid, mesenteric, renal, and femoral arteries from dogs and pigs, although there do appear to be species differences in the magnitude of this response (Angus et al., 1986). Furthermore, it has been suggested that endothelial alpha-2 adrenergic receptors mediate release of EDRF in coronary microvessels (Angus et al., 1986). Thus, alpha-2 adrenergic receptor agonists appear to have the capability of modulating vascular responsiveness via stimulation of the release of EDRF in both large arteries and the microcirculation, but this effect does not occur in all vessels.

2.3. Peripheral Venous Circulation

2.3.1. SAPHENOUS VEIN

In contrast to arterial alpha-2 adrenergic receptors, vascular alpha-2 adrenergic receptors in isolated veins can be studied easily in vitro, at least in certain tissues; indeed most of the characterization of venous alpha-2 adrenergic receptors has been performed in vitro.

The most commonly studied vein is the canine saphenous vein. DeMey and Vanhoutte (1981) first reported the potent vasoconstrictor activity of clonidine in this tissue. Additional studies have shown that highly selective alpha-2 adrenergic receptor agonists, such as B-HT 920, B-HT 933, and UK 14,304, will produce a vasoconstrictor response that is resistant to antagonism by prazosin and sensitive to blockade by rauwolscine (Fowler et al., 1984; Alabaster et al., 1985; Ruffolo and Zeid, 1985). Receptor dissociation constants calculated for alpha-2 adrenergic receptor antagonists in the canine saphenous vein correlate well with the values obtained at prejunctional alpha-2 adrenergic receptors (Fowler et al., 1984; Hieble et al., 1986).

Although alpha-1 adrenergic receptors are present also in canine saphenous vein, the use of selective agonists and/or antagonists allow the postsynaptic vascular alpha-2 adrenergic receptor to be studied without interference from the alpha-1 adrenergic receptor. In the canine saphenous vein, norepinephrine, which can activate both alpha-1 and alpha-2 adrenergic receptors, appears to activate preferentially the alpha-2 subtype. Fowler et al. (1984) found the response to low concentrations of norepinephrine to be relatively unaffected by prazosin. Flavahan et al. (1984) observed qualitatively similar results, with the lower portion of the dose–response curve to norepinephrine being blocked by prazosin to a lesser extent than the higher concentration range. In contrast, Alabaster et al. (1985) and Sullivan and Drew (1980) observed competitive blockade of the norepinephrine response by prazosin at both the lower and higher concentration ranges. The inconsistencies between these observations have yet to be resolved.

Saphenous veins from other species also show alpha-2 adrenergic receptor responsiveness. Alabaster et al. (1985) suggest that the rabbit saphenous vein represents a more useful model tissue to study postsynaptic vascular alpha-2 adrenergic receptors than the canine saphenous vein, based on a proposed smaller contribution of alpha-1 adrenergic receptor-mediated responses. Levitt and Hieble (1985), however, although observing potent blockade of B-HT 933-induced vasoconstriction by rauwolscine (K_B = 3.6 nM), found the response to norepinephrine to be sensitive to both prazosin (K_B = 13 nM) and rauwolscine (K_B = 7 nM), suggesting a significant contribution from alpha-1 adrenergic receptors as well as from alpha-2 adrenergic receptors in this tissue.

Experiments in isolated human saphenous vein (Muller-Schweinitzer, 1984) show similar results to those reported by Fowler et al. (1984) in the canine saphenous vein. The response to low concentrations of norepinephrine was essentially unaffected by prazosin, but potently antagonized by yohimbine (K_B = 25 nM). Inhibition of the vasoconstrictor response to field stimulation of adrenergic nerve terminals in both canine (Sullivan and Drew, 1980; Flavahan et al., 1984) and human (Docherty and Hyland, 1984; Gothert et al., 1984) saphenous vein is mediated via prejunctional alpha-2 adrenergic receptors, based on blockade of this response by yohimbine, but not by prazosin.

The venous circulation, in particular the canine saphenous vein, resembles the arterial circulation in that postsynaptic vascular alpha-1 and alpha-2 adrenergic receptors coexist, with each alpha-adrenergic receptor subtype mediating vasoconstriction (Constantine et al., 1982; DeMey and Vanhoutte, 1981; Flavahan et al., 1984; Docherty and Hyland, 1984). In contrast to the arterial circulation, however, postsynaptic vascular alpha-2 adrenergic receptors in the canine saphenous vein appear to be preferentially innervated, with postsynaptic vascular alpha-1 adrenergic receptors being innervated to a lesser degree and possibly located predominantly extrajunctionally (Fig. 1) (Flavahan et al., 1984).

In the canine saphenous vein, the alpha-1 adrenergic receptor occupancy–response relationship (obtained for cirazoline) is approximately four-fold more favorable than the alpha-2 adrenergic receptor occupancy–response relationship (obtained for B-HT 933), although both agonists are associated with a significant receptor reserve (Ruffolo and Zeid, 1985). Both the alpha-1 adrenergic receptor occupancy–response relationship of cirazoline and the alpha-2 adrenergic receptor occupancy–response relationship of B-HT 933 are rectangular hyperbolas, which suggests that both compounds have high intrinsic efficacy at their respective alpha-adrenergic receptor subtypes (Ruffolo and Zeid, 1985). The results indicate that the alpha-adrenergic receptor reserve may be significantly larger for postsynaptic vascular alpha-1 adrenergic receptors than for postsynaptic vascular alpha-2 adrenergic receptors in canine saphenous vein, as also appears to be the case in the peripheral arterial circulation (Ruffolo and Zeid, 1985).

The maximum contractile response observed to B-HT 933 is significantly less than that obtained with cirazoline in the canine saphenous vein (Ruffolo and Zeid, 1985). This observation has been made previously for several alpha-2 adrenergic receptor agonists that routinely produce lower maximum responses than alpha-1 adrenergic receptor agonists in the canine saphenous vein (Flavahan et al., 1984). The lower maximum responses commonly observed with alpha-2 adrenergic receptor agonists in the canine saphenous vein could result from four factors: (1) the alpha-2 adrenergic receptor agonists studied to date could be partial agonists (i.e., low intrinsic efficacy characterized by relatively linear occupancy–response relationships), (2) a relative deficiency in alpha-2 adrenergic receptor number, (3) poor coupling between alpha-2 adrenergic receptor activation and vasoconstrictor response, or (4) limitations in the alpha-2 adrenergic receptor excitation–contraction coupling mechanism. The first possibility may be eliminated because the alpha-2 adrenergic receptor occupancy–response relationship of B-HT 933 in canine saphenous vein is hyperbolic, characteristic of a full agonist. In fact, the alpha-2 adrenergic receptor occupancy–response relationship for B-HT 933 is as favorable for alpha-2 adrenergic receptors as cirazoline is for alpha-1 adrenergic receptors (Ruffolo and Zeid, 1985). In other words, B-HT 933 has as high an efficacy at alpha-2 adrenergic receptors as cirazoline has at alpha-1 adrenergic receptors, at least in canine saphenous vein. The second and third possibilities of limited alpha-2 adrenergic receptor number or poor coupling between alpha-2 adrenergic receptor occupancy and vasoconstrictor response may be eliminated because in the canine saphenous vein there exists an alpha-2 adrenergic receptor reserve for B-HT 933, and this could not occur if the number of alpha-2 adrenergic receptors was low or if coupling was poor. This reserve exists because of the high efficacy of the agonist, such that only 60% of the alpha-2 adrenergic receptor population is required by B-HT 933 to produce a maximum response (Ruffolo and Zeid, 1985). It is unlikely, therefore, that with 40% of the alpha-2 adrenergic receptor population representing excess or spare alpha-2 adrenergic receptors, that limited alpha-2 adrenergic receptor number could contribute to the lower maximum response observed for B-HT 933 relative to the alpha-1 adrenergic receptor agonist cirazoline in canine saphenous vein. In addition, if there were poor coupling between alpha-2 adrenergic receptor occupancy and vasoconstrictor response, there would not exist a reserve in alpha-2 adrenergic receptors, and the occupancy–response relationship for B-HT 933 would be linear. At present, evidence supports the contention that the lower maximum response observed with alpha-2 adrenergic receptor agonists relative to alpha-1 adrenergic receptor agonists in the canine saphenous vein results from fundamental differences in the excitation–contraction coupling mechanisms utilized by alpha-1 and alpha-2 adrenergic receptors, such that alpha-2 adrenergic receptor-mediated vasoconstriction is not capable of producing the degree of vasoconstriction observed with alpha-1 adrenergic receptor agonists. Although not completely understood, the excitation–contraction coupling mechanisms for postsynaptic vascular alpha-1 and alpha-2 adrenergic receptors in the canine saphenous vein appear to be different in terms of electrophysiological activity (Matthews et al., 1984a) and possibly calcium utilization (Langer and Shepperson, 1982a,b), and these factors may selectively limit the degree of vasoconstriction that can be produced by an alpha-2 adrenergic receptor agonist in this tissue.

There is one important difference between alpha-2 adrenergic receptor reserves in arteries and veins that must be emphasized. In the arterial circulation of the rat, there is no reserve in alpha-2 adrenergic receptors, the latter being characterized by a linear occupancy–response relationship for B-HT 933 (Ruffolo and Yaden, 1984). In contrast, in the canine saphenous vein, there is a significant alpha-2 adrenergic receptor reserve, such that a nonlinear, hyperbolic occupancy–response relationship is obtained for B-HT 933. The fact that there exists an alpha-2 adrenergic receptor reserve in the canine saphenous vein, but not in the arterial circulation of the rat, may have profound implications concerning fundamental differences between arteries and veins. As indicated previously, it has been suggested that the venous circulation is significantly more dependent on postsynaptic vascular alpha-2 adrenergic receptors than is the arterial circulation (Flavahan et al., 1984; Ruffolo, 1985b), and that this may be a direct consequence of the presence of spare alpha-2 adrenergic receptors in veins (at least the canine saphenous vein), but not arteries (Ruffolo and Yaden, 1984). In fact, it may be argued that the difficulty in studying the postsynaptic vascular alpha-2 adrenergic receptor in isolated arteries in vitro, and the ease with which these receptors may be studied in isolated venous preparations (such as the canine saphenous vein), may be attributed also to the spare alpha-2 adrenergic receptors found in larger veins, but not in larger arteries. Indeed, studies in vivo also tend to indicate that the postsynaptic vascular alpha-2 adrenergic receptor may play a more prominent role in venous capacitance vessels than in arterial resistance and conduit (Fig. 1) vessels (see below; Appleton et al.,

1984; Greenway and Innes, 1981; Patel et al., 1981; Segstro and Greenway, 1986).

As indicated previously, in the canine saphenous vein, the lower portion of the dose–response curve to norepinephrine is mediated predominantly by alpha-2 adrenergic receptors, whereas responses at higher concentrations are mediated predominantly by alpha-1 adrenergic receptors (Matthews et al., 1984b). The alpha-1 adrenergic receptor-mediated vasoconstrictor response in the canine saphenous vein results from electromechanical coupling in which the contractile response parallels electrophysiologic changes in membrane potential (Matthews et al., 1984a). In contrast, the alpha-2 adrenergic receptor-mediated response in the canine saphenous vein results from pharmacomechanical coupling, in which the contractile response is not paralleled by electrophysiologic changes.

Vasoconstriction in canine saphenous vein mediated by both the postsynaptic vascular alpha-1 and alpha-2 adrenergic receptors appears to be dependent on the translocation of extracellular calcium. In the canine saphenous vein, vasoconstrictor responses mediated by the alpha-1 adrenergic receptor agonist phenylephrine and by the alpha-2 adrenergic receptor agonist M-7 are both inhibited by calcium slow-channel antagonists, such as diltiazem and verapamil (Langer and Shepperson, 1981). Although alpha-1 and alpha-2 adrenergic receptor-mediated vasoconstriction of the canine saphenous vein is dependent predominantly upon the translocation of extracellular calcium, evidence exists to suggest that in this tissue, alpha-1 adrenergic receptor agonists may also trigger, to a smaller and limited extent, the release of intracellular calcium (Langer and Shepperson, 1981; Matthews et al., 1984b).

2.3.2. PULMONARY VEIN

Assessment of postjunctional alpha-adrenergic receptor activity in the pulmonary vasculature in vitro provides some interesting correlates to what is observed in canine saphenous vein. Intralobar pulmonary veins have been reported to contract to the selective alpha-2 adrenergic receptor agonist B-HT 933, and this response is sensitive to inhibition by the selective alpha-2 adrenergic receptor antagonist rauwolscine (Shebuski et al., 1987; Ohlstein et al., 1986). In contrast, intralobar pulmonary arteries are relatively unresponsive to B-HT 933 in vitro. These results indicate that postjunctional vascular alpha-2 adrenergic receptors may be preferentially located on the venous side of the pulmonary circulation, as also appears to be the case in the peripheral circulation (Ruffolo, 1985a).

2.3.3. HEPATIC PORTAL SYSTEM

A similar situation to that described in the canine saphenous vein also exists in vivo in the intestinal venous circulation (Patel et al., 1981). In addition, in the hepatic venous circulation of the cat in vivo, blood volume responses to norepinephrine are mediated by postsynaptic vascular alpha-2 adrenergic receptors, as is the hepatic venous response to sympathetic nerve stimulation (Segstro and Greenway, 1986). These results are suggestive of a dominance of alpha-2 over alpha-1 adrenergic receptors in the hepatic venous circulation, as well as a preferential, if not exclusive, junctional location of postsynaptic vascular alpha-2 adrenergic receptors (Fig. 1) (Segstro and Greenway, 1986). Furthermore, alpha-2 adrenergic receptor-mediated responses in the venous circulation appear to be more marked than those in the arterial circulation, consistent with the notion that postsynaptic vascular alpha-2 adrenergic receptors may play a more important functional role in venous relative to arterial blood vessels (Ruffolo, 1985a).

2.3.4. OTHER VEINS

Most other veins have less of an alpha-2 adrenergic receptor contribution relative to that observed in the saphenous vein. Shoji et al. (1983) compared the responsiveness of many canine veins to norepinephrine, phenylephrine, and clonidine. The saphenous and cephalic veins have the greatest response to clonidine, followed by the femoral vein. Analysis of the response to clonidine confirmed the presence of both alpha-1 and alpha-2 adrenergic receptors in the saphenous vein. Interestingly, longitudinal, but not helical, strips of portal vein, mesenteric vein, and vena cava readily respond to clonidine. Analysis of the response in the portal vein revealed only alpha-1 adrenergic receptor activation, however. Evidence for postjunctional alpha-2 adrenergic receptors in human femoral vein has been provided by the failure of prazosin to antagonize the response to low concentrations of norepinephrine in this tissue and by the potent contractile effect observed with guanfacine, a moderately selective alpha-2 adrenergic receptor agonist (Glusa and Markwardt, 1983a).

In a quantitative analysis of alpha-1 and alpha-2 adrenergic receptor characteristics in femoral and saphenous veins, the selective alpha-2 adrenergic receptor agonist UK 14,304 was much less effective in inducing contraction in the femoral vein (22% of norepinephrine maximum) compared to the saphenous vein (86% of norepinephrine maximum) (Flavahan and Vanhoutte, 1986a). As seen in some arteries, the response to norepinephrine in certain veins, such as the canine splenic vein, may be sensitive to blockade by both rauwolscine and prazosin, even though the tissue is unresponsive to highly selective alpha-2 adrenergic receptor agonists (Hieble and Woodward, 1984).

2.3.5. PHYSIOLOGIC SIGNIFICANCE OF VENOUS ALPHA-2 ADRENERGIC RECEPTORS

The physiologic significance of venous alpha-2 adrenergic receptors is unclear. In vivo studies with alpha-2 adrenergic receptor agonists in the rat (Gerold and Haesler, 1983) or dog (Zandberg et al., 1984) cannot demonstrate a significant hemodynamic effect clearly attributable to effects on venous capacitance vessels. Since venous alpha-2 adrenergic receptors are sensitive to temperature changes (McAdams and Waterfall, 1984; Flavahan and Vanhoutte, 1986b) and alpha-2 adrenergic receptors are most prominent in cutaneous vessels (Flavahan et al., 1984), the venous alpha-2 adrenergic receptor may be involved in blood flow redistribution to optimize the thermoregulatory process. It has been reported recently that alpha-2 adrenergic receptor-mediated venoconstriction can significantly reduce venous capacitance and thereby increase venous return to the heart resulting in an increase in cardiac output (Kalkman et al., 1984).

2.4. Myocardium

The cardiac adrenergic neuroeffector junction is in many respects similar to neuroeffector junctions in other peripheral tissues as far as alpha-adrenergic receptors are concerned. Presynaptic alpha-2 adrenergic receptors on postganglionic sympathetic nerve terminals have been identified in isolated hearts from many species. As in other organs, the presynaptic alpha-2 adrenergic receptors in myocardium, when activated, mediate an inhibitory effect on neurotransmitter release (Doxey and Roach, 1980; Drew, 1976; Hieble and Pendleton, 1979). As such, alpha-adrenergic receptor antagonists that are nonselective or selective alpha-2 adrenergic receptor antagonists have the capacity to produce positive inotropic and chronotropic responses (Benfey and Varma, 1962) by enhancing neurotransmitter liberation resulting from loss of the autoinhibition mediated by presynaptic alpha-2 adrenergic receptors (Starke et al., 1971a,b).

The predominant adrenergic receptor located postsynaptically in the heart is the beta-1 adrenergic receptor, which mediates a positive inotropic and chronotropic response (Broadley, 1982). Postsynaptic alpha-adrenergic receptors also exist in the hearts of many mammalian species, however, including humans, and mediate a positive inotropic response with little or no change in heart rate (Govier, 1967; Osnes, 1976; Schumann and Brodde, 1979; Schumann and Endoh, 1976). Most physiologic and radioligand binding data indicate that the postsynaptic alpha-adrenergic receptor in myocardium is exclusively of the alpha-1 subtype (Hoffman and Lefkowitz, 1980; Raisman et al., 1979; Schumann and Brodde, 1979). The mechanism by which cardiac alpha-1 adrenergic receptors increase force of myocardial contraction has not been established, but it appears not to be associated with the accumulation of cyclic-AMP or stimulation of adenylate cyclase (Brodde et al., 1978), and in this respect, alpha-1 adrenergic receptors differ from beta-1 adrenergic receptors in the myocardium. Other differences between alpha-1 and beta-1 adrenergic receptor-mediated effects in the heart include the rate of onset and duration of action, which are particularly long for alpha-1 adrenergic receptor-mediated inotropic effects (Schumann et al., 1975). Furthermore, although beta-1 adrenergic receptor-mediated inotropic responses occur at all frequencies of stimulation, the effect mediated by myocardial alpha-1 adrenergic receptors is apparent only at low frequencies (Broadley, 1982).

2.5. alpha-2 Adrenergic Receptors in the Kidney

The existence of alpha-adrenergic receptors in the kidney has been known for many years, since alpha-adrenergic drugs produce a variety of renal effects. The functions and locations of the renal alpha-adrenergic receptors are only now beginning to be understood. Radioligand binding studies indicate that alpha-1 and alpha-2 adrenergic receptors coexist in the kidneys of a variety of mammalian species including humans; the number, proportion, and distribution of each alpha-adrenergic receptor subtype may vary from one species to another, however (Jarrott et al., 1979; McPherson and Summers, 1981; Summers and McPherson, 1982; Summers, 1984).

The kidney receives a dense noradrenergic innervation that extends not only to the afferent and efferent arterioles (Barajas and Wang, 1979), but also to all portions of the nephron including the collecting duct (Barajas et al., 1984). In addition, alpha-adrenergic receptors are known to be present with an approximate two-fold greater density of alpha-2 over alpha-1 adrenergic receptors in crude membrane fractions of rat kidney (Sanchez and Pettinger, 1981). Although it is generally accepted that alpha-1 adrenergic receptors are most important in mediating vasoconstriction (Schmitz et al., 1981) and tubular sodium reabsorption (Osborn et al., 1982,1983), the precise functional role of the predominant alpha-2 adrenergic receptors is less clearly understood. In the rat, alpha-2 adrenergic receptors of the juxtaglomerular apparatus have been proposed to inhibit renin release (Pettinger, 1987). Both alpha-1 and alpha-2 adrenergic receptors have been proposed to alter electrolyte and fluid balance, but their exact roles are not fully understood.

Radioligand binding studies in rat kidney reveal that the major concentration of phentolamine-displacable [^{3}H]rauwolscine binding sites is found in the renal cortex, with a particularly high density associated with the proximal tubules, blood vessels, and glomeruli (Stephenson and Summers, 1985). In contrast to this predominant proximal tubular location of alpha-2 adrenergic receptors, as assessed by radioligand binding techniques, physiologic studies suggest a more important functional role for renal alpha-2 adrenergic receptors in the distal tubule and collecting duct. alpha-2 Adrenergic receptor activation weakly attenuates parathyroid hormone-induced activation of adenylate cyclase in isolated rat proximal convoluted tubule, but more effectively inhibits vasopressin-evoked stimulation of adenylate cyclase in the medullary and cortical collecting tubules with no effect in the medullary and cortical thick ascending limb (Umemura et al., 1985). In addition, alpha-2 adrenergic receptor stimulation antagonizes the vasopressin-induced reduction in sodium and water excretion in isolated rat perfused kidney (Smyth et al., 1985a) and water reabsorption in rabbit isolated cortical collecting tubules (Krothapalli et al., 1983).

The effects of the selective alpha-1 adrenergic receptor agonist cirazoline and the selective alpha-2 adrenergic receptor agonist B-HT 933 were assessed on renal hemodynamics and on water and solute excretion in conscious, chronically instrumented rats (Gellai and Ruffolo, 1987). Infusion of equipressor doses of cirazoline and B-HT 933 decreased renal plasma flow without changing glomerular filtration rate. Cirazoline infusion did not affect urinary excretion of water, electrolytes, or total solutes. In marked contrast, B-HT 933 increased urine flow and sodium excretion significantly, but did not significantly alter potassium and urea excretion. Urine osmolality decreased to hyposmotic levels (from 613 ± 86 to 172 ± 8 mOsm/kg/H_2O) during the infusion of B-HT 933, suggesting a possible interaction between the alpha-2 adrenergic receptor and the vasopressin system. This unique diuretic action of the selective alpha-2 adrenergic receptor agonist was also observed following the infusion of subpressor doses of B-HT 933 (Gellai and Ruffolo, 1987). In rats treated with the ganglionic blocker hexamethonium (10 mg/kg, i.v.), the B-HT

933-induced diuresis was not affected, confirming an action in the periphery, most likely at the level of the kidney. These results suggest that stimulation of renal alpha-2 adrenergic receptors mediates the inhibition of water and sodium reabsorption at the site of the distal renal nephron, most likely the cortical collecting duct, and is thereby responsible for producing diuresis and natriuresis (Gellai and Ruffolo, 1987).

As indicated above, the diuretic action of B-HT 933 is associated with sustained low levels of urine osmolality, maintained at 200 mOsm/kg H_2O or less, substantially less than plasma osmolality (290 mOsm/kg H_2O) (Gellai and Ruffolo, 1987). The formation of urine that is hypotonic to plasma is indicative of a reduction in vasopressin-associated renal water reabsorption. Infusions of other alpha-2 adrenergic receptor agonists (clonidine or guanabenz) have been reported to decrease secretion of vasopressin in anesthetized rats and dogs (Reid et al., 1979; Roman et al., 1979; Strandhoy et al., 1982). Olsen (1976) did not observe inhibition of vasopressin secretion by clonidine in conscious dogs, however. Strandhoy and coworkers (1982) emphasized that although infusion of guanabenz in anesthetized dogs decreased the plasma concentration of vasopressin, this alone could not account for the diuretic effect of guanabenz. Inhibition of the tubular action of vasopressin on water reabsorption by stimulation of tubular alpha-2 adrenergic receptors was suggested. Such inhibition has been demonstrated in studies of isolated rabbit cortical collecting tubules (Krothapalli et al., 1983; Chabardes et al., 1984) and in studies of isolated toad urinary bladder (Kinter et al., 1985). In these studies, alpha-2 adrenergic receptor stimulation inhibited vasopressin-dependent increases in water permeability by attenuating vasopressin-stimulated adenylate cyclase activity. Whether B-HT 933 impairs the secretion of vasopressin or inhibits its tubular action, or both, cannot be inferred at present. The formation of hypotonic urine, however, strongly implicates inhibition of vasopressin-dependent epithelial functions in the mechanism of action of B-HT 933 and supports the proposed collecting tubule/collecting duct site of action.

Smyth et al. (1985b) have proposed separate roles of the alpha-adrenergic receptor subtypes in the handling of water and electrolytes. These authors suggest that renal nerve stimulation potentiates tubular water and sodium reabsorption via alpha-1 adrenergic receptor stimulation, a concept that is generally accepted. In contrast, because of their extrajunctional location, renal alpha-2 adrenergic receptors may not be activated by renal nerve stimulation. Thus, renal alpha-2 adrenergic receptors may be stimulated by circulating catecholamines and play an important role in the regulation of water and sodium excretion, possibly by modulating the actions of vasopressin and other hormones in the distal nephron. Although this proposal by Smyth et al. (1985b) suggests a delicately balanced interaction of the two alpha-adrenergic receptor subtypes with neuronally released and circulating catecholamines, respectively, in the handling of water and electrolytes by the kidneys, the role of the alpha-2 adrenergic receptor in modulating sodium reabsorption under basal conditions has not been completely addressed. Smyth et al. (1985b) propose a relatively minor role for renal alpha-2 adrenergic receptors based on experimental data showing opposite effects by epinephrine on sodium reabsorption in alpha-1 adrenergic receptor blocked furosemide- or vasopressin-infused isolated rat kidneys, and no effect at all when epinephrine was infused under basal conditions in the same preparation (Smyth et al., 1984,1985a). In contrast, under normal physiologic conditions, renal alpha-2 adrenergic receptors may play an important role in the control of sodium and fluid reabsorption. It has been proposed that in a state of normal fluid balance (at least in the rat), the well-demonstrated role of alpha-1 adrenergic receptors to stimulate the increase in sodium reabsorption in the proximal tubule and thick ascending limb is balanced by the proposed action of the extrajunctional alpha-2 adrenergic receptor to inhibit sodium and water reabsorption in the collecting tubules and ducts. Accordingly, changing the activity of one of the alpha-adrenergic receptor subtypes or changing the balance in supply of catecholamine (i.e., neuronal vs. circulating) could unmask the actions of the other subtype or source of catecholamine. Thus, the natriuresis observed during the decrease in renal efferent sympathetic nerve activity, as would also occur upon stimulating the cardiopulmonary vagal afferent limb (left atrial stretch receptor stimulation), could result, at least partially, from the stimulating effect of circulating catecholamines on the extrajunctional alpha-2 adrenergic receptors; a selective alpha-2 adrenergic receptor antagonist would therefore be expected to attenuate the natriuresis. Application of highly selective alpha-1 and alpha-2 adrenergic receptor agonists and antagonists should assist in further elucidation of the potentially important role of the alpha-2 adrenergic receptor in the control of body fluid balance by the kidneys.

To summarize, it appears that the major function of alpha-2 adrenergic receptors in the kidney is to mediate the inhibition of vasopressin-induced sodium and water reabsorption in the cortical, and possibly medullary, collecting tubule and duct, leading to a natriuresis and diuresis, while possibly sparing potassium (Fig. 1). The diuresis is characterized by a hyposmotic urine. The role, if any, of the dense population of alpha-2 adrenergic receptors associated with the proximal convoluted tubule remains to be elucidated.

3. Functions Mediated by alpha-2 Adrenergic Receptors in Various Noncardiovascular Tissues

3.1. Inhibition of Neurotransmitter Release

The first physiologic action described for alpha-2 adrenergic receptors, and perhaps the most important role of the alpha-2 adrenergic receptor, is to mediate the inhibition of neurotransmitter release. Presynaptic alpha-2 adrenergic receptors mediate the inhibition of norepinephrine release from sympathetic nerves, acetylcholine release from parasympathetic nerves, as well as inhibition of the release of various central neurotransmitters (for reviews, see Starke, 1981; Doxey and Roach, 1980; Langer, 1973,1974,1979,1981; Langer and Arbilla, 1981; Dubocovich, 1984; Langer et al., 1985).

3.1.1. INHIBITION OF NOREPINEPHRINE RELEASE FROM SYMPATHETIC NERVE TERMINALS

The prejunctional alpha-2 adrenergic receptor serves as a key element in a local negative feedback system modulating neurotransmitter release. Activation of these alpha-2 adrenergic receptors, either by norepinephrine, epinephrine, or synthetic molecules having selectivity for the alpha-2 adrenergic receptor, such as B-HT 920, B-HT 933, UK-14,304, and clonidine (Langer et al., 1985), will inhibit stimulation-evoked neurotransmitter release. Conversely, alpha-2 adrenergic receptor antagonists, such as phentolamine, idazoxan, yohimbine, rauwolscine, and SK&F 86466, (Langer et al., 1985; Hieble et al., 1986), will potentiate stimulation-evoked norepinephrine release. This potentiation demonstrates that the prejunctional alpha-2 adrenergic receptor is normally under active tone as a result of endogenously released norepinephrine. The activation or blockade of prejunctional alpha-2 adrenergic receptors can be demonstrated in vivo in both animals (Langer et al., 1985) and in humans (Brown et al., 1985), as well as in isolated tissues in vitro. Since prejunctional alpha-2 adrenergic receptors have been found in all sympathetically innervated tissues thus far examined (Starke, 1981), this neuromodulatory system appears to be ubiquitous and to play an important role in the overall control of sympathetic outflow.

The magnitude of alpha-2 adrenergic receptor-mediated neuroinhibitory effects observed is dependent on stimulation parameters employed. As the amount of norepinephrine in the synaptic cleft is increased by greater frequency and/or duration of nerve stimulation, the prejunctional alpha-2 adrenergic receptor is activated to a greater extent by the endogenous norepinephrine. Hence, the ability of an exogenously administered alpha-2 adrenergic receptor agonist to inhibit sympathetic neurotransmission will decrease as stimulation frequency or duration of stimulation is increased. Conversely, potentiation of neurotransmitter overflow elicited by an alpha-2 adrenergic receptor antagonist will be enhanced as the stimulation parameters are increased to yield more intense stimulation of presynaptic alpha-2 adrenergic receptors (Auch-Schwelk et al., 1983). In this regard, an alpha-2 adrenergic receptor antagonist would not be expected to potentiate neurotransmitter release induced by a single pulse of nerve stimulation, since there should be little or no inhibitory tone mediated by the prejunctional alpha-2 adrenergic receptor under these conditions. Although there is controversy regarding this point (Kalsner, 1979,1980; Kalsner et al., 1980), most investigators report that alpha-2 adrenergic receptor antagonists do not potentiate the response of an innervated tissue to nerve stimulation with a single pulse or a train of widely separated pulses (Rand et al., 1973; Markiewicz et al., 1980; Baker and Marshall, 1982; Grant et al., 1980). It appears that a minimum interval of no greater than a few seconds is required between pulses for activation of the presynaptic alpha-2 adrenergic receptor to be of sufficient magnitude such that blockade by alpha-2 adrenergic receptor antagonist leads to enhanced norepinephrine efflux (Story et al., 1981; Auch-Schwelk et al., 1983).

Although the mechanism by which alpha-2 adrenergic receptor stimulation inhibits neurotransmitter release has not been established, an influence on the availability of intracellular calcium to promote exocytotic release is probably involved (Starke, 1977; Westfall, 1977; Langer, 1979). Evidence for the involvement of calcium derives from a greater effectiveness of alpha-2 adrenergic receptor-mediated inhibition of neurotransmitter release in low calcium media (Westfall, 1977) and the failure of alpha-2 adrenergic receptor agonists to inhibit tyramine-induced release, which does not require calcium ion (Starke and Montel, 1973). An alternative hypothesis has been suggested by Stjarne, however, involving an alpha-2 adrenergic receptor-mediated suppression of impulse transmission between varicosities (Stjarne, 1978,1979), since neurotransmitter release induced by veratridine in the guinea pig vas deferens, although requiring calcium ion, is unaffected by alpha-2 adrenergic receptor blockade with phentolamine (Stjarne et al., 1979).

Selective alpha-2 adrenergic receptor agonists will inhibit carbachol-induced catecholamine release from isolated bovine adrenal medullary cells, with a simultaneous inhibition of $^{45}Ca^{2+}$ uptake (Sakurai et al., 1983).

3.1.2. INHIBITION OF ACETYLCHOLINE RELEASE FROM CHOLINERGIC NERVE TERMINALS

alpha-Adrenergic receptor stimulation has been shown to inhibit nerve-evoked acetylcholine release from parasympathetic nerve terminals in guinea pig ileum (Paton and Vizi, 1969) and rabbit intestine (Vizi and Knoll, 1971), and from isolated superior cervical ganglia (Dawes and Vizi, 1973). These studies were performed before the alpha-1/alpha-2 adrenergic receptor subdivision was established; hence the nature (i.e., alpha-1 vs. alpha-2) of the prejunctional alpha-adrenergic receptor on cholinergic terminals was not originally established. More recently, several investigators have compared the prejunctional alpha-adrenergic receptors on sympathetic and parasympathetic nerve terminals and have concluded that both were of the alpha-2 subtype (Drew, 1978; Cambridge and Davey, 1980; Wikberg, 1978,1979a,b; Starke, 1981).

Prejunctional alpha-adrenergic receptors are also present at the skeletal neuromuscular junction. In contrast to autonomic neuroeffector junctions, activation of these receptors causes an enhancement of acetylcholine release (Malta et al., 1979). Based on the pharmacologic characterization of this receptor, however, it cannot be considered as an alpha-2 adrenergic receptor, since similar effects are produced by methoxamine, phenylephrine, norepinephrine, and oxymetazoline (Malta et al., 1979).

3.1.3. INHIBITION OF NEUROTRANSMITTER RELEASE FROM CENTRAL NEURONAL TERMINALS

As in the case of peripheral nerves, stimulation of presynaptic alpha-2 adrenergic receptors mediates the inhibition of neurotransmitter release from central neurons. Presynaptic alpha-2 adrenergic receptors are found on a variety of central neurons, including noradrenergic, dopaminergic, cholinergic, and serotonergic.

Most experiments studying central presynaptic alpha-2 adrenergic receptors have been performed using either brain slices or synaptosomes. Release of neurotransmitter from brain slices can be induced by either field stimulation or by exposure to high potassium ion concentration. alpha-2 Adrenergic receptors modulating central norepinephrine release were first demonstrated by Farnebo and Hamberger (1971), who showed that clonidine would depress the release of [³H]norepinephrine induced by field stimulation of rat cortical slices. Conversely, phentolamine and phenoxybenzamine were found to potentiate neurotransmitter overflow. In this study, the effect of both clonidine and the alpha-adrenergic receptor antagonists were small; other studies, however, have shown marked inhibition of stimulation-evoked release of [³H]norepinephrine from rat cortical slices (Pelayo et al., 1980; Hedler et al., 1981) and rabbit cortical slices (Reichenbacher et al., 1982) by clonidine and alpha-methyl-norepinephrine, as well as marked potentiation of neurotransmitter release by yohimbine. The variability in magnitude of the effects of alpha-2 adrenergic receptor agonists and antagonists is likely to be a result of differences in stimulation parameters and experimental conditions. Dismukes et al. (1977) reported that the release of norepinephrine induced by high potassium ion concentration (56 mM) was relatively unaffected by alpha-2 adrenergic receptor stimulation or blockade, in contrast to significant effects of these agents when release was induced by lower potassium ion concentrations.

As in the periphery, the magnitude of alpha-2 adrenergic receptor-mediated inhibition of neurotransmitter release from central noradrenergic neurons is dependent on calcium concentration (Dismukes et al., 1977), suggesting an alpha-2 adrenergic receptor-mediated alteration in the availability of intracellular calcium for exocytotic release of norepinephrine. It has also been suggested that alpha-2 adrenergic receptor-mediated modulation of cortical norepinephine release affects a step in the release process subsequent to calcium influx (Schoffelmeer and Mulder, 1983), perhaps via activation of an adenylate cyclase system (Wemer et al., 1982).

Tetrodotoxin treatment does not influence alpha-2 adrenergic receptor-mediated inhibition of release from cortical slices (Dismukes et al., 1977), indicating that neuronal transmission through intact neurons is not required. This is supported by the demonstration of an inhibitory effect of alpha-2 adrenergic receptor agonists, such as clonidine and oxymetazoline, in synaptosomes (De Langen et al., 1979). In contrast to the results in brain slices, alpha-2 adrenergic receptor antagonists do not potentiate synaptosomal release, presumably because of insufficient accumulation of endogenously released norepinephrine and, therefore, insufficient tone at the presynaptic alpha-2 adrenergic receptor.

The presynaptic alpha-2 adrenergic receptor on central noradrenergic neurons appears to play a physiologic role, since yohimbine will enhance the synthesis, release, and utilization of norepinephrine in the brain via removal of the presynaptic inhibitory pathway (Dubocovich, 1984). Of course, in vivo data on the effect of alpha-2 adrenergic receptor blockade undoubtedly reflects the summation of presynaptic and postsynaptic effects on a variety of interconnecting neuronal systems.

alpha-2 Adrenergic receptor stimulation will inhibit the field stimulation-induced release of [^{3}H]dopamine from superfused sections of rabbit retina (Dubocovich, 1984). This effect results from activation of presynaptic alpha-2 adrenergic receptors on dopaminergic amacrine cells. Although there are also dopamine autoreceptors present on these cells, studies with selective antagonists have clearly demonstrated discrete alpha-2 adrenergic receptor and dopamine DA$_2$-receptor mediated neuroinhibitory systems.

In isolated hippocampal and cortical slices, alpha-2 adrenergic receptor activation will inhibit stimulation-evoked serotonin release (Gothert and Huth, 1980; Frankhuyzen and Mulder, 1980). As in peripheral systems, alpha-2 adrenergic receptor stimulation can also inhibit acetylcholine release as evidenced by inhibition of stimulation-evoked release from superfused guinea pig cerebral cortex (Beani et al., 1978). alpha-2 Adrenergic receptor agonists will also inhibit ouabain-induced release of acetylcholine from rat cortical slices. The potency of a series of alpha-adrenergic receptor agonists in this system is consistent with an alpha-2 adrenergic receptor-mediated effect (Vizi, 1979), and release can be enhanced by phentolamine (Vizi, 1972), suggesting that activation of these presynaptic alpha-2 adrenergic receptors by endogenously released norepinephrine is physiologically relevant. alpha-2 Adrenergic receptor-mediated inhibition of central cholinergic tone has also been demonstrated in vivo (Myers and Waller, 1975; Beani et al., 1978).

Interestingly, alpha-2 adrenergic receptors do not seem to play a functional role in the striatum, since clonidine does not modify the release of acetylcholine from striatal slices (Vizi et al.,

1977) and norepinephrine does not affect ^{3}H-overflow from dopaminergic synaptosomes prelabeled with [^{3}H]norepinephrine (De Langen et al., 1979).

3.2. Functional Role of Platelet alpha-2 Adrenergic Receptors

Epinephrine induces aggregation of human platelets and potentiates aggregation induced by other agents such as ADP, collagen, and thrombin (O'Brien, 1963; Mills and Roberts, 1967; Thomas, 1967). Platelet aggregation induced by epinephrine is normally biphasic, consisting of a reversible partial aggregation, followed by a rapid, irreversible aggregation. When alpha-1 and alpha-2 adrenergic receptors were first differentiated, selective alpha-2 adrenergic receptor antagonists, such as yohimbine, were shown to block platelet aggregation induced by epinephrine (Hsu et al., 1979), suggesting an alpha-2 adrenergic receptor-mediated mechanism. Norepinephrine and alpha-methylnorepinephrine, which are also capable of stimulating alpha-2 adrenergic receptors, can mimic the response of epinephrine (Lasch and Jakobs, 1979). When synthetic alpha-2 adrenergic receptor agonists, such as clonidine or other imidazolidines, were shown to be incapable of producing an epinephrine-like aggregatory response, however, the platelet receptor was postulated to be distinct from either the alpha-1 or alpha-2 adrenergic receptor subtypes (Jakobs, 1978; Jakobs et al., 1978). The failure of most synthetic, nonphenethylamine alpha-2 adrenergic receptor agonists to induce platelet aggregation has subsequently been attributed to result from their relative lack of intrinsic efficacy at the alpha-2 adrenergic receptor. In most experiments, clonidine, which is a partial agonist at alpha-2 adrenergic receptors (Medgett et al., 1978; Ruffolo et al., 1982c,1984c,1985), will block epinephrine-induced platelet aggregation, a typical result expected for a partial agonist (Ruffolo, 1982). Several investigators have observed a small aggregatory response to clonidine (Barnes et al., 1982; Hsu et al., 1979), usually corresponding in magnitude to the initial phase of epinephrine-induced aggregation, suggesting that clonidine is also a partial alpha-2 adrenergic receptor agonist in platelet. UK 14,304, a highly potent and selective alpha-2 adrenergic receptor agonist (Cambridge, 1981), will produce an aggregatory effect in platelets comparable to that seen with epinephrine (Grant and Scrutton, 1980), which is consistent with the observation that UK 14,304 has higher efficacy than clonidine and most other imidazolidines at the alpha-2 adrenergic receptor (van Meel et al., 1983; Ruffolo et al., 1985). Several close structural analogs of UK 14,304 are also capable of inducing aggregation of human platelets (Clare et al., 1984).

Variability in sensitivity to aggregatory agents is a problem commonly observed in studying alpha-2 adrenergic receptor-mediated platelet aggregation. There appear to be subgroups of the general population with either high or low sensitivity to ephinephrine (Rossi and Louis, 1975; Swart et al., 1984a). In addition, variability is often observed in multiple determinations of the ephinephrine sensitivity of platelets from a single individual (O'Brien, 1963; Gaxiola et al., 1984). Platelet aggregatory responsiveness to ephinephrine appears to be an inherited trait in humans (Scrutton et al., 1981).

One important biochemical consequence resulting from the stimulation of platelet alpha-2 adrenergic receptors is an inhibition of adenylate cyclase activity. This inhibition is generally demonstrable only following activation with an agent that stimulates adenylate cyclase, such as PGE (Robinson et al., 1969; Jakobs et al., 1976). Although it is possible that the aggregatory response is a physiologic consequence of this biochemical event (Jakobs et al., 1978), several recent observations suggest that it may not be: (1) the inter-individual variation in adenylate cyclase responsiveness does not correlate with variations in sensitivity to ephinephrine-induced platelet aggregation (Swart et al., 1985), (2) alpha-2 adrenergic receptor agonists are available that can inhibit adenylate activity without inducing platelet aggregation (Clare et al., 1984), (3) other agents, such as certain adenosine analogs, can inhibit PGE_1-induced adenylate cyclase activity without inducing platelet aggregation (Haslam et al., 1978), and (4) preincubation of platelets with epinephrine results in a desensitization of the epinephrine-induced aggregatory response with no effect on the ability of epinephrine to decrease cyclic AMP accumulation (Motulsky et al., 1986). In addition, epinephrine-provoked platelet secondary aggregation and secretion can be eliminated by removal of extraplatelet Na^+, although epinephrine-mediated decreases in cyclic-AMP are not influenced by this manipulation (Connolly and Limbird, 1983). Further studies have suggested that this effect of Na^+ removal is a manifestation of the involvement of Na^+/H^+ exchange in the alpha-2 adrenergic receptor-induced mobilization of the arachidonic acid necessary as a prerequisite for secondary aggregation and secretion (Sweatt et al., 1985, 1986a,b; see below). Interestingly, alpha-2 adrenergic receptor–ligand interactions are modulated by Na^+ ions, even in detergent-solubilized preparations (Limbird and Speck, 1983), suggesting that a Na^+-binding regulatory site is either part of, or closely associated with, the alpha-2 adrenergic receptor.

The functional role of alpha-2 adrenergic receptor-mediated platelet aggregation is not clearly established. Many substances normally present in blood can induce platelet aggregation, and the in vitro concentrations of epinephrine required to induce aggregation are generally higher than those normally occurring in vivo. The most likely explanation is that physiologic control of platelet aggregation involves the action of multiple aggregatory hormones, each present at levels below those necessary for induction of aggregation individually. It is also possible that the high circulating levels of epinephrine seen during stress may have a direct effect on platelet aggregation (Rossi, 1978; Ardlie et al., 1984).

During the secondary phase of platelet aggregation, the contents of the platelet are released. These contents include several compounds that may themselves induce platelet aggregation or potentiate the aggregatory effects of other agents. Among the compounds released by platelets as they aggregate are catecholamines, which could potentiate aggregation induced by other agents, such as thrombin (Thomas, 1967).

Individuals whose platelets are insensitive to aggregation by epinephrine in vitro do not show any clinical evidence of bleeding diatheses (Swart et al., 1984a). Intradermal injection of phentolamine, a nonselective alpha-1/alpha-2 adrenergic receptor antagonist, has been shown to produce a localized increase in bleeding time in the area adjacent to the injection (O'Brien, 1963). Phenoxybenzamine, a relatively selective alpha-1 adrenergic receptor antagonist, had no effect in this assay.

Changes in cholesterol content of platelet membranes can affect the in vitro responsiveness to epinephrine. Cholesterol-rich platelets require an 18-fold lower concentration of epinephrine to induce aggregation, compared to normal platelets; conversely, cholesterol depletion decreased the sensitivity to epinephrine by eightfold (Insel et al., 1978). These changes in sensitivity were not accompanied by changes in binding affinity or capacity for [^{3}H]dihydroergocryptine or adenylate cyclase activity.

Enhanced sensitivity to epinephrine-induced platelet aggregation has been observed in young individuals suffering from stroke and other cerebral ischemic conditions (Gaxiola et al., 1984), and it has been hypothesized that susceptibility to stroke is inversely proportional to platelet epinephrine sensitivity. Platelets from patients suffering from variant angina, in which the coronary circulation is compromised, also show a highly significant enhancement in in vitro epinephrine sensitivity (Yokoyama et al., 1981). A positive correlation between epinephrine sensitivity and age has also been observed (Yokoyama et al., 1984), and platelets from newborns do not show the typical biphasic aggregation response to epinephrine (Corby and O'Barr, 1981). These age-related changes appear to correlate with alpha-2 adrenergic receptor number as determined from radioligand binding studies. It should be emphasized that in these studies, in vitro epinephrine sensitivity is assessed and that a pathologic effect attributable to alterations in epinephrine-induced platelet aggregation has not as yet been demonstrated in vivo.

alpha-2 Adrenergic receptor-mediated platelet aggregation is not as prominent in animal platelets. This seems to be related both to alpha-2 adrenergic receptor density and to coupling between receptor occupancy and triggering of the aggregatory response. Typical biphasic platelet aggregation is observed only in primates and occasionally in the cat (Hallam et al., 1981). In other species, such as dog and rabbit, alpha-2 adrenergic receptor agonists will potentiate aggregation induced by other agents, such as ADP (Myers et al., 1983; Grant and Scrutton, 1980; Hallam et al., 1981; Glusa and Markwardt, 1983b). This potentiating or pro-aggregatory action can be blocked by alpha-2 adrenergic receptor antagonists. Addition of subthreshold concentrations of the calcium ionophore A23187 will allow a full aggregatory response to occur to epinephrine and UK 14,304 in rabbit platelets (Grant and Scrutton, 1980). Platelets from the rat and guinea pig are unresponsive to epinephrine, both for induction of aggregation or potentiation of ADP-induced aggregation. This lack of activity correlates with the failure to demonstrate specific [^{3}H]yohimbine binding sites in platelets from these species (Glusa and Markwardt, 1983b).

3.3. Sedation Mediated by alpha-2 Adrenergic Receptors

Stimulation of presynaptic alpha-2 adrenergic receptors in both the periphery and central nervous system inhibits neurotransmitter release (see above). Since increased noradrenergic function generally leads to central excitation (Leppavuori and Putkonen, 1980; Jouvet, 1972), selective alpha-2 adrenergic receptor stimulation would be expected to produce sedation.

Sedation was observed as a common side effect during the initial clinical trials with clonidine (Michel et al., 1966; Bock et al., 1966; Davidov et al., 1967) and continues to be one of the principal limitations preventing wider use of this drug. Sedative effects of clonidine and other alpha-2 adrenergic receptor agonists have also been observed in a variety of animal models (Laverty and Taylor, 1969; Timmermans et al., 1981b; Hoefke et al., 1975; Drew et al., 1979; Delini-Stula et al., 1979; Delbarre and Schmitt, 1973; Virtanen and Nyman, 1985; Leppavuori, 1980; Ruffolo et al., 1984c). In addition to decreases in motor activity and potentiation of anesthesia, clonidine and other alpha-2 adrenergic receptor agonists can induce sleep (Holman et al., 1971).

Pharmacologic studies using both selective agonists and antagonists have established that alpha-adrenergic receptor agonist-induced sedation is produced by an alpha-2 adrenergic receptor-mediated mechanism (Timmermans et al., 1981b; Drew et al., 1979). Penetration into the central nervous system is required, since peripherally acting alpha-2 adrenergic receptor agonists are ineffective in producing sedation or sleep (Hoefke et al., 1975). Although the anatomical location of the alpha-2 adrenergic receptors responsible for sedation has not been conclusively established, a presynaptic site has been proposed. The principal evidence for this conclusion is the observation that clonidine and other alpha-2 adrenergic receptor agonists produce behavior stimulation, rather than sedation, following central catecholamine depletion by reserpine or destruction of central noradrenergic nerve terminals by 6-hydroxydopamine pretreatment (Zebrowska-Lupina et al., 1977). Furthermore, alpha-2 adrenergic receptor-mediated effects on catecholamine turnover correlate with their effects on sleep patterns (Leppavuori, 1980). Neuronal uptake blockers and monoamine oxidase inhibitors will attenuate clonidine-induced sedation, presumably by increasing synaptic norepinephrine concentrations that were decreased by clonidine acting presynaptically on alpha-2 adrenergic receptors (Gower and Marriott, 1980).

There have been extensive efforts to develop centrally acting alpha-2 adrenergic receptor agonists that are more selective in producing antihypertensive effects relative to sedation. Although some comparative studies suggest that less sedation may result with certain alpha-2 adrenergic receptor agonists in animals (Clough and Hatton, 1981; Simon et al., 1975; Kleinlogel et al., 1975; Laverty, 1969) and humans (Kho et al., 1975; Spiegel and Devos, 1980; Ashton and Rawlins, 1978), clinical experience as well as animal studies suggest that antihypertensive efficacy and sedative activity produced by alpha-2 adrenergic receptor agonists cannot be separated clearly (Reid et al., 1983a; Ruffolo et al., 1984c).

Central alpha-2 adrenergic receptor stimulation is known to decrease REM or paradoxical sleep, both in animals (Leppavuori, 1980; Jarrott et al., 1984; Kleinlogel et al., 1975) and in humans (Spiegel and Devos, 1980). Interestingly, this property has been implicated to explain the paradoxical efficacy of clonidine in the treatment of narcolepsy (Salin-Pascual et al., 1985).

3.4. Interaction of alpha-2 Adrenergic Receptors and Opiate Receptors in the Central Nervous System

3.4.1. ANTIHYPERTENSIVE EFFECT

alpha-2 Adrenergic receptors often appear to be localized in regions of the central nervous system containing high densities of opiate receptors, including the locus ceruleus, trigeminal nucleus, dorsal medial thalamus, and substantia gelatinosa of the spinal cord (Kuhar, 1982; Pert et al., 1976; Atweh and Kuhar, 1977). Furthermore, many central noradrenergic nerves appear to have an anatomic and functional interaction with nerves containing opioid peptides (Pickel, 1982; Pickel et al., 1979). Hence, it is not surprising to find pharmacologic interactions between alpha-2 adrenergic and opiate receptors.

Farsang and Kunos (1979) reported that the antihypertensive action of clonidine in spontaneously hypertensive rats could be reversed by the opiate receptor antagonist naloxone. This finding has been observed by several other groups (Baum and Becker, 1982; Valdman et al., 1981), and may apply also to the antihypertensive activity of alpha-methyldopa (Farsang et al., 1980; Petty and de Jonge, 1982; Ramirez-Gonzalez et al., 1983). Conflicting results have been reported regarding the ability of opiate receptor antagonists to block alpha-2 adrenergic receptor agonist-induced hypotension in normotensive rats, with Farsang et al. (1980) finding no effect in Wistar-Kyoto rats, whereas Bennett et al. (1982) demonstrated blockade in this strain.

In patients with essential hypertension, some investigators have reported attenuation of the antihypertensive activity of clonidine by opiate receptor blockade (Farsang et al., 1982,1984a,b,c), whereas other groups have found no effect (Bramnert and Hokfelt, 1983; Rogers and Cubeddu, 1983). Perhaps the inconsistencies can be partially explained by reports showing a heterogeneous response to naloxone in clonidine-treated hypertensive patients, with division of the patients into responders and nonresponders (Farsang et al., 1984b,c). It appears that the naloxone-sensitive patients have higher sympatho-adrenal tone than the naloxone-resistant group.

Blockade of the antihypertensive response to centrally acting alpha-2 adrenergic receptor agonists by naloxone has been attributed to the release of an endogenous opioid peptide resulting from central alpha-2 adrenergic receptor stimulation. Clonidine and alpha-methylnorepinephrine have been shown to induce the release of beta-endorphin-like immunoreactivity in the superfused brainstem of spontaneously hypertensive rats (Kunos et al., 1981), as well as in the anterior pituitary of normotensive rats (Pettibone and Mueller, 1981a), and to elevate plasma endorphin-like immunoreactivity in both rats (Pettibone and Mueller, 1981b) and patients with essential hypertension, but not in normotensive subjects (Farsang et al, 1983). The ability of naloxone to block clonidine-induced antihypertensive responses in a group of hypertensive patients correlated well with changes in plasma beta-endorphin immunoreactivity induced by clonidine (Farsang et al., 1984b). These findings suggest that in at least some patients, alpha-2 adrenergic receptor agonists can induce the release of an endogenous opioid peptide which may mediate part of the antihypertensive response. Since centrally active alpha-2 adrenergic receptor agonists are effective antihypertensive agents in almost all patients, however, and can lower blood pressure in normotensive individuals (Rubin et al., 1982), this opiate involvement cannot explain totally the cardiovascular response to central alpha-2 adrenergic receptor stimulation. It is likely that the sympatholytic effect mediated by a direct action of alpha-2 adrenergic receptor agonists on brainstem neurons occurs in all subjects, with an additional opiate-dependent pathway producing a further reduction in sympathetic outflow in a subgroup of the population. This opiate involvement may be dependent on the alpha-2 adrenergic receptor agonist employed, since guanfacine produces an antihypertensive effect that is insensitive to inhibition by naloxone in a group of patients in which the response to clonidine was previously shown to be attenuated by opiate receptor blockade (Farsang et al., 1984a).

3.4.2. ANALGESIA

Clonidine will produce analgesia in the rat (Paalzow, 1974; Paalzow and Paalzow, 1976; Fielding et al., 1978; Yasuoka and Yaksh, 1983). If administered centrally, other alpha-2 adrenergic receptor agonists that do not cross the blood–brain barrier, such as St 91, will also produce analgesia (Yasuoka and Yaksh, 1983). This effect is not mediated by endogenous opioid peptides, since it can be blocked by yohimbine, but not by naloxone (Fielding et al., 1978; Paalzow and Paalzow, 1976; Hynes et al., 1983). Most evidence supports a presynaptically mediated alpha-2 adrenergic receptor mechanism for this effect (Chance, 1983), although participation of an alpha-1 adrenergic receptor mediated mechanism on spinal neurons may also contribute (Zemlan et al., 1980). It is possible that alpha-2 adrenergic receptors may participate in the physiologic analgesia produced by stress (Chance, 1983). Anecdotal evidence suggests the potential clinical efficacy of clonidine as an analgesic (Goldstein, 1983).

3.4.3. SUPPRESSION OF OPIATE WITHDRAWAL

The interaction of alpha-2 adrenergic and opiate receptor systems has led to a useful clinical alternative in the treatment of opiate addiction. alpha-Adrenergic receptor agonists are known to inhibit firing of the locus ceruleus (Graham and Aghajanian, 1971; Cedarbaum and Aghajanian, 1977). The ability of piperoxan to block this response suggests an alpha-2 adrenergic receptor-mediated action (Aghajanian, 1978,1982). The alpha-2 adrenergic and opiate receptors mediate similar inhibitory responses in this nucleus, which sends projections throughout the central nervous system. In opiate-dependent rats, naloxone will potentiate locus ceruleus firing rate. Although under these conditions, acute administration of morphine is ineffective in suppressing firing rate, clonidine will still inhibit the firing rate in the locus ceruleus (Aghajanian, 1978). alpha-2 Adrenergic receptor-mediated inhibition of locus ceruleus firing is thought to be mediated by recurrent axons synapsing on the noradrenergic cell bodies located in the nucleus (Aghajanian, 1982). Recent evidence suggests that the opiate and alpha-2 adrenergic receptors on these cells may be linked through a common pathway to a potassium channel, activation of which results in hyperpolarization (Aghajanian and VanderMaelen, 1982; Aghajanian, 1985).

Since alpha-2 adrenergic and opiate receptors produce similar effects in the locus ceruleus, it was suggested that alpha-2 adrenergic receptor agonists, such as clonidine, may be effective in the treatment of opiate withdrawal. Animal studies support this hypothesis (Meyer and Sparber, 1976; Vetulani and Bednarczyk, 1977; Fielding et al., 1978) and show that clonidine can ameliorate the symptoms produced by either abstinence- or naloxone-precipitated opiate withdrawal in rats. The initial clinical trial of clonidine in methadone-maintained addicts showed significant clonidine-induced reductions in both physiological and psychological signs and symptoms of withdrawal (Gold et al., 1978a,b). The utility of clonidine in the treatment of opiate withdrawal has been confirmed in many clinical trials, both in an inpatient (Gold et al., 1982) and outpatient (Washton and Resnick, 1982; Kleber et al., 1985; Cami et al., 1985) setting, and clonidine has become the accepted alternative for methadone in the treatment of opiate withdrawal (Hughes and Morse, 1985). Initial clinical trials with lofexidine, another alpha-2 adrenergic receptor agonist having a similar pharmacologic profile to clonidine, have also demonstrated efficacy in suppression of opiate withdrawal symptoms (Washton et al, 1981; Washton and Resnick, 1982), confirming animal studies with this alpha-2 adrenergic receptor agonist (Shearman et al., 1980).

All evidence supports the validity of the initial hypothesis that clonidine is acting on the locus ceruleus to inhibit the symptoms of opiate withdrawal. Studies in primates show that clonidine will block the behavioral effects induced by electrical stimulation of this nucleus (Redmond and Huang, 1982), with piperoxan producing the opposite effect to clonidine. Biochemical studies in rats and primates, in which cortical levels of norepinephrine metabolites were measured, show elevations in MHPG during naloxone-precipitated opiate withdrawal. These biochemical effects can be blocked by clonidine (Roth et al., 1982). Such findings, both in animals and humans, suggest that central noradrenergic systems become hyperactive during opiate withdrawal, and this elevated activity can be suppressed by clonidine. Central noradrenergic hyperfunction may be associated also with benzodiazepine withdrawal (Cowen and Nutt, 1982), and preliminary data suggest that clonidine may be effective in relieving the symptoms in this condition (Ashton, 1984; Keshavan and Crammer, 1985).

3.5. Suppression of Insulin Release by alpha-2 Adrenergic Receptors

The endocrine pancreas of most mammalian species is innervated by the sympathetic nervous system (Smith and Porte, 1976). Activation of this sympathetic innervation by stimulation of either the splanchnic nerves (Porte et al., 1973) or the ventromedial hypothalamus (Frohman and Benardis, 1971) produces a reduction in insulin secretion in a manner similar to that produced by epinephrine infusion in vivo. Similarly, epinephrine inhibits glucose-stimulated insulin release from isolated rat pancreatic islets in vitro (Nakaki et al., 1980). This adrenergic inhibition of insulin release from the pancreas is blocked by phentolamine and dihydroergocryptine and is unaffected by propranolol, thus demonstrating that it is mediated via alpha-adrenergic receptors on the islet beta cells (Smith and Porte, 1976). These alpha-adrenergic receptors have been characterized subsequently as being of the alpha-2 subtype by radioligand bindings studies (Cherksey et al., 1982) and physiologic studies using selective agonists and antagonists (Nakaki et al., 1980).

The physiologic importance of the adrenergic regulation of insulin release mediated via alpha-2 adrenergic receptors in vivo has been studied by investigating the effect produced by selective alpha-2 adrenergic receptor blockade. Nakadate et al. (1980) found that phentolamine, dihydroergocryptine, or yohimbine markedly increase plasma immunoreactive insulin concentration in mice, whereas phenoxybenzamine or prazosin were found to have no effect. In accord with this, yohimbine increases plasma insulin concentration in the rat (Ahren et al., 1984). These results suggest that insulin secretion in vivo is under tonic inhibition by activation of alpha-2 adrenergic receptors in islet beta cells. The precise role of neurogenic and hormonal influences in this tonic activation of islet alpha-2 adrenergic receptors is unclear. It is possible that this effect is mediated partially, or even totally, by circulating epinephrine and norepinephrine released from the adrenal medulla, since inhibition of insulin release produced by electrical stimulation of the ventromedial hypothalamus utilizes a pathway involving catecholamine secretion from the adrenal medulla (Frohman and Benardis, 1971). There is also evidence from studies in the *in situ* saline-perfused canine pancreas that the alpha-2 adrenergic receptors on islet betacells are directly innervated. Thus, it is possible that circulating catecholamines released from the adrenal medulla, as well as neuronally released norepinephrine in islet betacells, may both contribute to the alpha-2 adrenergic receptor-mediated tonic inhibition of insulin release in vivo. Not until detailed studies are performed utilizing adrenalectomized and/or sympathectomized animals will it be possible to determine which mechanism predominates in the normal physiologic regulation of insulin release in vivo.

3.6. alpha-2 Adrenergic Receptors in the Gastrointestinal Tract

alpha-2 Adrenergic receptors have been shown to mediate several responses at different levels of the gastrointestinal tract, including regulation of gastric and intestinal motility and secretions.

Adrenergic regulation of gastric motility via an indirect mechanism was demonstrated by Jansson and Martinson (1966), who showed that sympathetic nerve stimulation inhibits excitatory gastric smooth muscle responses produced by vagal stimulation, but not those produced by exogenously administered acetylcholine in the cat stomach *in situ*. More recently, alpha-adrenergic receptor-mediated inhibition of neurogenic contraction of isolated gastric fundus from dog (Lefebvre et al., 1984) and rat (Verplanken et al., 1984) has been shown to be mediated by presynaptic alpha-2 adrenergic receptors located on the postganglionic parasympathetic neurons in the stomach, which, when activated, lead to a decrease in acetylcholine release. Prejunctional alpha-2 adrenergic receptors also produce an inhibition of vagally mediated gastric acid secretion from parietal cells in the rat (Yamaguchi et al., 1977; Cheng et al., 1981), by inhibiting acetylcholine release.

These peripheral gastric inhibitory alpha-2 adrenergic receptors may be located on the intramural parasympathetic ganglia (Wikberg, 1977), in addition to being present on the prejunctional terminal of postganglionic cholinergic neurons (Seno et al., 1978). Furthermore, it has been suggested that alpha-2 adrenergic receptors located in the central nervous system may be partially responsible for the inhibition of gastric secretion in pylorus-ligated rats (Yamaguchi et al., 1977). Thus, vagally mediated increases in motility and secretion may be inhibited by activation of alpha-2 adrenergic receptors in more than one location.

alpha-2 Adrenergic receptors may play a role in the etiology of stress and reserpine-induced gastric ulceration (Taylor and Mir, 1982). It has been postulated that, as a result of chronic stress or reserpine pretreatment, there is a reduction in neuronal levels of norepinephrine in the gastrointestinal tract and, therefore, a progressive failure of adrenergic neurotransmission, which leads to a reduction of the inhibitory alpha-2 adrenergic receptors located prejunctionally on the parasympathetic neurons. As a consequence, alpha-2 adrenergic receptor-mediated inhibition of gastric acid secretion is reduced or abolished, thus leading to gastric ulceration. These results suggest that presynaptic alpha-2 adrenergic receptors on cholinergic nerve terminals in the stomach may function physiologically to provide gastric protection from ulceration.

alpha-2 Adrenergic receptors have been shown to mediate effects on motility, as well as electrolyte and fluid transport, in the small and large intestine. As in the stomach, stimulation of the sympathetic nerves supplying the ileum inhibits the contractile response to parasympathetic nerve stimulation by an alpha-adrenergic receptor mechanism (Kroneberg and Oberdorf, 1974). This alpha-adrenergic receptor has been classified subsequently as being of the alpha-2 subtype by Drew (1978) and Andrejak et al. (1980). The alpha-2 adrenergic receptor-mediated inhibition of cholinergic-induced small intestinal contraction is accompanied by a reduction in acetylcholine release (Wikberg, 1977), thus indicating a prejunctional locus for the alpha-2 adrenergic receptors.

A similar role for prejunctional alpha-2 adrenergic receptors has been demonstrated further down the intestine in the rabbit colon (Gillespie and Khoyi, 1977). In addition, Fontaine et al. (1984) have provided evidence that prejunctional alpha-2 adrenergic receptors are present on nonadrenergic, noncholinergic (NANC) inhibitory neurons in the mouse isolated colon, activation of which inhibits the release of the NANC transmitter. Thus, alpha-2 adrenergic receptor activation paradoxically induces a contractile response in mouse colon. It would appear, therefore, that despite the apparent universal presence of inhibitory alpha-2 adrenergic receptors on neurons in the small intestine, the overall function mediated by alpha-2 adrenergic receptor activation in the colon is species dependent.

Field and McColl (1973) demonstrated that stimulation of alpha-adrenergic receptors in rabbit ileal mucosa promotes net Na^+ and Cl^- absorption. In addition, net HCO_3^- secretion is abolished, and the membrane potential across the mucosal cell is decreased (Field and McColl, 1973). The nature of the alpha-adrenergic receptor involved in this response has been examined in detail and has been shown to be of the alpha-2 subtype (Chang et al., 1982; Donowitz et al., 1982; Durbin et al., 1982; Dharmsathaphorn et al., 1984). alpha-2 Adrenergic receptors also mediate the inhibition of intestinal fluid secretion induced by PGE_1, VIP, dibutyryl cyclic AMP, and cholera toxin in rat isolated jejunum (Nakaki et al., 1982a,b), presumably by increasing net mucosal absorption. Further support for the presence of alpha-2 adrenergic receptors on intestinal mucosal cells has come from radioligand binding studies (Nakaki et al., 1983). The effects of alpha-2 adrenergic receptor activation on intestinal ion and water transport is not species-dependent, occurs in both the ileum and colon (Dharmsathaphorn et al., 1984), and has been shown to occur in humans where clonidine administration produces an inhibition of watery diarrhea (McArthur et al., 1982).

3.7. alpha-2 Adrenergic Receptors in Uterus

The uterus contains functional alpha-adrenergic receptors, activation of which by either exogenous catecholamines (Ahlquist, 1962) or sympathetic nerve stimulation (Miller and Marshall, 1965) produces uterine contraction. The presence and functional significance of alpha-2 adrenergic receptors in the uterus, however, has been investigated only recently. By using radioligand binding techniques, Hoffman et al. (1979) demonstrated the presence of both alpha-1 and alpha-2 adrenergic receptors in the rabbit uterus. Subsequent studies revealed that, despite a predominance of alpha-2 over alpha-1 adrenergic receptors in rabbit uterus, the contractile response to exogenous norepinephrine is mediated solely by the alpha-1 adrenergic receptor subtype (Hoffman et al., 1981). The increase, however, in total alpha-adrenergic receptor number known to occur in response to estrogen treatment (Roberts et al., 1977; Williams and Lefkowitz, 1977) was found by Hoffman et al. (1981) to be caused by a selective increase in alpha-2 adrenergic receptor density. A similar increase in alpha-2 adrenergic receptor number was observed with elevated estrogen levels in human myometrium (Bottarie et al., 1983). This estrogen-induced up-regulation of uterine alpha-2 adrenergic receptors is restricted to this target organ, since human platelet alpha-2 adrenergic receptor density does not increase with elevated estrogen levels (Rosen et al., 1984). Uterine alpha-2 adrenergic receptors also appear to be under the control of progesterone, since elevated plasma levels of progesterone reduce human myometrial alpha-2 adrenergic receptor density in both normal and estrogen-primed uteri (Bottarie et al., 1983).

Despite this regulation of uterine alpha-2 adrenergic receptor number by gonadal steroids, no functional role has yet been established for these receptors in this tissue. These alpha-2 adrenergic receptors do not appear to be located predominantly presynaptically, since surgical and chemical sympathectomy do not markedly affect alpha-2 adrenergic receptor density in rabbit uterus (Hoffman et al., 1981). It is possible that the alpha-2 adrenergic receptors in uterus regulate some aspect of cellular metabolism important for uterine function and that the steroid-induced changes in alpha-2 adrenergic receptor number mediate changes in the metabolic activity of the uterus during the menstrual cycle and pregnancy. No studies have yet been performed, however, to investigate the effect of alpha-2 adrenergic receptor activation on uterine metabolic function.

4. Correlation Between alpha-2 Adrenergic Receptor Occupancy (Binding) and Function

4.1. Occupancy–Response Relationships

The relationship between alpha-1 adrenergic receptor occupancy and binding has been studied in detail. Occupancy–response relationships for alpha-2 adrenergic receptor-mediated responses have received surprisingly little attention, however.

The first alpha-2 adrenergic receptor occupancy-response studies were performed in the cardiovascular system. In contrast to the ease with which alpha-1 adrenergic receptor-mediated responses are elicited in isolated arterial preparations in vitro, it is exceedingly difficult to demonstrate alpha-2 adrenergic receptor-mediated vasoconstrictor responses in similar preparations. On the other hand, alpha-2 adrenergic receptor-mediated arterial vasoconstriction may be readily demonstrated in vivo. Alkylation of alpha-2 adrenergic receptors in conscious rabbits and pithed rats with high doses of phenoxybenzamine produces a depressed maximum response to alpha-2 adrenergic receptor agonists with only small rightward shifts in the dose–response curves (Hamilton et al., 1983; Reid et al., 1983b; Ruffolo and Yaden, 1984). Detailed analysis of this phenomenon using B-HT 933 as the alpha-2 adrenergic receptor agonist revealed that the relationship between the maximum pressor response and the proportion of alpha-2 adrenergic receptors available for interaction with the agonist is linear (Ruffolo and Yaden, 1984), a situation highly characteristic of a lack of receptor reserve (Ruffolo, 1982). This lack of alpha-2 adrenergic receptor reserve in the arterial circulation may explain the difficulty experienced by many investigators in studying postjunctional vascular alpha-2 adrenergic receptors in isolated arteries in vitro.

In contrast to the arterial circulation in vivo, the canine saphenous vein in vitro possesses an alpha-2 adrenergic receptor reserve for B-HT 933, such that a hyperbolic occupancy–response relationship is observed for this agonist, and approximately 10% of the alpha-2 adrenergic receptors need be occupied to produce a half-maximal response (Fig. 2) (Ruffolo and Zeid, 1985). It would appear that there is also a reserve for presynaptic alpha-2 adrenergic receptors present on the postganglionic cholinergic neurons in the guinea pig ileum. Low concentrations of the irreversible alpha-adrenergic receptor antagonist benextramine initially produce rightward shifts in the dose–response curves of the alpha-2 adrenergic receptor agonist alpha-methylnorepinephrine, with no depression of the maximum response being observed. At higher concentrations of benextramine, reductions of the maximum response to alpha-methylnorepinephrine begin to occur (Mottram and Thakar, 1984). Analysis of these data by the method of Furchgott (1966) to yield the dissociation constant for alpha-methylnorepinephrine and the subsequent construction of an occupancy–response relationship, demonstrates that there exists a receptor reserve for alpha-methylnorepinephrine at pre-

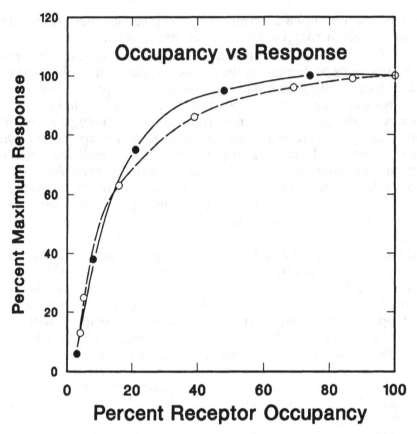

Fig. 2   alpha-2 Adrenergic receptor occupancy–response relationships for B-HT 933 interacting with postsynaptic vascular alpha-2
adrenergic receptors in canine saphenous vein (○) and alpha-
methylnorepinephrine interacting with presynaptic alpha-2 adrenergic
receptors on cholinergic neurons in guinea pig ileum (●).

junctional alpha-2 adrenergic receptors on postganglionic neurons in the guinea pig ileum, such that only approximately 10% of
the receptors need be occupied by alpha-methylnorepinephrine
to produce a half-maximal response (*see* Fig. 2). It is interesting to
note that the pre- and postsynaptic alpha-2 adrenergic receptor
occupancy–response relationships are nearly identical (Fig. 2).
    An alpha-2 adrenergic receptor reserve has been proposed to
exist in the human platelet for one of the postulated intermediate
responses between receptor activation and platelet aggregation,
namely inhibition of adenylate cyclase. Lenox et al. (1985) found
that the agonist dissociation constant for (−)-epinephrine (5.2
μM) was 22-fold greater than the concentration required to produce 50% inhibition of adenylate cyclase activity (0.24 μM). From

their binding and functional studies, Lenox et al. (1985) calculated that only approximately 10% of the alpha-2 adrenergic receptors need be occupied by ($-$)-epinephrine to produce a half-maximal response. The value obtained for the dissociation constant of ($-$)-epinephrine (i.e., 5.2 μM) is somewhat higher, however, than that value reported previously for alpha-2 adrenergic receptors on human platelets (100 nM, Swart et al., 1984b; 156 nM, Kawahara and Bylund, 1985), values that are, in fact, lower than the EC_{50} for inhibition of adenylate cyclase. Thus, the existence of an alpha-2 adrenergic receptor reserve for ($-$)-epinephrine in human platelet is open to some doubt. Furthermore, the relationship between ($-$)-epinephrine binding to alpha-2 adrenergic receptors in human platelet and induction of platelet aggregation has been studied by Swart et al. (1984b) who found that the agonist dissociation constant is five times lower than the EC_{50} for aggregation, which is clearly inconsistent with a receptor reserve (Ruffolo, 1982). These data would suggest that there may be no alpha-2 adrenergic receptor reserve in human platelets, and this would also account for the fact that alpha-2 adrenergic receptor partial agonists, which themselves are characterized by poor occupancy–response relationships, are relatively ineffective in producing platelet aggregation.

4.2. Second Messengers

4.2.1. INHIBITION OF ADENYLATE CYCLASE

As briefly alluded to above, alpha-2 adrenergic receptor activation in platelets induces an inhibition of adenylate cyclase. This biochemical response has been postulated to be a component of the transducing mechanism between receptor activation and response in most, if not all, cells that possess functional alpha-2 adrenergic receptors (Fain and Garcia-Sainz, 1980). alpha-2 Adrenergic receptor-mediated inhibition of adenylate cyclase enzyme activity has been observed to occur in human platelets (Mills, 1975), human adipocytes (Burns et al., 1981), human HT29 cells (Turner et al., 1985), neuroblastoma × glioma hybrid cells (Sabol and Nirenberg, 1979), rat renal cortex (Woodcock and Johnston, 1982), rabbit renal cortical collecting tubule (Krothapalli et al., 1983), porcine thyroid (Muraki et al., 1984), and rat pancreatic islet cells (Yamazaki et al., 1982), and has been postulated to occur as a result of prejunctional neuronal alpha-2 adrenergic receptor activation in rat neocortex (Schoffelmeer and Mulder, 1983) and following postjunctional vascular alpha-2 adrenergic receptor activation in pithed rats (Boyer et al., 1983). The actual importance of this effect in the transduction of alpha-2 adrenergic receptor activation into effector response in all systems is open to some doubt, however. Clare et al. (1984) observed that, although (−)-epinephrine induces human platelet aggregation and inhibits platelet adenylate cyclase, some alpha-2 adrenergic receptor agonists induce platelet aggregation without producing inhibition of adenylate cyclase. In addition, other agonists that induce platelet aggregation are, in fact, antagonists of (−)-epinephrine-induced inhibition of adenylate cyclase activity. Further evidence against inhibition of adenylate cyclase as being a necessary intermediate in alpha-2 adrenergic receptor-mediated platelet aggregation is provided by the work of Haslam (1975), in which it was shown that the aggregatory response produced by (−)-epinephrine is not accompanied by a decrease in intracellular levels of cyclic AMP in resting platelets and that a reduction in intracellular cyclic AMP levels is observed only after first elevating cyclic AMP levels by PGE_1. Another alpha-2 adrenergic receptor-mediated response in which inhibition of adenylate cyclase activity would not appear to be involved is the inhibition of cholera toxin-induced intestinal fluid accumulation (Nakaki et al., 1982b). The role of inhibition of adenylate cyclase activity vs other effector mechanisms in mediating alpha-2 adrenergic receptor-induced physiological effects will be dealt with in further detail in Chapter 6.

The mechanisms by which alpha-2 adrenergic receptor-mediated inhibition of adenylate cyclase results in the inhibition of neurohormonal secretion is also not known with certainty. An important role for intracellular cyclic AMP in regulating neurotransmitter release is suggested by the observation that elevation of intracellular cyclic AMP by dibutyryl cyclic AMP, 8-bromo-cyclic AMP, the direct adenylate cyclase activator forskolin, or the cyclic AMP phosphodiesterase inhibitor ZK62771 all increase norepinephrine release from rat neocortical slices (Schoffelmeer and Mulder, 1983). Nonetheless, although it would be expected that an alpha-2 adrenergic receptor-mediated reduction of intracellular cyclic AMP by inhibition of adenylate cyclase would lead to an inhibition of neurotransmitter release, rigorous proof of such an effector mechanism has not been established.

The mechanism for cyclic AMP-induced neurotransmitter release presumably involves, at least in part, the phosphorylation of proteins involved in the secretion process. For example, in central neurons, cyclic AMP-dependent protein kinase is believed to mediate phosphorylation of a variety of intracellular proteins, some of which may be involved in neurotransmitter release [e.g., sympathin I (Nestler and Greengard, 1982)]. Sulahke and St.

Louis (1980) have also suggested that Ca^{2+} channels in the nerve terminal may be opened after phosphorylation by cyclic AMP-dependent protein kinase. Thus, it may be hypothesized that prejunctional alpha-2 adrenergic receptor activation produces an inhibition of adenylate cyclase leading to a reduction in intracellular cyclic AMP levels. This reduction of intracellular cyclic AMP could cause a decrease in the degree of phosphorylation of membrane Ca^{2+} channels and intracellular proteins involved in the secretion process, and thus produce an inhibition of neurotransmitter release.

4.2.2. CALCIUM TRANSLOCATION

Another possible link between alpha-2 adrenergic receptor activation and effector organ response may involve calcium translocation. This phenomenon has been studied in great detail in vascular tissue in vivo and in vitro. van Meel et al. (1981b,c) first suggested that an influx of extracellular calcium ions is necessary for arteriolar vasoconstriction mediated by postsynaptic vascular alpha-2 adrenergic receptors in vivo based on the sensitivity of this response to inhibition by a variety of organic and inorganic slow calcium channel antagonists. The involvement of an influx of extracellular calcium ions in alpha-2 adrenergic receptor-mediated vasoconstriction has been confirmed in one of the few in vitro arterial preparations in which alpha-2 adrenergic receptor-mediated vasoconstriction has been demonstrated, namely the rat isolated tail artery (Medgett and Rajanayagam, 1984). In vitro studies in isolated veins, particularly the canine saphenous vein, have provided convincing evidence for a similar involvement of calcium translocation in alpha-2 adrenergic receptor-mediated vasoconstriction in the venous circulation. alpha-2 Adrenergic receptor-mediated venoconstriction is reduced by organic and inorganic slow calcium channel antagonists and is virtually abolished in calcium-free medium containing EGTA (Matthews et al., 1984b; Cooke et al., 1985). Furthermore, the alpha-2 adrenergic receptor agonist B-HT 920 produces an influx of $^{45}Ca^{2+}$, which is inhibited by the slow calcium channel antagonists verapamil and nifedipine (Matthews et al., 1984b). In this preparation, B-HT 920 does not induce intracellular calcium release from cytosolic stores (Jim and Matthews, 1985). The influx of calcium produced by alpha-2 adrenergic receptor agonists in the canine saphenous vein is not likely to be caused by activation of voltage-dependent calcium channels in the smooth muscle plasma membrane, since little change in membrane potential occurs during smooth muscle contraction elicited by alpha-2 adrenergic receptor agonists (Matthews et al., 1984a). In addition, much higher concentrations of the organic calcium channel antagonist nitrendipine than those that abolish depolarization-induced contraction of the canine saphenous vein are required to inhibit alpha-2 adrenergic receptor-mediated contraction (Cooke et al., 1985). In contrast to the dog, alpha-2 adrenergic receptor-mediated contraction of the rat saphenous vein is accompanied by membrane depolarization (Cheung, 1985a), and this response would appear to result from calcium ion influx initiated by opening of voltage sensitive calcium channels (Cheung, 1985). Thus, there is no uniform mechanism by which activation of alpha-2 adrenergic receptors induces calcium ion influx in vascular smooth muscle.

In contrast to vascular smooth muscle, human blood platelet alpha-2 adrenergic receptors do not seem to evoke an influx of extracellular calcium in order to produce aggregation. Clare and Scrutton (1983) have shown that, after inhibition of cyclooxygenase, concentrations of (−)-epinephrine that induce platelet aggregation do not promote $^{45}Ca^{2+}$ uptake. Another alpha-2 adrenergic receptor-mediated response in which enhanced calcium ion influx would appear to play no role is the inhibition of neurotransmitter release. It is believed that neurotransmitter release is induced by an influx of calcium ions through voltage-sensitive calcium channels in the presynaptic neuronal elements (Schoffelmeer et al., 1981), and it is unlikely, therefore, that increased calcium influx could lead to the reduction in transmitter release produced by alpha-2 adrenergic receptor agonists. Indeed, if alpha-2 adrenergic receptors had the capacity to increase calcium ion influx in nerve terminals, it would be expected that alpha-2 adrenergic receptor agonists would increase neurotransmitter release. As discussed earlier, prejunctional alpha-2 adrenergic receptor activation may actually inhibit Ca^{2+} influx by inhibiting phosphorylation of membrane Ca^{2+} channels.

4.2.3. Na^+/H^+ EXCHANGE

Because Ca^{2+} influx, on the one hand, and decreases in cyclic AMP levels, on the other, can often not fully account for the physiological effects of alpha-2 adrenergic receptors, it is possible that another intracellular second messenger system is involved in these responses. Such a novel pathway of intermediates between alpha-2 adrenergic receptor activation and the second wave of aggregation in human platelets has been recently proposed by

Limbird and coworkers (Limbird, 1984; Limbird and Speck, 1983; Sweatt et al., 1985). In this scheme, it is proposed that alpha-2 adrenergic receptor activation stimulates the plasma membrane-bound Na^+/H^+ exchange system, which leads to an increase in intracellular pH, resulting from increased extrusion of intracellular H^+, and a concomitant increase in the accessibility of membrane bound Ca^{2+}. Elevation of intracellular pH and $[Ca^{2+}]$, acting together, sufficiently increases the activity of phospholipase A_2 to release small amounts of arachidonic acid from the phospholipids in the cell membrane. This release of arachidonic acid leads to the production of the endoperoxide intermediates, PGG_2 and PGH_2, and subsequently thromboxane A_2 (TXA_2), which produce activation of membrane-bound phospholipase C. Activation of phospholipase C increases the conversion of phosphatidylinositol biphosphate into inositol triphosphate and diacylglycerol. Diacylglycerol is hydrolyzed sequentially by a di- and monoacylglycerol lipase or can be phosphorylated to phosphatidic acid, which can serve as a substrate for phospholipase A_2 to release large amounts of arachidonic acid, which are then converted into the potent aggregatory prostanoids PGG_2, PGH_2, and TXA_2 (Sweatt et al., 1986b). This scheme is summarized in Fig. 3. In Chapter 6, the possible role of Na^+/H^+ exchange in mediating the effects of alpha-2 adrenergic receptors in other target organs is explored.

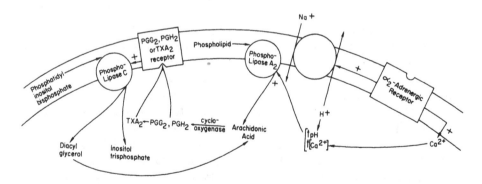

Fig. 3   Scheme for the relationship between alpha-2 adrenergic receptor activation and the ultimate mobilization of membrane-bound arachidonic acid resulting from alpha-2 adrenergic receptor-mediated Na^+/H^+ exchange. This scheme has been proposed to be involved in alpha-2 adrenergic receptor-mediated platelet aggregation.

5. Comparison of Drug Potencies in Biochemical and Physiologic Assay Systems for alpha-2 Adrenergic Receptors

When comparing drug potencies assessed in different systems, it is important to bear in mind the definition of potency and its physical meaning. The potency of an agonist in producing a physiologic response is usually given as the concentration of that agonist that is required to produce a given response, generally the half-maximal response. There are two independent parameters of agonist activity, however, namely affinity and intrinsic efficacy, that combine with a tissue property, receptor density, to give rise to the observed physiologic potency (Ruffolo, 1982; Kenakin, 1984).

The affinity of a ligand, in its simplest terms, is simply a measure of the tightness with which the ligand binds to the receptor; it depends on the ratio of the rates of association with, and dissociation from, the receptor. Affinity does not determine whether or not a ligand will activate the receptor, and hence be an agonist or an antagonist, when it is bound. Intrinsic efficacy is the parameter that determines whether a ligand–receptor interaction will lead to receptor activation and, thus, the production of a biological stimulus and, ultimately, a response (Ruffolo, 1982). As with affinity, intrinsic efficacy can vary widely from drug to drug. Once a ligand has bound to and activated a receptor, the final physiologic response will depend not only upon the intrinsic efficacy of the agonist, but also upon the receptor density in the tissue (Furchgott, 1966). Even an agonist with high intrinsic efficacy may still produce only a submaximal response when all the receptors are occupied if the number of receptors on the effector cells is sufficiently small that full receptor activation cannot elicit the maximum biological stimulus to which the tissue is physically capable of responding (Furchgott, 1966).

Since all three parameters may differ between various drugs, and from tissue to tissue, it is important that when agonist potencies are compared, it is only the affinities and/or efficacies that are used in such comparisons. Radioligand binding studies measure only one of these drug-related parameters, namely affinity. Although manipulation of the incubation medium (e.g., addition of ions or nucleotides) may also give some information as to whether a ligand is likely to be an agonist, partial agonist, or antagonist, such studies do not at present provide good estimates of intrinsic efficacy. Thus when comparing drug potencies in radioligand binding and physiologic test systems, only the affinity terms should be used.

There is an abundance of literature reporting affinities of various ligands at alpha-2 adrenergic receptors determined in radioligand binding studies. There are relatively few reports, however, of pharmacologically determined affinities of alpha-2 adrenergic receptor agonists in physiologic systems. An estimate of agonist affinity at alpha-2 adrenergic receptors has been made for B-HT 933 in canine saphenous vein (Ruffolo and Zeid, 1985). The dissociation constant obtained for B-HT 933 in this test system (5 μM) is fivefold greater than that obtained in radioligand binding studies in rat cerebral cortical membranes (Decker et al., 1984). Similarly, Jannsens and Vanhoutte (1978) found the dissociation constant of low concentrations of norepinephrine, which interact selectively at postjunctional alpha-2 adrenergic receptors in canine saphenous vein (360 nM), to be fourfold greater than that obtained in radioligand binding studies in rat cerebral cortex and human platelet (Cheung et al., 1982). A further comparison of the dissociation constants for B-HT 920 (467 nM) and alpha-methylnorepinephrine (524 nM) at prejunctional alpha-2 adrenergic receptors on postganglionic cholinergic neurons in guinea pig ileum, calculated from the data of Mottram and Thaker (1984), with those obtained in radioligand binding studies in rat cerebral cortex by Decker et al. (1984), again reveals that the affinities of these alpha-2 adrenergic receptor agonists are three- to fivefold lower when assessed by the physiologic assay techniques. Thus, it appears that the affinity of agonists estimated by physiologic assays are consistently lower than those obtained in radioligand binding assays by a factor of three- to fivefold.

In contrast to the consistent difference between agonist affinity in radioligand binding and physiologic assays, there is excellent agreement between the affinity of antagonists estimated in the two types of assay. For example, the affinities for alpha-2 adrenergic receptors of idazoxan and a series of its analogs as assessed in radioligand binding assays in rat cerebral cortical membranes (Gadie et al., 1984) are remarkably consistent with their affinities determined pharmacologically at prejunctional alpha-2 adrenergic receptors in rat vas deferens (Doxey et al., 1984). Similarly, the affinity of rauwolscine at postjunctional alpha-2 adrenergic receptors determined pharmacologically in canine saphenous vein (Ruffolo and Zeid, 1985) is the same as that found in radioligand binding studies in human platelets (Cheung et al., 1982).

A possible reason for the difference in agonist affinities determined in physiologic and radioligand binding assays, and the lack of difference in antagonist affinities, may lie in the differential effect of guanine nucleotides on agonist and antagonist binding. It is widely believed that alpha-2 adrenergic receptors utilize a GTP binding protein in the transduction of alpha-2 adrenergic receptor activation to a biological response (Limbird, 1984). For alpha-2 adrenergic receptor activation to be converted into a biological response, GTP must bind to a guanine nucleotide binding protein. In this state, the affinity of the agonist for the receptor is reduced, which is clearly demonstrated when the nonhydrolyzable guanine nucleotide Gpp(NH)p is included in the incubation medium in alpha-2 adrenergic receptor binding studies. Gpp(NH)p produces an approximate 10-fold decrease in the affinity of agonists for the alpha-2 adrenergic receptor in human platelets, whereas antagonist affinity is not affected (Smith and Limbird, 1981). In radioligand binding studies, Gpp(NH)p is not routinely included in the incubation medium, whereas in physiologic systems, GTP is present and likely binds to the guanine nucleotide binding protein in order to produce a physiologic response. Thus, it would be expected that agonist affinity in a physiologic assay system will be lower than that assessed in a binding assay, but antagonist affinity should not, and does not, differ.

6. Summary and Conclusions

Like the diversity of structural requirements for binding at the alpha-2 adrenergic receptor described in the previous chapter, the biochemical correlates of alpha-2 adrenergic receptor stimulation are also heterogeneous and resistant to broad generalizations. In some systems, alpha-2 adrenergic receptor activation promotes the translocation of extracellular Ca^{2+} in order to induce a biological response, such as in the vasculature, although in other circumstances, such as in the inhibition of neurotransmitter release, alpha-2 adrenergic receptor activation may inhibit Ca^{2+} influx. Likewise, in some systems, effector organ response to alpha-2 adrenergic receptor stimulation relies on receptor-mediated inhibition of adenylate cyclase, such as in the suppression of lipolysis, whereas in other systems, such as the platelet or beta cells of the pancreas, the significance of inhibition of adenylate cyclase is questionable. As discussed in the final chapter of the volume, however, it is hoped that integrative experimental approaches will be fruitful in understanding the physiological, biochemical, and pharmacological mechanisms of the alpha-2 adrenergic receptor function in its diverse target organs.

REFERENCES

Aghajanian, G. K. (1982) Central noradrenergic neurons: A locus for the functional interplay between α_2 adrenoceptors and opiate receptors. *J. Clin. Psychiat.* **43**, 20–24.

Aghajanian, G. K. (1985) The neurobiology of opiate withdrawal: Receptors, second messengers and ion channels. *Psychiatry Lett.* **111**, 57–60.

Aghajanian, G. K. (1978) Tolerance of locus ceruleus neurons to morphine and suppression of withdrawal response by clonidine. *Nature* **276**, 186–188.

Aghajanian, G. K. and VanderMaelen, C. P. (1982) α_2-Adrenoceptor-mediated hyperpolarization of locus ceruleus neurons: Intracellular studies in vivo. *Science* **215**, 1394–1396.

Ahlquist, R. P. (1962) The adrenoceptor receptor-detector. *Arch. Int. Pharmacodyn. Ther.* **139**, 38–41.

Ahren, B., Ludquist, I., and Jarhult, J. (1984) Effects of α_1-, α_2- and β-adrenoceptor blockers on insulin secretion in the rat. *Acta Endocrinol.* **105**, 78–82.

Alabaster, V. A., Keir, R. F., and Peters, C. J. (1985) Comparison of activity of α-adrenoceptor agonists and antagonists in dog and rabbit saphenous vein. *Naunyn Schmiedebergs Arch. Pharmacol.* **330**, 33–36.

Andrejak, M., Pommier, Y., Mouille, P., and Schmitt, H. (1980) Effects of some α-adrenoceptor agonists and antagonists on the guinea pig ileum. *Naunyn Schmiedebergs Arch. Pharmacol.* **314**, 83–87.

Angus, J. A., Cocks, T. M., and Satoh, K. (1986) The α-adrenoceptors on endothelial cells. *Fed. Proc.* **45**, 2355–2359.

Appleton, C., Martin, G. V., Algeo, S., Olajos, M., and Goldman, S. (1984) α_1- and α_2-adrenergic venoconstriction in intact dogs. *Circulation* **70** (suppl. II), 232.

Ardlie, N. G., Cameron, H. A., and Garrett, J. (1984) Platelet activation by circulating levels of hormones: A possible link in coronary heart disease. *Thrombosis Res.* **36**, 315–322.

Ariens, E. J. and van Rossum, J. M. (1957) pD_x, pA_x and pD_x' values in the analysis of pharmacodynamics. *Arch. Int. Pharmacodyn. Ther.* **110**, 275–299.

Ashton, H. (1984) Benzodiazepine withdrawal: An unfinished story. *Br. Med. J.* **288**, 1135–1140.

Ashton, H. and Rawlins, M. D. (1978) Central nervous system depressant actions of clonidine and UK-14,304: Partial dissociation of EEG and behavioral effects. *Br. J. Clin. Pharmacol.* **5**, 135–140.

Atweh, S. F. and Kuhar, M. J. (1977) Autoradiographic localization of opiate receptors in rat brain. I. Spinal cord and lower medulla. *Brain Res.* **142**, 53–67.

Auch-Schwelk, W., Starke, K., and Steppeler, A. (1983) Experimental conditions required for the enhancement by α-adrenoceptor antagonists of noradrenaline release in the rabbit ear artery. *Br. J. Pharmacol.* **78**, 543–551.

Awad, R., Payne, R., and Deth, R. C. (1983) α-Adrenergic receptor subtype associated with receptor binding, calcium influx, calcium release and contractile events in the rabbit aorta. *J. Pharmacol. Exp. Ther.* **227**, 60–67.

Baker, S. and Marshall, I. (1982) The effect of RX781094, a selective α_2-adrenoceptor antagonist, on [^{3}H]-noradrenaline release in the mouse vas deferens. *Br. J. Pharmacol.* **76**, 212P.

Barajas, L. and Wang, P. (1979) Localization of tritiated norepinephrine in the renal arteriolar nerves. *Anat. Res.* **195**, 525–534.

Barajas, L., Powers, K., and Wang, P. (1984) Innervation of the renal cortical tubules: A quantitative study. *Am. J. Physiol.* **247**, F50–F60.

Barnes, J. S., Dubocovich, M. L., Nies, A. S., and Gerber, J. G. (1982) Lack of interaction between tricyclic antidepressants and clonidine at the α_2-adrenoceptor on human platelets. *Clin. Pharmacol. Ther.* **32**, 744–748.

Baum, T. and Becker, F. T. (1982) Alpha-adrenergic and 5-hydroxytryptaminergic receptor stimulants as new antihypertensive drugs, with observations on involvement of opiate receptors. *Clin. Exp. Hyper.* **A4**, 235–248.

Beani, L., Bianchi, C., Giacomelli, A., and Tamperi, F. (1978) Noradrenaline inhibition of acetylcholine release from guinea pig brain. *Eur. J. Pharmacol.* **48**, 179–193.

Beckeringh, J. J., Thoolen, M. J., M. C., de Jonge, A, Wilffert, B., Timmermans, P. B. M. W. M., and van Zwieten, P. A. (1984) Differential effects of the calcium entry blocker D600 on contractions of rat and guinea pig aortas elicited by various α-adrenoceptor agonists. *J. Pharmacol. Exp. Ther.* **229**, 515–521.

Benfey, B. G. and Varma, D. R. (1962) Studies on the cardiovascular actions of antisympathomimetic drugs. *Int. J. Neuropharmacol.* **1**, 9–12.

Bennett, D. A., de Feo, J. J., Elko, E. E., and Lal, H. (1982) Naloxone-induced reversal of clonidine, but not hydralazine, hypotension. *Drug. Devel. Res.* **2**, 175–179.

Bentley, S. M., Drew, G. M., and Whiting, S. B. (1977) Evidence for two distinct types of postsynaptic α-adrenoceptor. *Br. J. Pharmacol.* **61**, 116P–117P.

Berne, R. M. (1958) Effect of epinephrine and norepinephrine on coronary circulation. *Circ. Res.* **6**, 644–655.

Besse, J. C. and Furchgott, R. F. (1976) Dissociation constants and relative efficacies of agonists acting on alpha adrenergic receptors in rabbit aorta. *J. Pharmacol. Exp. Ther.* **197**, 66–78.

Bevan, J. A. (1979) Sites of transition between functional systemic and cerebral arteries of rabbits occur at embryological junctional sites. *Science* **204**, 635–637.

Bevan, J. A. (1984) Autonomic pharmacologist's guide to the cerebral circulation. *Trends Pharmacol. Sci.* **5**, 234–236.

Bevan, J. A., Bevan, R. D., and Laher, I. (1985) Role of α-adrenoceptors in vascular control. *Clin. Sci.* **68** (suppl. 10), 83s–85s.

Bock, K. D., Heimsoth, V., Merguet, P., and Schoenermark, J. (1966) Klinische and kinische-experimentelle Untersuchungen mit einer neuen blutdrucksenkender Substanz: Dicholorphenylaminoimidazolin. *Deutsch. Med. Wschr.* **91**, 1761–1770.

Bolli, P., Erne, P., Ji, B. H., Block, L. H., Kiowski, W., and Buhler, F. R. (1984) Adrenaline induces vasoconstriction through postjunctional alpha₂-adrenoceptors and this response is enhanced in patients with essential hypertension. *J. Hyperten.* **2** (suppl. 3), 115–118.

Bottarie, S. P., Vokaer, A., Kaivez, E., Lescrainier, J.-P., and Vauquelin, G. P. (1983) Differential regulation of α-adrenergic receptor subclasses by gonadal steriods in human myometrium. *J. Clin. Endocrinol. Metab.* **57**, 937–941.

Bousquet, P. and Guertzenstein, P. G. (1973) Localization of the central cardiovascular action of clonidine. *Br. J. Pharmacol.* **49**, 573–579.

Boyer, J. L., Cardenas, C., Posadas, C., and Garcia-Sainz, J. A. (1983) Pertussis toxin induces tachycardia and impairs the increase in blood pressure produced by α₂-adrenergic agonists. *Life Sci.* **33**, 2627–2633.

Bramnert, M. and Hokfelt, B. (1983) Failure of naloxone to reduce the clonidine induced reduction of blood pressure and plasma noradrenaline in patients with essential hypertension. *Acta. Physiol. Scand.* **118**, 379–383.

Broadley, K. J. (1970) An analysis of the coronary vascular responses to catecholamines using a modified Langendorff heart preparation. *Br. J. Pharmacol.* **40**, 617–629.

Broadley, K. J. (1982) Cardiac adrenoceptors. *J. Auton. Pharmacol.* **2**, 119–145.

Brodde, O.-E., Motomura, S., Endoh, M., and Schumann, H. J. (1978) Lack of correlation between the positive inotropic effect evoked by α-adrenoceptor stimulation and the levels of cyclic AMP and/or cyclic GMP in the isolated ventricle strip of the rabbit. *J. Mol. Cell. Cardiol.* **10**, 207–219.

Brown, M. J., Struthers, A. D., Di Silvio, L., Yeo, T., Ghatei, M., and Burrin, J. M. (1985) Metabolic and haemodynamic effects of α₂-adrenoceptor stimulation and antagonism in man. *Clin. Sci.* **68** (suppl 10), 137s–139s.

Burns, T. W., Langley, P. E., Terry, B. E., Bylund, D. B., Hoffman, B. B., Lefkowitz, R. J., Garcia-Sainz, J. A., and Fain, J. W. (1981) Pharmacological characterization of adrenergic receptors in human adipocytes. *J. Clin. Invest.* **67**, 467–475.

Calvette, J. A., Hayes, R. J., Oates, N. S., Sever, P. S., and Thom, S. (1984) Alpha₁ and α₂ adrenoceptor responses in isolated human arteries. *Br. J. Pharmacol.* **83**, 364P.

Cambridge, D. (1981) UK-14,304, a potent and selective α₂-agonist for the characterization of α-adrenoceptor subtypes. *Eur. J. Pharmacol.* **72**, 413–415.

Cambridge, D., and Davey, M. J. (1980) Comparison of the α-adrenoceptors located on sympathetic and parasympathetic nerve terminals. *Br. J. Pharmacol.* **69**, 345P–346P.

Cami, J., de Torres, S., San, L., Sole, A., Guerra, D., and Ugena, B. (1985) Efficacy of clonidine and of methadone in the rapid detoxification of patients dependent on heroin. *Clin. Pharmacol. Ther.* **38**, 336–341.

Campbell, B. C., Elliot, H. L., Hamilton, C. A., and Reid, J. L. (1980) Changes in blood pressure, heart rate and sympathetic activity on abrupt withdrawal of tiamenidine (Hoe440) in essential hypertension. *Eur. J. Clin. Pharmacol.* **18**, 449–454.

Cavero, I., Shepperson, N., Lefevre-Borg, F., and Langer, S. Z. (1983) Differential inhibition of vascular smooth muscle responses to α_1- and α_2-adrenoceptor agonists by diltiazem and verapamil. *Circ. Res.* **52**, (suppl. 1), 69–76.

Cedarbaum, J. M. and Aghajanian, G. K. (1977) Catecholamine receptors on locus ceruleus neurons: Pharmacological characterization. *Eur. J. Pharmacol.* **44**, 375–385.

Chabardes, D., Montegut, M., Imbert-Teboul, M., and Morel, F. (1984) Inhibition of α_2-adrenergic agonist on AVP-induced cAMP accumulation in isolated collecting tubule of the rat kidney. *Mol. Cell. Endocrinol.* **37**, 263–275.

Chance, W. T. (1983) Clonidine analgesia: Tolerance and cross-tolerance to autoanalgesia. *Life Sci.* **33**, 2241–2246.

Chang, E. B., Field, M., and Miller, R. J. (1982) α-Adrenergic receptor regulation of ion transport in rabbit ileum. *Am. J. Physiol.* **242**, G237–G242.

Cheng, H. C., Gleason, E. M., Nathan, B. A., Lachman, P. J., and Woodward, J. K. (1981) Effects of clonidine on gastric secretion in the rat. *J. Pharmacol. Exp. Ther.* **217**, 121–126.

Cherksey, B., Mendelsohn, S., Zadunaisky, J., and Altszuler, N. (1982) Demonstration of α_2-adrenergic receptors in rat pancreatic islets using radioligand binding. *Proc. Soc. Exp. Biol. Med.* **171**, 196–200.

Cheung, D. W. (1985) The effect of Bay K 8644 on contraction mediated by α-adrenoceptors in the rat saphenous vein. *Br. J. Pharmacol.* **85**, 317–319.

Cheung, Y. D., Barnett, D. B., and Nahorski, S. R. (1982) [^{3}H]Rauwolscine and [^{3}H] yohimbine binding to rat cerebral and human platelet membranes; Possible heterogeneity of α_2-adrenoceptors. *Eur. J. Pharmacol.* **84**, 79–85.

Chiu, A. T., McCall, D. E., Thoolen, M. J. M. C., and Timmermans, P. B. M. W. M. (1986) Ca^{++} utilization in the contraction of rat aorta to full and partial α_1-adrenoceptor agonists. *J. Pharmacol. Exp. Ther.* **238**, 224–231.

Clare, K. A. and Scrutton, M. C. (1983) The properties of $^{45}Ca^{2+}$ uptake into human blood platelets induced by PAF and adrenaline. *Thromb. Haemost.* **50**, 41.

Clare, K. A., Scrutton, M. C., and Thompson, N. T. (1984) Effects of α_2-adrenoceptor agonists and of related compounds on aggregation of, and on adenylate cyclase activity in, human platelets. *Br. J. Pharmacol.* **82**, 467–476.

Clifton, C. G., O'Neill, W. M., and Wallin, J. D. (1981) Tiamenidine, a new antihypertensive agent: Efficacy, safety and rebound hypertension. *Curr. Ther. Res.* **30**, 397–404.

Clough, D. P. and Hatton, R. (1981) Hypotensive and sedative effects of α-adrenoceptor agonists: Relationship to α_1 and α_2-adrenoceptor potency. *Br. J. Pharmacol.* **73**, 595–604.

Cocks, T. M. and Angus, J. A. (1983) Endothelium-dependent relaxation of coronary arteries by noradrenaline and serotonin. *Nature* **305**, 627–630.

Cohn, J. N., Levine, T. B., Olivari, M. T., Garberg, V., Lura, D., Francis, G. S., Simon, A. B., and Rector, T. (1984) Plasma norepinephrine as a guide to prognosis in patients with chronic congestive heart failure. *N. Engl. J. Med.* **311**, 819–823.

Connolly, T. M. and Limbird, L. E. (1983) The influence of Na$^+$ on the α_2-adrenergic system of human platelets. *J. Biol. Chem.* **258**, 3907–3912.

Connor, H. E., Drew, G. M., and Finch, L. (1982) Clonidine-induced potentiation of reflex vagal bradycardia in anesthetized cats. *J. Pharm. Pharmacol.* **34**, 22–26.

Constantine, J. W., Gunnell, D., and Weeks, R. A. (1980) Alpha$_1$ and α_2 vascular adrenoceptors in the dog. *Eur. J. Pharmacol.* **66**, 281–286.

Constantine, J. W., Level, W., and Archer, R. (1982) A dog saphenous vein preparation with functional postsynaptic α_2- but not α_1-adrenoceptors. *Eur. J. Pharmacol.* **85**, 325–329.

Conway, E. L., Louis, W. J., and Jarrott, B. (1979) The effect of acute α-methyldopa administration on catecholamine levels in anterior hypothalamic and medullary nuclei in rat brain. *Neuropharmacology* **18**, 279–286.

Cooke, J. H., Johns, E. J., MacLeod, J. H., and Singer, B. (1972) Effect of renal nerve stimulation, renal blood flow and adrenergic blockade on plasma renin activity in the cat. *J. Physiol.* **226**, 15–36.

Cooke, J. P., Rimele, T. J., Flavahan, N. A., and Vanhoutte, P. M. (1985) Nimodipine and inhibition of α-adrenergic activation of the isolated canine saphenous vein. *J. Pharmacol. Exp. Ther.* **234**, 598–602.

Corby, D. G. and O'Barr, T. G. (1981) Decreased α-adrenergic receptors in newborn platelets: Cause of abnormal response to epinephrine. *Dev. Pharmacol. Ther.* **2**, 215–225.

Cowen, P. J. and Nutt, D. J. (1982) Abstinence symptoms after withdrawal of tranquilizing drugs: Is there a common neurological mechanism? *Lancet* **ii**, 360–362.

Cutter, W. E., Bier, D. M., Shah, S. D., and Cryer, P. E. (1980) Epinephrine plasma clearance rates and physiologic thresholds for metabolic and hemodynamic actions in man. *J. Clin. Invest.* **66**, 94–101.

Davidov, M., Kakaviatos, N., and Finnerty, F. A. (1967) The antihypertensive effects of an imidazoline compound. *Clin. Pharmacol. Ther.* **8**, 810–816.

Davis, D. S., Wing, L. M. H., Reid, J. L., Neill, E., Tippett, P., and Dollery, C. T. (1977) Pharmacokinetics and concentration-effect relationships of intravenous and oral clonidine. *Clin. Pharmacol. Ther.* **21**, 593–600.

Dawes, P. M. and Vizi, E. S. (1973) Acetylcholine release from the rabbit isolated superior cervical ganglion preparation. *Br. J. Pharmacol.* **48**, 225–232.

Decker, N., Ehrhardt, J. D., Leclerc, G., and Schwartz, J. (1984) Postjunctional α-adrenoceptors. *Naunyn Schmiedebergs Arch. Pharmacol.* **326**, 1–6.

deJonge, A., Timmermans, P. B. M. W. M., and van Zwieten, P. A. (1981) Participation of cardiac presynaptic α_2-adrenoceptors in the bradycardic effects of clonidine and analogues. *Naunyn Schmiedebergs Arch. Pharmacol.* **317**, 8–12.

De Langen, C. D. J., Hogenboom, F., and Mulder, A. H. (1979) Presynaptic noradrenergic α-receptors and modulation of ^{3}H-noradrenaline release from rat brain synaptosomes. *Eur. J. Pharmacol.* **60**, 79–89.

Delbarre, B. and Schmitt, H. (1973) A further attempt to characterize sedative receptors activated by clonidine in chickens and mice. *Eur. J. Pharmacol.* **22**, 355–359.

Delini-Stula, A., Baumann, P., and Buch, O. (1979) Depression of exploratory activity by clonidine in rats as a model for the detection of relative pre- and postsynaptic central noradrenergic receptor selectivity of α-adrenolytic drugs. *Naunyn Schmiedebergs Arch. Pharmacol.* **307**, 115–122.

DeMey, J. and Vanhoutte, P. M. (1981) Uneven distribution of postjunctional α_1 and α_2-like adrenoceptors in canine arterial and venous smooth muscle. *Circ. Res.* **48**, 875–884.

Dharmsathaphorn, K., Yamashiro, D. J., Lindeborg, D., Mandel, K. G., McRoberts, J., and Ruffolo, R. R., Jr. (1984) Effects of structure-activity relationships of α-adrenergic compounds on electrolyte transport in the rabbit ileum and rat colon. *Gastroenterology* **86**, 120–128.

Dismukes, K., De Boer, A. A., and Mulder, A. H. (1977) On the mechanism of α-receptor mediated modulation of ^{3}H-noradrenaline release from slices of rat brain neocortex. *Naunyn Schmiedebergs Arch. Pharmacol.* **299**, 115–122.

Docherty, J. R. and Hyland, L. (1984) Neuro-effector transmission through postsynaptic α_2-adrenoceptors in human saphenous vein. *Br. J. Pharmacol.* **83**, 362P.

Docherty, J. R., MacDonald, A., and McGrath, J. C. (1979) Further subclassification of α-adrenoceptors in the cardiovascular system, vas deferens and anococcygeus muscle of the rat. *Br. J. Pharmacol.* **67**, 421P–422P.

Dollery, C. T. and Reid, J. L. (1973) Central noradrenergic neurones and the cardiovascular actions of clonidine in the rabbit. *Br. J. Pharmacol.* **47**, 206–216.

Donowitz, M., Cusolito, S., Battisti, L., and Fogel, R. (1982) Dopamine stimulation of active Na^+ and Cl^- absorption in rabbit ileum. Interaction with α-adrenergic and specific dopamine receptors. *J. Clin. Invest.* **69**, 1008–1016.

Doxey, J. C. and Roach, A. G. (1980) Presynaptic α-adrenoceptors; in vitro methods and preparations utilised in the evaluation of agonists and antagonists. *J. Auton. Pharmacol.* **1**, 73–99.

Doxey, J. C., Roach, A. G., Strachan, D. A., and Virdee, N. K. (1984) Selectivity and potency of 2-alkyl analogues of the alpha-2 adrenoceptor antagonist idazoxan (RX781094) in peripheral systems. *Br. J. Pharmacol.* **83**, 713–722.

Drew, G. M. (1976) Effects of alpha-adrenoceptor agonists and antagonists on pre- and postsynaptically located alpha-adrenoceptors. *Eur. J. Pharmacol.* **36**, 313–320.

Drew, G. M. (1978) Pharmacological characterization of the presynaptic α-adrenoceptors regulating cholinergic activity in the guinea pig ileum. *Br. J. Pharmacol.* **64**, 293–300.

Drew, G. M. and Whiting, S. B. (1979) Evidence for two distinct types of postsynaptic α-adrenoceptor in vascular smooth muscle in vivo. *Br. J. Pharmacol.* **67**, 207–215.

Drew, G. M., Gower, A. J., and Marriott, A. S. (1979) α_2-adrenoceptors mediate clonidine-induced sedation in the rat. *Br. J. Pharmacol.* **67**, 133–141.

Dubocovich, M. L. (1984) Presynaptic α-adrenoceptors in the central nervous system. *Ann. NY Acad. Sci.* **430**, 7–25.

Duckles, S. P. (1980) Functional activity of the noradrenergic innervation of large cerebral arteries. *Br. J. Pharmacol.* **69**, 193–199.

Duckles, S. P., Bevan, R. D., and Bevan, J. A. (1976) An *in vitro* study of prolonged vasospasm of a rabbit cerebral artery. *Stroke* **7**, 174–178.

Durbin, T., Rosenthal, L., McArthur, K., Anderson, D., and Dharmsathaphorn, K. (1982) Clonidine and lidamidine (WHR-1142) stimulate sodium and chloride absorption in the rabbit intestine. *Gastroenterology* **82**, 1352–1358.

Egleme, C., Godfraind, T., and Miller, R. C. (1984) Enhanced responsiveness of isolated rat aorta to clonidine after removal of the endothelial cells. *Br. J. Pharmacol.* **81**, 16–18.

Elliott, H. L., and Reid, J. L. (1983) Evidence for postjunctional vascular α_2-adrenoceptors in peripheral vascular regulation in man. *Clin. Sci.* **65**, 237–241.

Elsner, D., Saeed, M., Sommer, O., Holtz, J., and Bassenge, E. (1984) Sympathetic vasoconstriction sensitive to α_2-adrenergic receptor blockade. *Hypertension* **6**, 915–925.

Engberg, G., Elam, M., and Svensson, T. H. (1982) Clonidine withdrawal: Activation of brain noradrenergic neurons with specifically reduced α_2-receptor sensitivity. *Life Sci.* **30**, 235–243.

Fain, J. N. and Garcia-Sainz, J. A. (1980) Role of phosphatidylinositol turnover in α_1- and of adenylate cyclase inhibition in α_2-effects of catecholamines. *Life Sci.* **26**, 1183–1194.

Farnebo, L-O and Hamberger, B. (1971) Drug-induced changes in the release of ^{3}H-monoamines from field stimulated rat brain slices. *Acta Physiol. Scand.* (suppl. 371), 35–44.

Farsang, C. and Kunos, G. (1979) Naloxone reverses the antihypertensive effect of clonidine. *Br. J. Pharmacol.* **67**, 161–164.

Farsang, C., Varga, K., Vajda, L., Alfoldi, S., and Kapocsi, J. (1984a) Effects of clonidine and guanfacine in essential hypertension. *Clin. Pharmacol. Ther.* **36**, 588–594.

Farsang, C., Varga, K., Vajda, L., Kapocsi, J., Balas-Eltes, A., and Kunos, G. (1984b) beta-Endorphin contributes to the antihypertensive effect of clonidine in a subset of patients with essential hypertension. *Neuropeptides* **4**, 293–302.

Farsang, C., Kapocsi, J., Vajda, L., Varga, K., Malisak, Z., Fekete, M., and Kunos, G. (1984c) Reversal by naloxone of the antihypertensive action of clonidine: Involvement of the sympathetic nervous system. *Circulation* **69**, 461–467.

Farsang, C., Ramirez-Gonzalez, M. D., Mucci, L. and Kunos, G. (1980) Possible role of an endogenous opioid in the cardiovascular effects of central α-adrenoceptor stimulation in spontaneously hypertensive rats. *J. Pharmacol. Exp. Ther.* **214**, 203–208.

Farsang, C., Kapocsi, J., Juhasz, I., and Kunos, G. (1982) Possible involvement of an endogenous opioid in the antihypertensive effect of clonidine in patients with essential hypertension. *Circulation* **66**, 1268–1272.

Farsang, C., Vajda, L., Kopocsi, J., Malisal, Z., Alfoldi, S., Varga, K., Juhasz, I., and Kunos, G. (1983) Diurnal rhythm of beta endorphin in normotensive and hypertensive patients: The effect of clonidine. *J. Clin. Endocrinol. Metab.* **56**, 865–867.

Feigl, E. O. (1967) Sympathetic control of coronary circulation. *Circ. Res.* **20**, 262–270.

Feigl, E. O. (1975) Control of myocardial oxygen tension by sympathetic coronary vasoconstriction in the dog. *Circ. Res.* **37**, 88–95.

Field, M. and McColl, I. (1973) Ion transport in rabbit ileal mucosa. III. Effects of catecholamines. *Am. J. Physiol.* **225**, 852–857.

Fielding, S., Wilker, J., Hynes, M., Szewczak, M., Novick, W. J., and Lal, H. (1978) A comparison of clonidine with morphine for antinociceptive and antiwithdrawal actions. *J. Pharmacol. Exp. Ther.* **207**, 899–905.

Flavahan, N. A. and Vanhoutte, P. M. (1986a) Alpha$_1$ and α_2 adrenoceptor: Response coupling in canine saphenous and femoral veins. *J. Pharmacol. Exp. Ther.* **238**, 131–138.

Flavahan, N. A. and Vanhoutte, P. M. (1986b) The effect of cooling on α_1 and α_2-adrenergic responses in canine saphenous and femoral veins. *J. Pharmacol. Exp. Ther.* **238**, 131–138.

Flavahan, N. A., Rimele, T. J., Cooke, J. P., and Vanhoutte, P. M. (1984) Characterization of postjunctional alpha-1 and alpha-2 adrenoceptors activated by exogenous or nerve-released norepinephrine in the canine saphenous vein. *J. Pharmacol. Exp. Ther.* **230**, 699–705.

Fontaine, J., Grivegnee, A., and Reuse, J. (1984) Adrenoceptors and regulation of intestinal tone in the isolated colon of the mouse. *Br. J. Pharmacol.* **81**, 231–243.

Fowler, P. J., Grous, M., Price, W., and Matthews, W. D. (1984) Pharmacological differentiation of postsynaptic alpha-adrenoceptors in the dog saphenous vein. *J. Pharmacol. Exp. Ther.* **229**, 712–718.

Frankhuyzen, A. L. and Mulder, A. H. (1980) Noradrenaline inhibits depolarization induced ^{3}H-serotonin release from slices of rat hippocampus. *Eur. J. Pharmacol.* **63**, 179–182.

Frisk-Holmberg, M., Paalzow, L., and Wibell, L. (1984) Relationship between the cardiovascular effects and steady-state kinetics of clonidine in hypertension. *Eur. J. Clin. Pharmacol.* **26**, 309–313.

Frohman, L. A. and Benardis, L. L. (1971) Effect of hypothalamic stimulation on plasma glucose, insulin and glucagon levels. *Am. J. Physiol.* **221**, 1596–1603.

Furchgott, R. F. (1966) The use of beta-haloalkylamines in the differentiation of receptors and in the determination of dissociation constants of receptor-agonist complexes. *Adv. Drug Res.* **3**, 21–55.

Furchgott, R. F. (1983) Role of the endothelium in response of vascular smooth muscle. *Circ. Res.* **53**, 557–573.

Gaal, G., Kattus, A. A., Kolin, A., and Ross, G. (1966) Effects of adrenaline and noradrenaline on coronary blood flow before and after beta-adrenergic blockade. *Br. J. Pharmacol.* **26**, 713–722.

Gadie, B., Lane, A. C., McCarthy, P. S., Tulloch, I. F., and Walter, D. S. (1984) 2-Alkyl analogues of idazoxan (RX781094) with enhanced antagonist potency and selectivity at central α_2-adrenoceptors in the rat. *Br. J. Pharmacol.* **83**, 707–712.

Gaxiola, B., Friedl, W., and Propping, P. (1984) Epinephrine-induced platelet aggregation. A twin study. *Clin. Genet.* **26**, 543–548.

Gellai, M. and Ruffolo, R. R., Jr. (1987) Renal effects of selective α_1 and

α_2-adrenoceptor agonists in conscious, normotensive rats. *J. Pharmacol. Exp. Ther.* **240**, 723–728.

Gerold, M. and Haeusler, G. (1983) Alpha$_2$-adrenoceptors in rat resistance vessels. *Naunyn Schmiedebergs Arch. Pharmacol.* **322**, 29–33.

Giles, T. D., Thomas, M. G., Sander, G. E., and Quiroz, A. C. (1985) Central α-adrenergic agonists in chronic heart failure and ischemic heart disease. *J. Cardiovasc. Pharmacol.* **7** (suppl. 8), S51–S55.

Gillespie, J. S. and Khoyi, M. A. (1977) The site and receptors responsible for the inhibition by sympathetic nerves of intestinal smooth muscle and its parasympathetic motor nerves. *J. Physiol.* **267**, 767–789.

Gillis, R. A., Gatti, P. J., and Quest, J. A. (1985) Mechanism of the antihypertensive effect of α_2-agonists. *J. Cardiovasc. Pharmacol.* **7** (suppl. 8), S38–S44.

Glomstein, A., Kauge, A., Oye, I., and Sinclair, D. (1967) Effects of adrenaline on coronary flow in isolated perfused rat hearts. *Acta Physiol. Scand.* **69**, 102–110.

Glusa, E. and Markwadt, F. (1983a) Characterization of postjunctional α-adrenoceptors in isolated human femoral veins and arteries. *Naunyn Schmiedebergs Arch. Pharmacol.* **323**, 101–105.

Glusa, E. and Markwadt, F. (1983b) Characterization of α_2-adrenoceptors on blood platelets from various species using, ^{3}H-yohimbine. *Haemostasis* **13**, 96–101.

Gold, M. S., Redmond, D. E., and Kleber, H. D. (1978a) Clonidine in opiate withdrawal. *Lancet* i, 929–930.

Gold, M. S., Redmond, D. E., and Kleber, H. D. (1978b) Clonidine blocks acute opiate withdrawal symptoms. *Lancet* ii, 599–602.

Gold, M. S., Pottach, A. L. C., and Extein, I. (1982) Clonidine: Inpatient studies from 1978 to 1981. *J. Clin. Psychiatry* **43**, 35–38.

Goldberg, M. R. and Robertson, D. (1984) Evidence for the existence of vascular α_2-adrenergic receptors in humans. *Hypertension* **6**, 551–556.

Goldstein, J. A. (1983) Clonidine as analgesic. *Biol. Psychiatry* **18**, 1339–1340.

Gothert, M. and Huth, H. (1980) Alpha-adrenoceptor mediated modulation of 5-hydroxytyrptamine release from rat brain cortex slices. *Naunyn Schmiedebergs Arch. Pharmacol.* **313**, 21–26.

Göthert, M., Schlicker, E., Hentrich, F., Rohm, N., and Zerkowski, H-R. (1984) Modulation of noradrenaline release in human saphenous vein via presynaptic alpha-adrenoceptors. *Eur. J. Pharmacol.* **102**, 261–267.

Govier, W. C. (1967) A positive inotropic effect of phenylephrine mediated through α-adrenergic receptors. *Life Sci.* **6**, 1361–1365.

Gower, A. J. and Marriott, A. S. (1980) The inhibition of clonidine-induced sedation in the mouse by anti-depressant drugs. *Br. J. Pharmacol.* **69**, 287P–288P.

Graham, A. W. and Aghajanian, G. K. (1971) Effects of amphetamine on single-cell activity in a catecholamine nucleus, the locus ceruleus. *Nature* **234**, 100–102.

Granger, D. N., Richardson, P. D. I., Kvietys, P. R., and Mortillaro, N. A. (1980) Intestinal blood flow. *Gastroenterology* **78**, 837–863.

Grant, J. A. and Scrutton, M. C. (1980) Interaction of selective α-adrenoceptor agonists and antagonists with human and rabbit blood platelets. *Br. J. Pharmacol.* **71**, 121–134.

Grant, K., Marshall, I., and Nasmyth, P. A. (1980) Cocaine and presynaptic α-adrenoceptor regulation of noradrenaline release in response to one and two pulses. *Br. J. Pharmacol.* **69**, 344P–345P.

Greenway, C. V. and Innes, I. R. (1981) Effects of arteriolar vasodilators on hepatic venous compliance and cardiac output in anesthetized cats. *J. Cardiovasc. Pharmacol.* **3**, 1321–1331.

Gunnar Wallin, B. and Frisk-Holmberg, M. (1981) The antihypertensive mechanism of clonidine in man. *Hypertension* **3**, 340–346.

Haeusler, G. (1974) Clonidine-induced inhibition of sympathetic nerve activity; no indication for a central presynaptic or an indirect sympathomimetic mode of action. *Naunyn Schmiedebergs Arch. Pharmacol.* **286**, 97–111.

Hallam, T. J., Scrutton, M. C., and Wallis, R. B. (1981) Responses of rabbit platelets to adrenaline induced by other agonists. *Thromb. Res.* **20**, 413–424.

Hamilton, F. M. and Feigl, E. (1976) Coronary vascular sympathetic beta-receptor innervation. *Am. J. Physiol.* **230**, 1569–1576.

Hamilton, C. A. and Reid, J. L. (1982) A postsynaptic location of α_2-adrenoceptors in vascular smooth muscle: In vivo studies in the conscious rabbit. *Cardiovasc. Res.* **16**, 11–15.

Hamilton, C. A., Reid, J. L., and Sumner, D. J. (1983) Acute effects of phenoxybenzamine on α-adrenoceptor responses in vivo and in vitro: Relation of in vivo pressor responses to the number of specific binding sites. *J. Cardiovasc. Pharmacol.* **5**, 868–873.

Hannah, J. A. M., Hamilton, C. A., and Reid, J. L. (1983) RX 781094, a new potent α_2-adrenoceptor antagonist. *Naunyn Schmiedebergs Arch. Pharmacol.* **322**, 221–227.

Hansson, L. (1983) Clinical aspects of blood pressure crisis due to withdrawal of centrally acting antihypertensive drugs. *Br. J. Clin. Pharmacol.* **15**, 485S–489S.

Hansson, L., Hunyor, S. N., Julius, S., and Hoobler, S. W. (1973) Blood pressure crisis following withdrawal of clonidine, with special reference to arterial and urinary catecholamine levels and suggestions for acute management. *Am. Heart J.* **85**, 605–610.

Hashimto, K., Shigel, T., Imai, S., Saito, Yo., Yago, N., Uei, I., and Clark, R. E. (1960) Oxygen consumption and coronary vascular tone in the isolated fibrillating dog heart. *Am. J. Physiol.* **198**, 965–970.

Haslam, R. J. (1975) Role of Cyclic Nucleotides in Platelet Function, in *CIBA Foundation Symposium* no. 25, Elsevier/North Holland Biomedical, Amsterdam.

Haslam, R. J., Davidson, M. M. L., and Desjardins, J. V. (1978) Inhibition of adenylate cyclase by adenosine analogs in preparations of broken and intact human platelets: Evidence for unidirectional control of platelet function by cyclic AMP. *Biochem. J.* **176**, 83–95.

Hedler, L., Stamm, G., Weitzell, R., and Starke, K. (1981) Functional characterization of central α-adrenoceptors by yohimbine diastereomers. *Eur. J. Pharmacol.* **70**, 43–52.

Hesse, I. F. A. and Johns, E. J. (1984) An in vivo study of the α-adrenoceptor subtypes on the renal vasculature of the anaesthetised rabbit. *J. Autonom. Pharmacol.* **4**, 145–152.

Heusch, G. and Deussen, A. (1983) The effects of cardiac sympathetic nerve stimulation on perfusion of stenotic coronary arteries in the dog. *Circ. Res.* **53**, 8–15.

Heusch, G., Deussen, A., Schipke, J., and Thamer, V. (1984) Alpha$_1$- and alpha$_2$-adrenoceptor-mediated vasoconstriction of large and small canine coronary arteries *in vivo*. *J. Cardiovasc. Pharmacol.* **6**, 961–968.

Hicks, P. E. and Waldron, C. (1982) Selective post-junctional supersensitivity to α_2-adrenoceptor agonists after reserpine pretreatment in rats. *Br. J. Pharmacol.* **75**, 152P.

Hieble, J. P. and Pendleton, R. G. (1979) Effects of ring substitution on the pre- and postjunctional alpha-adrenergic activity of arylimino-imidazolidines. *Naunyn Schmiedebergs Arch. Pharmacol.* **309**, 217–224.

Hieble, J. P. and Woodward, D. F. (1984) Different characteristics of postjunctional α-adrenoceptors on arterial and venous smooth muscle. *Naunyn Schmiedebergs Arch. Pharmacol.* **328**, 44–50.

Hieble, J. P., DeMarinis, R. M., Matthews, W. D., and Fowler, P. J. (1986) Selective alpha-2 adrenoceptor blockade by SK&F 86466: In vitro characterization of receptor selectivity. *J. Pharmacol. Exp. Ther.* **236**, 90–96.

Hiley, C. R. and Thomas, G. R. (1987) Effects of alpha-adrenoceptor agonists on cardiac output and its regional distribution in the pithed rat. *Br. J. Pharmacol.* **90**, 61–70.

Hoefke, A., Kobinger, W., and Walland, A. (1975) Relationship between activity and structure in derivatives of clonidine. *Arzneimittelforsch.* **25**, 786–793.

Hoffman, B. B. and Lefkowitz, R. J. (1980) Radioligand binding studies of adrenergic receptors: New insights into molecular and physiological regulation. *Ann. Rev. Pharmacol. Toxicol.* **20**, 581–608.

Hoffman, B. B., De Lean, A., Wood, C. L., Schocken, D. D., and Lefkowitz, R. J. (1979) α-Adrenergic receptor subtypes: Quantitative assessment by ligand biding. *Life Sci.* **24**, 1739–1746.

Hoffman, B. B., Lavin, T. N., Lefkowitz, R. J., and Ruffolo, R. R., Jr. (1981) α-Adrenergic receptor subtypes in rabbit uterus: Mediation of myometrial contraction and regulation by estrogens. *J. Pharmacol. Exp. Ther.* **219**, 290–295.

Holman, R. B., Shillito, E., and Vogt, M. (1971) Sleep produced by clonidine (2-(2,6-dichlorophenylamino)-2-imidazoline hydrochloride). *Br. J. Pharmacol.* **43**, 685–695.

Holtz, J., Saeed, M., Sommer, O., and Bassenge, E. (1982) Norepinephrine constricts the canine coronary bed via postsynaptic α_2-adrenoceptors. *Eur. J. Pharmacol.* **82**, 199–202.

Horn, P. T., Kohli, J. D., Listinsky, J. J., and Goldberg, L. I. (1982) Regional variation in the α-adrenergic receptors in the canine resistance vessels. *Naunyn Schmiedebergs Arch. Pharmacol.* **318**, 166–172.

Hsu, C. Y., Knapp, D. R., and Halushka, P. V. (1979) The effects of α-adrenergic agents on human platelet aggregation. *J. Pharmacol. Exp. Ther.* **208**, 366–370.

Hughes, P. L. and Morse, R. M. (1985) Use of clonidine in a mixed-drug detoxification regimen: Possibility of masking of clinical signs of sedative withdrawal. *Mayo Clin. Proc.* **60**, 47–49.

Hyman, A. L. and Kadowitz, P. J. (1985) Evidence for existence of postjunctional α_1- and α_2-adrenoceptors in cat pulmonary vascular bed. *Am. J. Physiol.* **249**, H891–H898.

Hyman, A. L. and Kadowitz, P. J. (1986) Enhancement of α- and β-adrenoceptor responses by elevations in vascular tone in pulmonary circulation. *Am. J. Physiol.* **250**, H1109–H1116.

Hyman, A. L., Lippton, H. L., and Kadowitz, P. J. (1985) Autonomic regulation of the pulmonary circulation. *J. Cardiovasc. Pharmacol.* **7**, S80–S95.

Hynes, M. D., Atlas, D., and Ruffolo, R. R., Jr. (1983) Analgesic activity of HP-aminoclonidine, a novel analog of clonidine: Role of opioid receptors and α-adrenoceptors. *Pharmacol. Biochem. Behav.* **19**, 879–882.

Insel, P. A., Nirenberg, P., Turnbull, J., and Shattil, S. J. (1978) Relationships between membrane cholesterol, α-adrenergic receptors, and platelet function. *Biochemistry* **17**, 5269–5274.

Ito, T. and Chiba, S. (1985) Existence of two types of postjunctional α adrenoceptors in the isolated canine intermediate auricular artery. *J. Pharmacol. Exp. Ther.* **234**, 698–702.

Jakobs, K. H. (1978) Synthetic α-adrenergic agonists are potent α-adrenergic blockers in human platelets. *Nature* **274**, 819–820.

Jakobs, K. H., Saur, W., and Schultz, G. (1978) Characterization of α- and beta-adrenergic receptors linked to human platelet adenylate cyclase. *Naunyn Schmiedebergs Arch. Pharmacol.* **302**, 285–291.

Jakobs, K. H., Saur, W., and Schultz, G. (1976) Reduction of adenylate cyclase activity in lysates of human platelets by the α-adrenergic component of epinephrine. *J. Cycl. Nucl. Res.* **2**, 381–392.

Jain, A. K., Hiremath, A., Michael, R., Ryan, J. R., and McMahon, F. G. (1985) Clonidine and guanfacine in hypertension. *Clin. Pharmacol. Ther.* **37**, 271–276.

Kalsner, S. (1980) Limitations of presynaptic adrenoceptor theory: The characteristics of the effects of noradrenaline and phenoxybenzamine on stimulation-induced efflux of [^{3}H]-noradrenaline in vas deferens. *J. Pharmacol. Exp. Ther.* **212**, 232–239.

Kalsner, S., Suleiman, M., and Dobson, R. E. (1980) Adrenergic presynaptic receptors: An overextended hypothesis? *J. Pharm. Pharmacol.* **32**, 290–292.

Kawahara, R. S. and Bylund, D. B. (1985) Solubilization and characterization of putative α$_2$-adrenergic isoceptors from the human platelet and the rat cerebral cortex. *J. Pharmacol. Exp. Ther.* **233**, 603–610.

Kenakin, T. P. (1984) The classification of drugs and drug receptors in isolated tissues. *Pharmacol. Rev.* **36**, 165–222.

Keshavan, M. S. and Crammer, J. L. (1985) Clonidine in benzodiazepine withdrawal. *Lancet* **i**, 1325–1326.

Kho, T. L., Schalekamp, M. A. D. H., Zaal, G. A., Wester, A., and Birkenhager, W. H. (1975) Comparison between the effects of St 600 and clonidine. *Arch. Int. Pharmacodyn.* **217**, 162–169.

Kinter, L. B., Mann, W. A., and Naselsky, D. (1985) Site of alpha-2 agonist inhibition of epithelial water flux. *Fed. Proc.* **44**, 1014.

Kleber, H. D., Riordan, C. E., Rounsaville, B., Kosten, T., Charney, D., Gaspari, J., Hogan, I., and O'Connor, C. (1985) Clonidine in outpatient detoxification from methadone maintenance. *Arch. Gen. Psychiat.* **42**, 391–394.

Kleinlogel, H., Scholtysik, G., and Sayers, A. C. (1975) Effects of clonidine and BS 100-141 on the EEG sleep patterns in rats. *Eur. J. Pharmacol.* **33**, 159–163.

Kobinger, W. (1978) Central α-adrenergic systems as targets for antihypertensive drugs. *Rev. Physiol. Biochem. Pharmacol.* **81**, 39–100.

Kobinger, W. and Pichler, L. (1981) Alpha$_2$-adrenoceptor agonist effect of

B-HT 920 in isolated perfused hindquarters of rats. *Eur. J. Pharmacol.* **76**, 101–105.

Kobinger, W. and Pichler, L. (1976) Centrally induced reduction in sympathetic tone, a postsynaptic α-adrenoceptor stimulating action of imidazolidines. *Eur. J. Pharmacol.* **40**, 311–320.

Kobinger, W., Lillie, C., and Pichler, L. (1980) Central cardiovascular alpha-adrenoceptors. Relation to peripheral receptors. *Circ. Res.* **46**, 121–25.

Kopia, G. A., Kopaciewicz, L. J., and Ruffolo, R. R., Jr. (1986) α-Adrenoceptor regulation of coronary artery blood flow in normal and stenotic canine coronary arteries. *J. Pharmacol. Exp. Ther.* **239**, 641–647.

Kroneberg, G. and Oberdorf, A. (1974) Inhibition of Acetylcholine Release and Acetylcholine Action in the Guinea Pig Ileum by Sympathetic α- and β-Receptor Stimulation, in *Proceedings, First Congress of the Hungarian Pharmacological Society* (Knoll, J. and Vizi, E. S., eds.) Akademaiai Kiado, Budapest.

Krothapalli, R. K., Duffy, B., Senekjian, H., and Suki, W. (1983) Modulation of the hydro-osmotic effect of vasopressin on the rabbit cortical collecting tubule by adrenergic agents. *J. Clin. Invest.* **72**, 287–294.

Kuhar, M. J. (1982) Receptors for clonidine in brain: Insights into therapeutic actions. *J. Clin. Psychiat.* **43**, 17–19.

Kunos, G., Farsang, C., and Ramirez-Gonzalez, M. D. (1981) Beta-endorphin: Possible involvement in the antihypertensive effect of central α-receptor activation. *Science* **211**, 82–84.

Jannseens, W. J. and Vanhoutte, P. M. (1978) Instantaneous changes of α-adrenoceptor affinity by moderate cooling in canine cutaneous veins. *Am. J. Physiol.* **234**, H330–H337.

Jansson, G. and Martinson, J. (1966) Studies on the ganglionic site of action of sympathetic outflow to the stomach. *Acta Physiol. Scand.* **68**, 184–192.

Jarrott, B., Lewis, S., Conway, E. L., Summers, R., and Louis, W. J. (1984) The involvement of central α adrenoceptors in the antihypertensive actions of methyldopa and clonidine in the rat. *Clin. Exp. Hyperten.* **A6**, 387–400.

Jarrott, B., Louis, W. J., and Summers, R. J. (1979) The characteristics of ^{3}H-clonidine binding to an α-adrenoceptor in membranes from guinea pig kidney. *Br. J. Pharmacol.* **65**, 663–670.

Jauernig, R. A., Moulds, R. F. W., and Shaw, J. (1978) The action of prazosin on human vascular preparations. *Arch. Int. Pharmacodyn.* **231**, 81–89.

Jie, K., van Brummelen, P., Vermey, P., Timmermans, P. B. M. W. M., and van Zwieten, P. A. (1986) Alpha$_1$- and alpha$_2$-adrenoceptor mediated vasoconstriction in the forearm of normotensive and hypertensive subjects. *J. Cardiovasc. Pharmacol.* **8**, 190–196.

Jim, K. F. and Matthews, W. D. (1985) Role of extracellular calcium in contractions produced by activation of postsynaptic α$_2$-adrenoceptors in canine saphenous vein. *J. Pharmacol. Exp. Ther.* **234**, 161–165.

Jouvet, M. (1972) The role of monoamines and acetylcholine containing neurons in the regulation of the sleep-waking cycle. *Ergeb. Physiol.* **64**, 166–307.

Kalkman, H. O., Thoolen, M. J. M. C., Timmermans, P. B. M. W. M., and van Zwieten, P. A. (1984) The influence of α$_1$- and α$_2$-adrenoceptor agonists on cardiac output in rats and cats. *J. Pharm. Pharmacol.* **36**, 265–268.

Kalsner, S. (1979) Single pulse stimulation of guinea pig vas deferens and the pre-synaptic receptor hypothesis. *Br. J. Pharmacol.* **66**, 343–349.

Langer, S. Z. (1973) The Regulation of Transmitter Release Elicited by Nerve Stimulation Through a Presynaptic Feed-Back Mechanism, in *Frontiers in Catecholamine Research* (Usden, E. and Synder, S., eds.) Pergamon, New York.

Langer, S. Z. (1974) Presynaptic regulation of catecholamine release. *Biochem. Pharmacol.* **23**, 1783.1800.

Langer, S. Z. (1979) Presynaptic Adrenoceptors and Regulation of Release, in *The Release of Catecholamines from Adrenergic Neurons* (Paton, D. M., ed.) Pergamon, Oxford.

Langer, S. Z. (1981) Presynaptic regulation of the release of catecholamines. *Pharmacol. Rev.* **32**, 337–362.

Langer, S. Z. and Arbilla, S. (1981) Presynaptic receptors and modulation of the release of noradrenaline, dopamine and GABA. *Postgrad. Med. J.* **57** (suppl. 1), 18–29.

Langer, S. Z. and Hicks, P. E. (1984) Alpha-adrenoceptor subtypes in blood vessels: Physiology and pharmacology. *J. Cardiovasc. Pharmacol.* **6** (suppl. 4), S547–S558.

Langer, S. Z. and Shepperson, N. B. (1981) Antagonism of α-adrenoceptor mediated contractions of the isolated saphenous vein of the dog by diltizem and verapamil. *Br. J. Pharmacol.* **74**, 942P.

Langer, S. Z. and Shepperson, N. B. (1982a) Postjunctional α_1- and α_2-adrenoceptors: Preferential innervation of α_1-adrenoceptors and the role of neuronal uptake. *J. Cardiovasc. Pharmacol.* **4**, S8–S13.

Langer, S. Z. and Shepperson, N. B. (1982b) Recent developments in vascular smooth muscle pharmacology: The postsynaptic α_2-adrenoceptor. *Trends Pharmacol. Sci.* **3**, 440–444.

Langer, S. Z., Duval, N., and Massingham, R. (1985) Pharmacological and therapeutic significance of alpha adrenoceptor subtypes. *J. Cardiovasc. Pharmacol.* **7** (suppl. 8), S1–S8.

Langer, S. Z., Shepperson, N. B., and Massingham, R. (1981) Preferential noradrenergic innervation of α_1-adrenergic receptors in vascular smooth muscle. *Hypertension* **3** (suppl. 1), I112–I118.

Lasch, P. and Jakobs, K. H. (1979) Agonistic and antagonistic effects of various α-adrenergic agonists in human platelets. *Naunyn Schmiedebergs Arch. Pharmacol.* **306**, 119–125.

Laverty, R. (1969) A comparison of the behavioral effects of some hypotensive imidazoline derivatives in rats. *Eur. J. Pharmacol.* **9**, 163–169.

Laverty, R. and Taylor, K. M. (1969) Behavioral and biochemical effects of 2-(2,6-dichlorophenylamine)-2-imidazoline hydrochloride (St 155) on the central nervous system. *Br. J. Pharmacol.* **35**, 253–264.

Lefebvre, R. A., Willems, J. L., and Bogaert, M. G. (1984) Inhibitory effect of dopamine on canine gastric fundus. *Naunyn Schmiedebergs Arch. Pharmacol.* **326**, 22–28.

Lenox, R. H., Ellis, J., Riper, D. V., and Ehrlich, Y. H. (1985) α_2-Adrenergic receptor mediated regulation of adenylate cyclase in the intact human platelet. *Mol. Pharmacol.* **27**, 1–9.

Leppavuori, A. (1980) The effects of an α-adrenergic agonist or antagonist on sleep during blockade of catecholamine synthesis in the cat. *Brain Res.* **193**, 117–128.

Leppavuori, A. and Putkonen, P. T. S. (1980) Alpha-adrenoceptive influences on the control of the sleep waking cycle in the cat. *Brain Res.* **193**, 95–115.

Levine, T. B., Francis, G. S., and Goldsmith, S. R. (1982) Activity of the sympathetic nervous system assessed by plasma hormone levels and their relation to hemodynamic abnormalaties in congestive heart failure. *Am. J. Cardiol.* **49**, 1659–1666.

Levitt, B. and Hieble, J. P. (1985) Characterization of pre- and postjunctional α-adrenoceptors in rabbit lateral saphenous vein. *Fed. Proc.* **44**, 1465.

Lewis, S. J., Fennessy, M. R., and Taylor, D. A. (1981) Central aminergic mechanisms associated with the clonidine withdrawal syndrome. *Clin. Exp. Pharmacol. Physiol.* **8**, 489–495.

Limbird, L. E. (1984) GTP and Na^+ modulate receptor-adenyl cyclase coupling and receptor-mediated function. *Am. J. Physiol.* **247**, E59–E68.

Limbird, L. E. and Speck, J. L. (1983) N-ethylmaleimide, elevated temperature and digitonin solubilization eliminate guanine nucleotide but not sodium effects on human platelet α adrenergic receptor agonist interactions. *J. Cycl. Nucl. Phos. Res.* **9**, 191–201.

Lioy, F. (1967) An analysis of the mechanism of catecholamine effects on coronary circulation. *Am. J. Physiol.* **213**, 487–491.

Lopez, L. M. and Mehta, J. L. (1984) Comparative efficacy and safety of lofexidine and clonidine in mild to moderately severe systemic hypertension. *Am. J. Cardiol.* **53**, 787–790.

Lowenthal, D. T. (1980) Pharmacokinetics of clonidine. *J. Cardiovasc. Pharmacol.* **2**, S29–S37.

Madjar, H., Docherty, J. R., and Starke, K. (1980) An examination of pre- and postsynaptic α-adrenoceptors in the autoperfused rabbit hindlimb. *J. Cardiovasc. Pharmacol.* **2**, 619–627.

Malindzak, G. S., Kosinski, E. J., Green, H. D., and Yarborough, G. W. (1978) The effects of adrenergic stimulation on conductive and resistive segments of the coronary vascular bed. *J. Pharmacol. Exp. Ther.* **206**, 248–258.

Malta, E., McPherson, G. A., and Raper, C. (1979) Comparison of prejunctional α-adrenoceptors at the neuromuscular junction with vascular postjunctional α-receptors in cat skeletal muscle. *Br. J. Pharmacol.* **65**, 249–256.

Malta, E., Schini, V., and Miller, R. C. (1986) Role of efficacy in the assessment of the actions of α-adrenoceptor agonists in rat aorta with endothelium. *J. Pharm. Pharmacol.* **38**, 209–213.

Markiewicz, M., Marshall, I., and Nasmyth, P. A. (1980) Lack of feedback via pre-synaptic α-adrenoceptors by noradrenaline released by a single pulse. *Br. J. Pharmacol.* **69**, 343P–344P.

Martin, W., Furchgott, R. F., Villani, G. M., and Jothianandan, D. (1986) Depression of contractile responses in rat aorta by spontaneously released endothelium-derived relaxing factor. *J. Pharmacol. Exp. Ther.* **237**, 529–538.

Matsuda, H., Kuon, E., Holtz, J., and Busse, R. (1985) Endothelium-mediated dilations contribute to the polarity of the arterial wall in vasomotion induced by α_2-adrenergic agonists. *J. Cardiovasc. Pharmacol.* **7**, 680–688.

Matthews, W. D., McCafferty, G. P., and Grous, M. (1984a) Characterization of α-adrenoceptors on vascular smooth muscle: Electrophysiological differentiation in canine saphenous vein. *J. Pharmacol. Exp. Ther.* **231**, 355–360.

Matthews, W. D., Jim, K. F., Hieble, J. P., and DeMarinis, R. M. (1984b) Post-synaptic α-adrenoceptors on vascular smooth muscle. *Fed. Proc.* **43**, 2923–2928.

McAdams, R. P. and Waterfall, J. F. (1984) Functional subsensitivity of postjunctional α_2- but not α_1-adrenoceptors in the dog saphenous vein with reduction in temperature. *Br. J. Pharmacol.* **83**, 412P.

McArthur, K. E., Anderson, D. S., Durbin, T. E., Orloff, M. J., and Dharmsathaphorn, K. (1982) Clonidine and lidamidine to inhibit watery diarrhea in a patient with lung cancer. *Ann. Int. Med.* **96**, 323–325.

McCalden, T. A. (1981) Sympathetic control of the cerebral circulation. *J. Auton. Pharmacol.* **1**, 421–431.

McGrath, J. C., Flavahan, N. A., and McKean, C. E. (1982) Alpha$_1$ and α_2-adrenoceptor mediated pressor and chronotropic effects in the rat and rabbit. *J. Cardiovasc. Pharmacol.* **4** (suppl. 1), S101–S107.

McPherson, G. A. and Summers, R. J. (1981) ^{3}H-Prazosin and ^{3}H-clonidine binding to α-adrenoceptors in membranes prepared from regions of rat kidney. *J. Pharm. Pharmacol.* **33**, 189–191.

Medgett, I. C. and Rajanayagam, M. A. S. (1984) Effects of reduced calcium ion concentration and of diltiazem on vasoconstrictor responses to noradrenaline and sympathetic nerve stimulation in rat isolated tail artery. *Br. J. Pharmacol.* **83**, 889–898.

Medgett, I. C., McCulloch, M. W., and Rand, M. J. (1978) Partial agonist action of clonidine on prejunctional and postjunctional alpha-adrenoceptors. *Naunyn Schmiedebergs Arch. Pharmacol.* **304**, 215–221.

Meyer, D. R. and Sparber, S. B. (1976) Clonidine antagonizes body weight loss and other symptoms used to measure withdrawal in morphine pelleted rats given naloxone. *Pharmacologist* **18**, 236.

Michel, D., Zimmerman, W., Nassehi, A., and Seraphim, P. (1966) Erste Beobachtungen uber einen antihypertensiven Effekt von 2-(2,6-Dichlorphenyl-amino)-2-imidazolin hydrochlorid am Menschen. *Deutsch. med. Woschr.* **91**, 1540–1547.

Michell, R. H. (1979) Inositol phospholipids in membrane function. *Trends Biochem. Sci.* **4**, 128–131.

Miller, M. D. and Marshall, J. M. (1965) Uterine responses to nerve stimulation: Relation to hormonal status and catecholamines. *Am. J. Physiol.* **209**, 859–865.

Mills, D. C. B. (1975) Initial Biochemical Responses of Platelets to Stimulation, in *CIBA Foundation Symposium* no. 35, Elsevier/North Holland Biomedical, Amsterdam.

Mills, D. C. B. and Roberts, G. C. K. (1967) Effects of adrenaline on human blood platelets. *J. Physiol.* (Lond.) **193**, 443–453.

Misu, Y. and Kubo, T. (1982) Central and peripheral cardiovascular responses of rats to guanabenz and clonidine. *Jpn. J. Pharmacol.* **32**, 925–928.

Mottram, D. R. and Thakar, Y. (1984) The action and interaction β-phenethylamines and imidazolines on prejunctional α_2-adrenoceptors of guinea pig ileum in the presence of the non-competitive antagonist benextramine. *J. Pharm. Pharmacol.* **36**, 668–672.

Motulsky, H. J., Shattil, S. J., Ferry, N., Rozansky, D., and Insel, P. A. (1986) Desensitization of epinephrine-initiated platelet aggregation does not alter binding to the α_2-adrenergic receptor or receptor coupling to adenylate cyclase. *Mol. Pharmacol.* **29**, 1–8.

Muller-Schweinitzer, E. (1984) Alpha-adrenoceptors, 5-hydroxytryptamine receptors and the action of dihydroergotamine in human venous preparations obtained during saphenectomy procedures for varicose veins. *Naunyn Schmiedebergs Arch. Pharmacol.* **327**, 299–303.

Muraki, T., Nakai, T., and Kato, R. (1984) Predominance of α_2-adrenoceptors in porcine thyroid: Biochemical and pharmacological correlations. *Endocrinology* **114**, 1645–1651.

Myers, R. D. and Waller, M. B. (1975) 5HT and norepinephrine-induced release of acetylcholine from hypothalamus and mesencephalon of the monkey during thermoregulation. *Brain Res.* **84**, 47–61.

Myers, K. M., Huston, L. Y., and Clemmons, R. M. (1983) Regulation of canine platelet function. II. Catecholamines. *Am. J. Physiol.* **245**, R100–R109.

Nakadate, T., Nakaki, T., Muraki, T., and Kato, R. (1980) Adrenergic regulation of blood glucose levels: Possible involvement of postsynaptic alpha-2 type adrenergic receptors regulating insulin release. *J. Pharmacol. Exp. Ther.* **215**, 226–230.

Nakaki, T., Nakadate, T., and Kato, R. (1980) α-Adrenoceptors modulating insulin release from isolated pancreatic islet cells. *Naunyn Schmiedebergs Arch. Pharmacol.* **313**, 151–153.

Nakaki, T., Nakadate, T., Yamamoto, S., and Kato, R. (1982a) α_2-Adrenoceptors inhibit the cholera-toxin-induced intestinal fluid accumulation. *Naunyn Schmiedebergs Arch. Pharmacol.* **318**, 181–184.

Nakaki, T., Nakadate, T., Yamamoto, S., and Kato, R. (1982b) α_2-Adrenergic inhibition of intestinal secretion induced by prostaglandin E_2, vasoactive intestinal peptide and dibutyryl cyclic AMP in rat jejunum. *J. Pharmacol. Exp. Ther.* **220**, 637–641.

Nakaki, T., Nakadate, T., Yamamoto, S., and Kato, R. (1983) α_2-Adrenergic receptor in intestinal epithelial cells. Identification by [^{3}H]yohimbine and failure to inhibit cyclic AMP accumulation. *Mol. Pharmacol.* **23**, 228–234.

Nestler, E. J. and Greegard, P. (1982) Nerve impulses increase the phosphorylation state of protein I in rabbit superior cervical ganglion. *Nature* **296**, 452–454.

Nichols, A. J. (1985) Pharmacological studies on the hepatosplanchnic circulation of the rat. PhD thesis, University of Cambridge, Cambridge.

Nichols, A. J. and Hiley, C. R. (1985) Identification of adrenoceptors and dopamine receptors mediating vascular responses in the superior mesenteric arterial bed of the rat. *J. Pharm. Pharmacol.* **37**, 110–115.

Nichols, A. J. and Ruffolo, R. R., Jr. (1986) The relationship between alterations in α_1-adrenoceptor reserve by phenoxybenzamine and benextramine and the sensitivity of cirazoline-induced pressor responses to inhibition by nifedipine. *Eur. J. Pharmacol.* **126**, 297–301.

O'Brien, J. R. (1963) Some effects of adrenaline and anti-adrenaline compounds on platelets in vitro and in vivo. *Nature* **200**, 763–764.

Ohlstein, E. H., Shebuski, R. J., and Ruffolo, R. R., Jr. (1986) Localization of α_2-adrenoceptors in the canine pulmonary vasculature. *Pharmacologist* **28**, 141.

Olsen, U. B. (1976) Clonidine-induced increase of renal prostaglandin activity and water diuresis in conscious dogs. *Eur. J. Pharmacol.* **36**, 95–101.

Onesti, G., Schwartz, A. B., and Kim, K. E. (1971) Antihypertensive effect of clonidine. *Circ. Res.* **28** (suppl. 2), 53–69.

Oriowo, M. A., Bevan, R. D., and Bevan, J. A. (1985) Abrupt decrease in sensitivity at the origin of the rabbit internal iliac artery to norepinephrine but not to histamine and 5-hydroxytryptamine. *Fed. Proc.* **44**, 883.

Osborn, J. L., DiBona, G. F., and Thames, M. D. (1982) Role of renal α-adrenoceptor mediating renin secretion. *Am. J. Physiol.* **242**, F620–F626.

Osborn, J. L., Holdaas, H., Thames, M. D., and Di Bona, G. F. (1983) Renal adrenoceptor mediation of antinatriuretic and renin secretion responses to low frequency renal nerve stimulation in the dog. *Circ. Res.* **53**, 298–305.

Osnes, J.-B. (1976) Positive inotropic effect without cyclic AMP elevation after α-adrenergic stimulation of perfused hearts from hypothyroid rats. *Acta Pharmacol. Tox.* **398**, 232–240.

Oswald, H. and Greven, J. (1981) Effects of Adrenergic Activators and Inhibitors on Kidney Function, in *Handbook of Experimental Pharmacology* vol. 54 (Szekeres, L., ed.) Springer Verlag, Berlin.

Owen, M. P. and Bevan, J. A. (1985) Influence of norepinephrine uptake on neurogenic vasoconstriction is greater in smaller compared with larger arteries of the rabbit ear. *Fed. Proc.* **44**, 1733.

Paalzow, L. (1974) Analgesia produced by clonidine in mice and rats. *J. Pharm. Pharmacol.* **26**, 361–363.

Paalzow, G. and Paalzow, L. (1976) Clonidine antinociceptive activity: Effects of drugs influencing central monoaminergic and cholinergic mechanisms in the rat. *Naunyn Schmiedebergs Arch. Pharmacol.* **292**, 119–126.

Parratt, J. R. (1969) The effect of adrenaline, noradrenaline, and propranolol on myocardial blood flow and metabolic heat production in monkeys and baboons. *Cardiovasc. Res.* **3**, 306–314.

Patel, P., Bose, D., and Greenway, C. (1981) Effects of prazosin and phenoxybenzamine on α- and β-receptor mediated responses in intestinal resistance and capacitance vessels. *J. Cardiovasc. Pharmacol.* **3**, 1050–1059.

Paton, W. D. M. and Vizi, E. S. (1969) The inhibitory action of noradrenaline and adrenaline on acetylcholine output by guinea pig ileum longitudinal muscle strip. *Br. J. Pharmacol.* **35**, 10–28.

Pelayo, F., Dubocovich, M. L., and Langer, S. Z. (1980) Inhibition of neuronal uptake reduces the presynaptic effects of clonidine but not of α-methylnoradrenaline on the stimulation-evoked release of ^{3}H-noradrenaline from rat occipital cortex slices. *Eur. J. Pharmacol.* **64**, 143–155.

Pert, C. B., Kuhar, M. J., and Snyder, S. H. (1976) Opiate receptor: Autoradiographic localization in rat brain. *Proc. Natl. Acad. Sci. USA* **73**, 3729–3733.

Pettibone, D. J. and Mueller, G. P. (1981a) Clonidine releases immunoreactive beta endorphin from rat pars distalis. *Brain Res.* **221**, 409–414.

Pettibone, D. J. and Mueller, G. P. (1981b) α-Adrenergic stimulation by clonidine increases plasma concentration of immunoreactive beta-endorphin in rats. *Endocrinology* **109**, 798–802.

Pettinger, W. (1987) Renal α_2-adrenergic receptors and hypertension. *Hypertension* **9**, 3–6.

Petty, M. A. and de Jonge, W. (1982) Does beta-endorphin contribute to the central antihypertensive action of α-methyldopa in rats? *Clin. Sci.* **63**, 293s–295s.

Pickel, V. (1982) Central noradrenergic neurons: Identification, distribution and synaptic interactions with axons containing morphine-like peptides. *J. Clin. Psychiatry* **43**, 13–16.

Pickel, V. M., Joh, T. S., Reis, D. J., Leeman, S. E., and Miller, R. J. (1979) Electron microscopic localization of substance P and enkephalin in axon terminals related to dendrites of catecholaminergic neurons. *Brain Res.* **160**, 387–400.

Pitt, B., Elliot, E. C., and Gregg, D. E. (1967) Adrenergic receptor activity in the coronary arteries of the unanesthetized dog. *Circ. Res.* **21**, 75–84.

Planitz, V. (1984) Crossover comparison of moxonidine and clonidine in mild to moderate hypertension. *Eur. J. Clin. Pharmacol.* **27**, 147–152.

Porte, D., Jr., Girardier, L., Seydoux, J., Kanazawa, Y., and Posternak, J. (1973) Neural regulation of insulin secretion in the dog. *J. Clin. Invest.* **52**, 210–214.

Proctor, E. (1968) The effects of physiological concentrations of noradrenaline on the coronary resistance of isolated perfused hearts of the cat, dog, and monkey. *J. Pharm. Pharmacol.* **20**, 36–40.

Purdy, R. E. and Stupecky, G. L. (1984) Characterization of the α-adrenergic properties of rabbit ear artery and thoracic aorta. *J. Pharmacol. Exp. Ther.* **229**, 459–468.

Raisman, R., Brile, M., and Langer, S. Z. (1979) Specific labeling of postsynaptic α_1-adrenoceptors in rat heart ventricle by ^{3}H-WB-4101. *Arch. Pharmacol.* **307**, 223–226.

Ramirez-Gonzalez, M. D., Tchakarov, L., Mosque-da-Garcia, R., and Kunos, G. (1983) β-Endorphin acting on the brainstem is involved in the antihypertensive action of clonidine and α-methyldopa in rats. *Circ. Res.* **53**, 150–157.

Rand, M. J., Story, D. F., Allen, G. S., Glover, A. B., and McCulloch, M. W. (1973) Pulse-to-Pulse Modulation of Noradrenaline Release Through a Prejunctional α-Receptor Auto-inhibitory Mechanism, in *Frontiers in Catecholamine Research* (Usdin, E. and Snyder, S. H., eds.) Pergamon, Oxford.

Redmond, D. E. and Huang, Y. H. (1982) The primate locus ceruleus and effects of clonidine on opiate withdrawal. *J. Clin. Psychiatry* **43**, 25–29.

Reese, J. B. and Matthews, W. D. (1986) α-Adrenergic agonists stimulate phosphatidylinositol hydrolysis in canine saphenous vein. *Pharmacologist* **28**, 161.

Reichenbacher, D., Reimann, W., and Starke, K. (1982) Alpha-adrenoceptor-mediated inhibition of noradrenaline release in rabbit brain cortex slices. *Naunyn Schmiedebergs Arch. Pharmacol.* **319**, 71–77.

Reid, J. L. (1985) Central α_2 receptors and the regulation of blood pressure in humans. *J. Cardiovasc. Pharmacol.* **7** (suppl. 8), S45–S50.

Reid, J. L., Dargie, H. J., Davies, D. S., Wing, L. M. H., Hamilton, C. A., and Dollery, C. T. (1977) Clonidine withdrawal in hypertension. *Lancet* i, 1171–1174.

Reid, J. L., Rubin, P. C., and Howden, C. W. (1983a) Central

α_2-adrenoceptors and blood pressure regulation in man: Studies with guanfacine (BS 100-141) and azepexole (B-HT 933). *Br. J. Clin. Pharmacol.* **15** (suppl. 4), 463–469.

Reid, J. L., Hamilton, C. A., and Hannah, J. A. M. (1983b) Peripheral α_1- and α_2-adrenoceptor mechanisms in blood pressure control. *Chest* **83**, 302–304.

Reid, I. A., Nolan, P. L., Wolf, J. A., and Keil, L. C. (1979) Suppression of vasopressin release by clonidine: Effect of α-adrenoceptor antagonists. *Endocrinology* **104**, 1403–1406.

Roberts, J. M., Insel, P. A., Goldfien, R. D., and Goldfien, A. (1977) α-Adrenoceptors but not β-adrenoceptors increase in rabbit uterus with oestrogen. *Nature* **270**, 624–625.

Robison, G. A., Arnold, A., and Hartmann, R. C. (1969) Cyclic AMP as a second messenger. *Pharmacol. Res. Commun.* **1**, 325–332.

Rogers, J. F. and Cubeddu, L. X. (1983) Naloxone does not antagonize the antihypertensive effect of clonidine in essential hypertension. *Clin. Pharmacol. Ther.* **34**, 68–73.

Roman, R. J., Cowley, A. W., Jr., and Lecherie, C. (1979) Water diuretic and natriuretic effect of clonidine in the rat. *J. Pharmacol. Exp. Ther.* **221**, 385–393.

Rosen, S. G., Berk, M. A., Popp, D. A., Serusclat, P., Smith, E. B., Shah, S. D., Ginsberg, A. M., and Clutter, W. E. (1984) β- and α-Adrenergic receptors and receptor coupling to adenylate cyclase in human mononuclear leukocytes and platelets in relation to physiological variations of sex steroids. *J. Clin. Endocrinol. Metab.* **58**, 1068–1076.

Rosendorff, C., Mitchell, G., Scriven, D. R., and Shapiro, C. (1976) Evidence for a dual innervation affecting local blood flow in the hypothalamus of the conscious rabbit. *Circ. Res.* **38**, 140–145.

Rossi, E. C. (1978) Interactions Between Epinephrine and Platelets, in: *Platelets: A Multidisciplinary Approach* (deGaetano, G. and Garattini, S., eds.) Raven, New York.

Rossi, E. C. and Louis, G. (1975) A time-dependent increase in the responsiveness of platelet rich plasma to epinephrine. *J. Lab. Clin. Med.* **85**, 300–306.

Roth, R. H., Elsworth, J. D., and Redmond, D. E. (1982) Clonidine suppression of noradrenergic hyperactivity during morphine withdrawal by clonidine: Biochemical studies in rodents and primates. *J. Clin. Psychiatry* **43**, 42–46.

Rubin, P. C., Howden, C. W., McLean, K., and Reid, J. L. (1982) Pharmacodynamic studies with a specific α_2-adrenoceptor agonist (BHT-933) in man. *J. Cardiovasc. Pharmacol.* **4**, 527–530.

Ruffolo, R. R. (1982) Important concepts of receptor theory. *J. Auton. Pharmacol.* **2**, 277–295.

Ruffolo, R. R., Jr. (1984a) α-Adrenoceptors. *Monogr. Neural Sci.* **10**, 224–253.

Ruffolo, R. R., Jr. (1984b) Interaction of agonists with peripheral α-adrenergic receptors. *Fed. Proc.* **43**, 2910–2916.

Ruffolo, R. R., Jr. (1984c) Stereochemical requirements for activation and blockade of α_1 and α_2-adrenoceptors. *Trends Pharmacol. Sci.* **5**, 160–164.

Ruffolo, R. R., Jr. (1985a) Relative agonist potency as a means of differentiating α-adrenoceptors and α-adrenergic mechanisms. *Clin. Sci.* **68** (suppl. 10), 9s–14s.

Ruffolo, R. R., Jr. (1985b) Distribution and function of peripheral α-adrenoceptors in the cardiovascular system. *Pharmacol. Biochem. Behav.* **22**, 827–833.

Ruffolo, R. R., Jr. and Waddell, J. E. (1982) Stereochemical requirement of α_2-adrenergic receptors for α-methyl substituted phenethylamines. *Life Sci.* **31**, 2999–3007.

Ruffolo, R. R., Jr. and Yaden, E. L. (1984) Existence of spare α_1-adrenoceptors, but not α_2-adrenoceptors, for respective vasopressor effects of cirazoline and B-HT 933 in the pithed rat. *J. Cardiovasc. Pharmacol.* **6**, 1011–1019.

Ruffolo, R. R., Jr. and Zeid, R. L. (1985) Relationship between α-adrenoceptor occupancy and response for the α_1-adrenoceptor agonist, cirazoline, and the α_2-adrenoceptor agonist, B-HT 933, in canine saphenous vein. *J. Pharmacol. Exp. Ther.* **235**, 636–643.

Ruffolo, R. R., Jr., Rosing, E. L., and Waddell, J. E. (1979) Receptor interactions of imidazolines. I. Affinity and efficacy for alpha adrenergic receptors in rat aorta. *J. Pharmacol. Exp. Ther.* **211**, 733–738.

Ruffolo, R. R., Yaden, E. L., and Waddell, J. E. (1982a) Stereochemical requirements of alpha-2 adrenergic receptors. *J. Pharmacol. Exp. Ther.* **222**, 645–651.

Ruffolo, R. R., Yaden, E. L., and Ward, J. S. (1982b) Receptor interactions of imidazolines. Influence of ionization constant on the diffusion of clonidine and a series of structurally related imidazolines into and out of the central nervous system. *Eur. J. Pharmacol.* **81**, 367–375.

Ruffolo, R. R., Yaden, E. L., Waddell, J. E., and Ward, J. S. (1982c) Receptor interactions of imidazolines. X1. α-Adrenergic and antihypertensive effects of clonidine and its methylene-bridged analog, St 1913. *Pharmacology* **25**, 187–201.

Ruffolo, R. R., Jr., Goldberg, M. R., and Morgan, E. L. (1984a) Interactions of epinephrine, dopamine, and their corresponding α-methyl-substituted derivatives with α- and β-adrenoceptors in the pithed rat. *J. Pharmacol. Exp. Ther.* **230**, 595–600.

Ruffolo, R. R., Jr., Morgan, E. L., and Messick, K. (1984b) Possible relationship between receptor reserve and the differential antagonism of α_1 and α_2-adrenoceptor mediated pressor responses by calcium channel antagonists. *J. Pharmacol. Exp. Ther.* **230**, 587–594.

Ruffolo, R. R., Jr., Yaden, E. L., Timmermans, P. B. M. W. M., van Zwieten, P. A., and Hynes, M. D. (1984c) Characterization of the α-adrenoceptor-mediated effects and antihypertensive activity of ICI 106270: Comparison with clonidine. *J. Pharmacol. Exp. Ther.* **229**, 58–66.

Ruffolo, R. R., Messick, K., and Horng, J. S. (1985) Interactions of dimethoxy-substituted tolazoline derivatives with alpha-1 and alpha-2 adrenoceptors *in vitro*. *J. Auton. Pharmacol.* **5**, 71–79.

Sabol, S. L. and Nirenberg, M. (1979) Regulation of adenylate cyclase of neuroblastoma × glioma hybrid cells by α-adrenergic receptors. I. Inhibition of adenylate cyclase mediated by α-receptors. *J. Biol. Chem.* **254**, 1913–1920.

Sakakibara, Y., Fujiwars, M., and Muramatsu, I. (1982) Pharmacological characterization of the α-adrenoceptors of the dog basilar artery. *Naunyn Schmiedebergs Arch. Pharmacol.* **319**, 1–7.

Sakurai, S., Wada, A., Izumi, F., Kobayashi, H., and Yanagihara, N. (1983)

Inhibition by α_2-adrenoceptor agonists of the secretion of catecholamines from isolated adrenal medullary cells. *Naunyn Schmiedebergs Arch. Pharmacol.* **324**, 15–19.

Salin-Pascual, R., de la Fuente, J., and Fernandez-Guardiola, A. (1985) Effects of clonidine in narcolepsy. *Clin. Psychiatry* **46**, 528–531.

Sanchez, A. and Pettinger, W. A. (1981) Dietary sodium regulation of blood pressure and renal α_1- and α_2-receptors in WKY and SHR. *Life Sci.* **29**, 2795–2802.

Sastre, A., Griendling, K. K., Rusher, M. M., and Milnor, W. R. (1984) Relation between α-adrenergic receptor occupation and contractile response: Radioligand and physiologic studies in canine aorta. *J. Pharmacol. Exp. Ther.* **229**, 887–896.

Sawyer, R., Warnock, P., and Docherty, J. R. (1985) Role of vascular α_2-adrenoceptors as targets for circulating catecholamines in the maintenance of blood pressure in anesthetized spontaneously hypertensive rats. *J. Cardiovasc. Pharmacol.* **7**, 809–812.

Schumann, H. J. and Brodde, O.-E. (1979) Demonstration of α-adrenoceptors in the rabbit heart by [^{3}H]-dihydroergocrytine binding. *Arch. Pharmacol.* **308**, 191–198.

Schumann, H. J. and Endoh, M. (1976) α-Adrenoceptors in the ventricular myocardium: Clonidine, naphazoline and methoxamine as partial α-agonists exerting a competitive dualism in action to phenylephrine. *Eur. J. Pharmacol.* **36**, 413–421.

Schumann, H. J., Endoh, M., and Brodde, O.-E. (1975) The time course of the effects of β- and α-adrenoceptor stimulation by isoprenaline and methoxamine on the contractile force and cAMP level of the isolated rabbit papillary muscle. *Arch. Pharmacol.* **289**, 291–302.

Scrutton, M. C., Clare, K. A., Hutton, R. A., and Bruckdorfer, K. R. (1981) Depressed responsiveness to adrenaline in platelets from apparently normal human donors: A familial trait. *Br. J. Haematol.* **49**, 303–341.

Segstro, R. and Greenway, C. (1986) α-Receptor subtype mediating sympathetic mobilization of blood from the hepatic venous system in anesthetized cats. *J. Pharmacol. Exp. Ther.* **236**, 224–229.

Seno, N., Nakazato, Y, and Ohga, A. (1978) Presynaptic inhibitory effects of catecholamines on cholinergic transmission in the smooth muscle of the chick stomach. *Eur. J. Pharmacol.* **51**, 229–237.

Schmitt, H. (1971) Action des α-sympathomimetiques sur les structures nerveuses. *Actual. Pharmacol.* **24**, 93–131.

Schmitt, H. and Schmitt, H. (1969) Localization of the hypotensive effect of 2-(2,6-dichlorophenylamino)-2-imidazoline hydrochloride. *Eur. J. Pharmacol.* **6**, 8–12.

Schmitz, J. M., Graham, K. M., Saglowsky, A., and Pettinger, W. A. (1981) Renal α_1- and α_2-adrenergic receptors: Biochemical and pharmacological correlations. *J. Pharmacol. Exp. Ther.* **219**, 400–406.

Schoffelmeer, A. N. M., Wemer, J., and Mulder, A. H. (1981) Comparison between electrically-evoked and potassium-induced ^{3}H-noradrenaline release from rat neocortex slices: Role of calcium ions and transmitter pools. *Neurochem. Int.* **3**, 129–136.

Schoffelmeer, A. N. M. and Mulder, A. H. (1983) ^{3}H-Noradrenaline release from rat neocortical slices in the absence of extracellular calcium and its presynaptic α_2-adrenergic modulation. A study on the possible role of cyclic AMP. *Naunyn Schmiedebergs Arch. Pharmacol.* **323**, 188–192.

Scholtysik, G. (1980) Pharmacology of guanfacine. *Br. J. Clin. Pharmacol.* **10**, 21S–24S.

Scholtysik, G., Lauener, H., Eichenberger, E., Burki, H., Salzmann, R., Muller-Schweinitzer, E., and Waite, R. (1975) Pharmacological actions of the antihypertensive agent N-amidino-2-(2,6-dichlorophenyl) acetamide hydrochloride (BS 100-141). *Arzneimittelforsch.* **25**, 1483–1491.

Schultz, H. S., Chertien, S. D., Brewer, D. D., Eltorai, M. T., and Weber, M. A. (1981) Centrally acting antihypertensive agents: A comparison of lofexidine with clonidine. *J. Clin. Pharmacol.* **21**, 65–71.

Shearman, G. T., Lal, H., and Ursillo, R. C. (1980) Effectiveness of lofexidine in blocking morphine-withdrawal signs in the rat. *Pharmacol. Biochem. Behav.* **12**, 573–575.

Shebuski, R. J., Fujita, T., and Ruffolo, R. R., Jr. (1986) Evaluation of α_1- and α_2-adrenoceptor-mediated vasoconstriction in the *in situ* autoperfused, pulmonary circulation of the anesthetized dog. *J. Pharmacol. Exp. Ther.* **238**, 217–223.

Shebuski, R. J., Ohlstein, E. H., Smith, J. M., Jr., and Ruffolo, R. R., Jr. (1987) Enhanced pulmonary α_2-adrenoceptor responsiveness under conditions of elevated pulmonary vascular tone. *J. Pharmacol. Exp. Ther.*, **242**, 158–165.

Shepperson, N. B., Duval, N., and Langer, S. Z. (1982) Dopamine decreases mesenteric blood flow in the anesthetized dog through stimulation of postsynaptic α_2-adrenoceptors. *Eur. J. Pharmacol.* **81**, 627–635.

Shoji, T., Tsuru, H., and Shigei, T. (1983) A regional difference in the distribution of postsynaptic α-adrenoceptor subtypes in canine veins. *Naunyn Schmiedebergs Arch. Pharmacol.* **324**, 246–255.

Simon, P., Chermat, R., and Boissier, J. R. (1975) Comparison des effets sedatifs de deux α-sympathomimetiques antihypertenseurs: La clonidine et la tiamenidine. *Therapie* **30**, 855–861.

Skarby, T. (1984) Pharmacological properties of prejunctional α-adrenoceptors in isolated feline middle cerebral arteries; comparison with the postjunctional α-adrenoceptors. *Acta Physiol. Scand.* **122**, 165–174.

Skarby, T. V. C., Anderson, K. E., and Edvinsson, L. (1983) Pharmacological characterization of postjunctional α-adrenoceptors in isolated feline cerebral and peripheral arteries. *Acta. Physiol Scand.* **117**, 63–73.

Smith, S. K. and Limbird, L. E. (1981) Solubilization of human platelet α-adrenergic receptors: Evidence that agonist occupancy of the receptor stabilizes receptor-effector interactions. *Proc. Natl. Acad. Sci. USA* **78**, 4026–4030.

Smith, P. H. and Porte, D., Jr. (1976) Neuropharmacology of the pancreatic islets. *Ann. Rev. Pharmacol. Toxicol.* **16**, 269–285.

Smyth, D. D., Umemura, S., and Pettinger, W. A. (1984) Alpha$_2$-adrenoceptors and sodium reabsorption in the isolated perfused rat kidney. *Am. J. Physiol.* **247**, F680–F685.

Smyth, D. D., Umemura, S., and Pettinger, W. A. (1985a) α_2-Adrenoceptor antagonism of vasopressin-induced changes in sodium excretion. *Am. J. Physiol.* **248**, F767–F772.

Smyth, D. D., Umemura, S., and Pettinger, W. A. (1985b) Renal nerve stimulation causes α_1-adrenoceptor-mediated sodium retention but not α_2-adrenoceptor antagonism of vasopressin. *Circ. Res.* **57**, 304–311.

Spiegel, R. and Devos, J. E. (1980) Central effects of guanfacine and clonidine during wakefullness and sleep in healthy subjects. *Br. J. Clin. Pharmacol.* **10**, 165S–168S.

Srimal, R. C., Gulati, K., and Dhawan, B. N. (1977) On the mechanism of the central hypotensive action of clonidine. *Can. J. Physiol. Pharmacol.* **55**, 1007–1014.

Starke, K. (1977) Regulation of noradrenaline release by presynaptic receptor systems. *Rev. Physiol. Biochem. Pharmacol.* **77**, 1–124.

Starke, K. (1981) Alpha-adrenoceptor subclassification. *Rev. Physiol. Biochem. Pharmacol.* **88**, 199–236.

Starke, K. and Montel, H. (1973) Involvement of alpha-receptors in clonidine-induced inhibition of transmitter release from central monoamine neurons. *Neuropharmacology* **12**, 1073–1080.

Starke, K., Montel, H., and Schumann, H. J. (1971a) Influence of cocaine and phenoxybenzamine on noradrenaline uptake and release. *Naunyn Schmiedebergs Arch. Pharmacol.* **270**, 210–214.

Starke, K., Montel, H., and Wagner, J. (1971b) Effect of phentolamine on noradrenaline uptake and release. *Naunyn Schmiedebergs Arch. Pharmacol.* **271**, 181–192.

Stephenson, J. A. and Summers, R. J. (1985) Light microscopic autoradiography of the distribution of [^{3}H]rauwolscine binding to α-adrenoceptors in rat kidney. *Eur. J. Pharmacol.* **116**, 271–278.

Stevens, M. J. and Moulds, R. F. W. (1982) Are the pre- and postsynaptic α-adrenoceptors in human vascular smooth muscle atypical? *J. Cardiovasc. Pharmacol.* **4** (suppl. 1), S129–S133.

Stjarne, L. (1978) Facilitation and receptor-mediated regulation of noradrenaline secretion by control of recruitment of varicosities as well as by control of electro-secretory coupling. *Neuroscience* **3**, 1147–1155.

Stjarne, L. (1979) Presynaptic α-receptors do not depress the secretion of ^{3}H-noradrenaline induced by veratridine. *Acta. Physiol. Scand.* **106**, 379–381.

Stjarne, L., Alberts, P., and Bartfai, T. (1979) Models of Regulation of Norepinephrine Secretion by Prejunctional Receptors and by Facilitation: Role of Calcium and Cyclic Nucleotides, in *Catecholamines: Basic and Clinical Frontiers* vol 1. (Usdin, E., Kopin, I. J., and Barchas, J., eds.) Pergamon, Oxford.

Story, D. F., McCulloch, M. W., Rand, M. J., and Standford-Starr, C. A. (1981) Conditions required for the inhibitory feed-back loop in noradrenergic transmission. *Nature* **293**, 62–65.

Strandhoy, J. W., Morris, M., and Buckalew, V. M., Jr. (1982) Renal effects of the antihypertensive, guanabenz, in the dog. *J. Pharmacol. Exp. Ther.* **221**, 347–352.

Sulahke, P. V. and St. Louis, P. J. (1980) Passive and active calcium fluxes across plasma membranes. *Prog. Biophys. Mol. Biol.* **35**, 135–195.

Sullivan, A. T. and Drew, G. M. (1980) Pharmacological characterization of pre- and postsynaptic α-adrenoceptors in dog saphenous vein. *Naunyn Schmiedebergs Arch. Pharmacol.* **314**, 249–258.

Sulpizio, A. C. and Hieble, J. P. (1985) The effect of calcium channel blockade on the dose–response relationship for B-HT 920 in the canine saphenous vein. *Pharmacologist* **27**, 205.

Sulpizio, A. C. and Hieble, J. P. (1987) Demonstration of α$_2$-adrenoceptor mediated contraction in the isolated canine saphenous artery treated with BAY k 8644. *Eur. J. Pharmacol.* **135**, 107–110.

Summers, R. J. (1984) Renal α-adrenoceptors. *Fed. Proc.* **43**, 2917–2922.

Summers, R. J. and McPherson, G. A. (1982) Radioligand studies of α-adrenoceptors in the kidney. *Trends Pharmacol. Sci.* **3**, 291–294.

Swart, S. S., Maguire, M., Wood, J. K., and Barnett, D. B. (1985) α_2-Adrenoceptor coupling to adenylate cyclase in adrenaline insensitive human platelets. *Eur. J. Pharmacol.* **116**, 113–119.

Swart, S. S., Pearson, D., Wood, J. K., and Barnett, D. B. (1984a) Functional significance of the platelet α_2-adrenoceptor: Studies in patients with myoproliferative disorders. *Thromb. Res.* **33**, 531–541.

Swart, S. S., Pearson, D., Wood, J. K., and Barnett, D. B. (1984b) Human platelet α_2-adrenoceptors: Relationship between radioligand binding studies and adrenaline-induced aggregation in normal individuals. *Eur. J. Pharmacol.* **103**, 25–32.

Sweatt, J. D., Johnson, S. L., Cragoe, E. J., and Limbird, L. E. (1985) Inhibitors of Na^+/H^+ exchange block stimulus-provoked arachidonic acid release in human platelets. *J. Biol. Chem.* **260**, 12910–12918.

Sweatt, J. D., Blair, I. A., Cragoe, E. J., and Limbird, L. E. (1986a) Inhibitors of Na^+/H^+ exchange block epinephrine- and ADP-induced stimulation of human platelet phospholipase C. by blockage of arachidonic acid release at a prior step. *J. Biol. Chem.* **261**, 8660–8666.

Sweatt, J. D., Connolly, T. M., Cragoe, E. J., and Limbird, L. E. (1986b) Evidence that Na^+/H^+ exchange regulates receptor-mediated phospholipase A_2 activation in human platelets. *J. Biol. Chem.* **261**, 8667–8673.

Taylor, J. A. and Mir, G. N. (1982) α-Adrenergic receptors and gastric function. *Drug Dev. Res.* **2**, 105–122.

Thomas, D. P. (1967) Effects of catecholamines on platelet aggregation caused by thrombin. *Nature* **215**, 298–299.

Thoolen, M. J. M. C., Timmermans, P. B. M. W. M., and van Zwieten, P. A. (1983) Cardiovascular effects of withdrawal of some centrally acting antihypertensive drugs in the rat. *Br. J. Clin. Pharmacol.* **15**, 491S–505S.

Timmermans, P. B. M. W. M. and Thoolen, M. J. M. C. (1987) Ca^{2+} *Utilization in Signal Transformation of α_1-Adrenergic Receptors* (Ruffolo, R. R., Jr., ed.) Humana, Clifton, New Jersey.

Timmermans, P. B. M. W. M. and van Zwieten, P. A. (1978) Dissociation constants of clonidine and structurally related imidazolines. *Arzneimittelforsch.* **28**, 1676–1681.

Timmermans, P. B. M. W. M., Brands, A., and van Zwieten, P. A. (1977) Lipophilicity and brain disposition of clonidine and structurally related imidazolidines. *Naunyn Schmiedebergs Arch. Pharmacol.* **300**, 217–226.

Timmermans, P. B. M. W. M., Kwa, H. Y., and van Zwieten, P. A. (1979) Possible subdivision of postsynaptic α-adrenoceptors mediating pressor responses in the pithed rat. *Naunyn Schmiedebergs Arch. Pharmacol.* **310**, 189–193.

Timmermans, P. B. M. W. M., Hoefke, W., Stahle, H., and van Zwieten, P. A. (1980) Structure-activity relationships in clonidine-like imidazolidines and related compounds. *Prog. Pharmacol.* **3**, 1–104.

Timmermans, P. B. M. W. M., deJonge, A., van Meel, J. C. A., Slothorst-Grisdijk, F. P., Lam, E., and van Zwieten, P. A. (1981a) Characterization of α-adrenoceptor populations. Quantitative relationships between cardiovascular effects initiated at central and peripheral α-adrenoceptors. *J. Med. Chem.* **24**, 502–507.

Timmermans, P. B. M. W. M., Schoop, A. M. C., Kwa, H. Y., and van
    Zwieten, P. A. (1981b) Characterization of alpha-adrenoceptors
    participating in the central hypotensive and sedative effects of clonidine
    using yohimbine, rauwolscine and corynanthine. Eur. J. Pharmacol. 70,
    7–15.
Timmermans, P. B. M. W. M., deJonge, A., van Meel, J. C. A., Mathy, M.
    J., and van Zwieten, P. A. (1983) Influence of nifedipine on functional
    responses in vivo initiated at α_2-adrenoceptors. J. Cardiovasc. Pharmacol.
    5, 1–11.
Toda, N. (1983) Alpha adrenergic receptor subtypes in human, monkey and
    dog cerebral arteries. J. Pharmacol. Exp. Ther. 226, 861–868.
Turner, J. T., Ray-Prenger, C., and Bylund, D. B. (1985) Alpha-2 adrenergic
    receptors in the human cell line, HT29: Characterization with the full
    agonist radioligand, [^{3}H]UK-14,304 and inhibition of adenylate cyclase.
    Mol. Pharmacol. 28, 422–430.
Umemura, S., Marver, D., Smyth, D. M., and Pettinger, W. A. (1985)
    α_2-Adrenoceptors and cellular cAMP levels in single nephron segments
    from the rat. Am. J. Physiol. 249, F28–F33.
Usui, H., Fujiwara, M., Tsukahara, T., Taniguchi, T., and Kurahashi, K.
    (1985) Differences in contractile responses to electrical stimulation and
    alpha-adrenergic binding sites in isolated cerebral arteries of humans,
    cows, dogs and monkeys. J. Cardiovasc. Pharmacol. 7 (suppl. 3), S47–S52.
Valdman, A. V., Medvedev, O. S., and Rozhanskaya, N. I. (1981) Naloxone
    blocks the antihypertensive effect of clonidine on hypertensive and
    normotensive animals. Bull. Exp. Biol. Med. 11, 560–562.
Van Breeman, C., Hwang, O., and Cauvin, C. (1982) Calcium Antagonist
    Inhibition of Norepinephrine Stimulated Calcium Influx in Vascular
    Smooth Muscle, in International Symposium on Calcium Modulators
    (Godfraind, T., ed.) Elsevier/North Holland, Amersterdam.
van Meel, J. C. A., deJonge, A., Timmermans, P. B. M. W. M., and van
    Zwieten, P. A. (1981a) Selectivity of some alpha adrenoceptor agonists
    for peripheral alpha-1 and alpha-2 adrenoceptors in the normotensive
    pithed rat. J. Pharmacol. Exp. Ther. 219, 760–767.
van Meel, J. C. A., de Jonge, A., Kalkman, H. O., Wilffert, B.,
    Timmermans, P. B. M. W. M., and van Zwieten, P. A. (1981b) Vascular
    smooth muscle contraction initiated by postsynaptic α_2-adrenoceptor
    activation is induced by an influx of extracellular calcium. Eur. J.
    Pharmacol. 69, 205–208.
van Meel, J. C. A., de Jonge, A., Kalkman, H. O., Wilffert, B.,
    Timmermans, P. B. M. W. M., and van Zweiten, P. A. (1981c) Organic
    and inorganic calcium antagonists reduce vasoconstriction in vivo me-
    diated by postsynaptic α_2-adrenoceptors. Naunyn Schmiedebergs Arch.
    Pharmacol. 316, 288–293.
van Meel, J. C. A., de Zoeten, K., Timmermans, P. B. M. W. M., and van
    Zwieten, P. A. (1982) Impairment by nifedipine of vasopressor re-
    sponses to stimulation of postsynaptic α_2-adrenoreceptors in ganglion-
    blocked rabbits. Further evidence for the selective inhibition of
    postsynaptic α_2-adrenoceptor induced pressor responses by calcium
    antagonists. J. Auton. Pharmacol. 2, 1320.
van Meel, J. C. A., Timmermans, P. B. M. W. M., and van Zwieten, P. A.
    (1983) Alpha$_1$ and α_2-adrenoceptor stimulation in the isolated perfused
    hindquarters of the rat: An in vitro model. J. Cardiovasc. Pharmacol. 5,
    580–585.

van Zwieten, P. A. (1980) Pharmacology of centrally acting hypotensive drugs. *Br. J. Clin. Pharmacol.* **10**, 13S–20S.

van Zwieten, P. A., Thoolen, M. J. M. C., and Timmermans, P. B. M. W. M. (1983) The pharmacology of centrally acting antihypertensive drugs. *Br. J. Clin. Pharmacol.* **15**, 455S–462S.

Verplanken, P. A., Lefebvre, R. A., and Bogaert, M. G. (1984) Pharmacological characterization of α-adrenoceptors in the rat gastric fundus. *J. Pharmacol. Exp. Ther.* **231**, 404–410.

Vetulani, J. and Bednarczyk, B. (1977) Depression by clonidine of shaking behavior elicited by nalorphine in morphine-dependent rats. *J. Pharm. Pharmacol.* **29**, 567–569.

Villalobos-Molina, R., Uc, M., Hong, E., and Garcia-Sainz, J. A. (1982) Correlation between phosphatidylinositol labeling and contraction in rabbit aorta: effect of α_1-adrenergic activation. *J. Pharmacol. Exp. Ther.* **222**, 258–261.

Virtanen, R. and Nyman, L. (1985) Evaluation of the alpha-1 and alpha-2 adrenoceptor effects of detomidine, a novel veterinary sedative analgesic. *Eur. J. Pharmacol.* **108**, 163–169.

Vizi, E. S. (1979) Presynaptic modulation of neurochemical transmission. *Prog. Neurobiol.* **12**, 181–190.

Vizi, E. S. (1972) Stimulation by inhibition of $(Na^+-K^+-Mg^{++})$activated ATP-ase, of acetylcholine release in cortical slices from rat brain. *J. Physiol.* (Lond) **226**, 95–117.

Vizi, E. S., Harsing, L. G., and Knoll, J. (1977) Presynaptic inhibition leading to disinhibition of acetylcholine release from interneurons of the caudate nucleus: Effects of dopamine, beta-endorphin and D-Ala^2-Pro^5-enkephalinamide. *Neuroscience* **2**, 953–961.

Vizi, E. S. and Knoll, J. (1971) The effects of sympathetic nerve stimulation and guanethidine on parasympathetic neuroeffector transmission; the inhibition of acetylcholine release. *J. Pharm. Pharmacol.* **23**, 918–25.

Walland, A. (1978) Inhibition of a somato-sympathetic reflex via peripheral presynaptic α-adrenoceptors. *Eur. J. Pharmacol.* **47**, 211–221.

Washton, A. M. and Resnick, R. B. (1982) Outpatient opiate detoxification with clonidine. *J. Clin. Psychiatry* **43**, 39–41.

Washton, A. M., Resnick, R. B., Perzel, J. F., and Garwood, J. (1981) Lofexidine, a clonidine analogue effective in opiate withdrawal. *Lancet* **i**, 991–992.

Weber, M. A. (1980) Discontinuation syndrome following cessation of treatment with clonidine and other antihypertensive agents. *J. Cardiovasc. Pharmacol.* **2**, (suppl. 1), S73–S89.

Wemer, J., Schoffelmeer, A. N. M., and Mulder, A. H. (1982) Effects of cyclic AMP analogues and phosphodiesterase inhibitors on K^+-induced release of ^{3}-H noradrenaline release from rat brain slices and on its presynaptic α-adrenergic modulation. *J. Neurochem.* **39**, 349–356.

Westfall, T. C. (1977) Local regulation of adrenergic neurotransmission. *Physiol. Rev.* **57**, 659–728.

Wikberg, J. E. S. (1978) Differentiation between pre- and postjunctional α-receptors in guinea pig ileum and rabbit aorta. *Acta Physiol. Scand.* **103**, 225–239.

Wikberg, J. (1977) Localization of adrenergic receptors in guinea pig ileum and rabbit jejunum to cholinergic neurons and to smooth muscle cells. *Acta Physiol. Scand.* **99**, 190–207.

Wikberg, J. E. S. (1979a) The pharmacological classification of adrenergic alpha-1 and alpha-2 receptors and their mechanisms of action. *Acta Physiol. Scand.* (suppl.) **468**, 1–89.

Wikberg, J. E. S. (1979b) Pre- and Postjunctional α-Receptors, in *Presynaptic Receptors* (Langer, S. Z., Starke, K., and Dubocovich, M. L., eds.) Pergamon, Oxford.

Wilffert, B., Timmermans, P. B. M. W. M., and van Zwieten, P. A. (1982) Extrasynaptic location of α_2- and noninnervated β_2-adrenoceptors in the vascular system of the pithed normotensive rat. *J. Pharmacol. Exp. Ther.* **221**, 762–768.

Wilkins, K. H., Wintermitz, S. R., Oparil, S., Smith, L. R., and Sustan, H. P. (1981) Lofexidine and clonidine in moderate essential hypertension. *Clin. Pharmacol. Ther.* **30**, 752–757.

Williams, L. T. and Lefkowitz, R. J. (1977) Regulation of rabbit myometrial receptors by estrogen and progesterone. *J. Clin. Invest.* **60**, 815–818.

Williams, D. O. and Most, A. S. (1981) Responsiveness of the coronary circulation to brief vs. sustained alpha-adrenergic stimulation. *Circulation* **63**, 11–16.

Woodcock, E. A. and Johnston, C. I. (1982) Selective inhibition by epinephrine of parathyroid hormone-stimulated adenylate cyclase in rat renal cortex. *Am. J. Physiol.* **242**, F721–F726.

Yamaguchi, I. and Kopin, I. J. (1980) Differential inhibition of α_1- and α_2-adrenoceptor-mediated pressor responses in pithed rats. *J. Pharmacol. Exp. Ther.* **214**, 275–281.

Yamaguchi, I., Hiroi, J., and Kumada, S. (1977) Central and peripheral adrenergic mechanisms regulating gastric secretion in the rat. *J. Pharmacol. Exp. Ther.* **203**, 125–131.

Yamazaki, S., Katada, T., and Ui, M. (1982) α_2-Adrenergic inhibition of insulin secretion via interference with cyclic AMP generation in rat pancreatic islets. *Mol. Pharmacol.* **21**, 648–653.

Yasuoka, S. and Yaksh, T. L. (1983) Effects on nociceptive threshold and blood pressure of intrathecally administered morphine and α-adrenergic agonists. *Neuropharmacology* **22**, 309–315.

Yokoyama, M., Kawashima, S., Sakamoto, S., Akita, H., Okada, T., Maekawa, K., Mizutani, T., and Fukuzaki, H. (1981) Platelet α-adrenergic receptor function in ischemic heart disease. *Jpn. Circ. J.* **45**, 873.

Yokoyama, M., Kusui, A., Sakamoto, S., and Fukuzaki, H. (1984) Age-associated increments in human platelet α-adrenoceptor capacity. Possible mechanism for platelet hyperactivity to epinephrine in aging man. *Thrombosis Res.* **34**, 287–295.

Zandberg, P., Timmermans, P. B. M. W. M., and van Zwieten, P. A. (1984) Hemodynamic profiles of methoxamine and B-HT 933 in spinalized ganglion-blocked dogs. *J. Cardiovasc. Pharmacol.* **6**, 256–262.

Zebrowska-Lupina, I., Przegalinski, E., Sloniec, M., and Kleinrok, Z. (1977) Clonidine induced locomotor hyperactivity in rats. *Naunyn Schmiedebergs Arch. Pharmacol.* **297**, 227–231.

Zemlan, F. P., Corrigan, S. A., and Pfaff, D. W. (1980) Noradrenergic and serotonergic mediation of spinal analgesia mechanisms. *Eur. J. Pharmacol.* **61**, 111–124.

SECTION 5
RECEPTOR REGULATION

Chapter 6

Regulation of alpha-2 Adrenergic Receptors

Paul A. Insel and Harvey J. Motulsky

1. Introduction

In this chapter, we will review available literature (as of January, 1986) on the regulation of expression of alpha-2 adrenergic receptors. We have somewhat arbitrarily divided regulation of alpha-2 adrenergic receptors into "physiologic," "pharmacologic," and "disease-related" changes in receptor expression. Although some may disagree with these distinctions, we believe they provide a framework to examine settings that may involve different mechanisms for receptor regulation.

Because we are interested in providing readers with mechanistic information, this chapter will focus on the use of radioligand binding and other biochemical approaches. The application of these types of techniques to the study of alpha-2 adrenergic receptors has lagged considerably behind study of several other classes of neurotransmitter and hormone receptors. Thus, studies in other ligand/receptor systems, such as nicotinic cholinergic receptors, certain peptide hormone receptors, and various transport protein receptors (e.g., lipoprotein, transferrin, and so on) have led to several general concepts regarding receptor "life cycles" (Brown et al., 1983; Pastan and Willingham, 1985). Such concepts can provide a framework for thinking about cellular and biochemical mechanisms that underlie observed changes in receptor expression. Unfortunately, a knowledge of the "life cycle" of alpha-2 adrenergic receptors is so rudimentary that it is difficult to do more than speculate as to whether settings with increased expression of receptor-mediated activity represent increases in receptor synthesis/insertion in the plasma membrane, decreases in receptor clearance, enhanced coupling to G_i or other system involved in transmembrane signaling, or changes in processes that serve to terminate receptor responses. A major challenge to those working in this area is to develop tools and approaches that will permit testing of such possibilities. A necessary first step is to define the appropriate settings in which expression of receptors is altered. Our goal is to provide readers with this type of information.

2. Physiologic Regulation of Alpha-2 Adrenergic Receptors

We will discuss four physiologic settings in which changes in alpha-2 adrenergic receptors have been described: development (ontogeny) and aging, states of altered nutrition, exercise and changes related to the menstrual (estrous) cycle, and pregnancy.

2.1 Developmental Changes in alpha-2 Adrenergic Receptors

Responses to catecholamines vary with the age of experimental animals and humans (Weiss et al., 1984b; Kelly et al., 1984; Whitsett et al., 1982). During fetal development, alpha-2 adrenergic receptors (radioligand binding sites) are present on some tissues (e.g., lung) that fail to express receptors in adult animals (Latifpour et al., 1982). Conversely, other tissues (e.g., brain) express more receptors as animals mature (Dausse et al., 1982). Studies in rats have recently indicated a biphasic ontogenic pattern in expression of alpha-2 adrenergic receptors in brain (a decrease in animals at age 7–14 d, but an increase at age 45–60 d) (Dausse et al., 1982). Expression of these receptors is enhanced at particular times of postnatal development in mice who have genetic impairment in differentiation of oligodendrocytes, accompanied by severe myelin deficiency and increased frequency of convulsions (Maurin et al., 1985). The cellular and molecular basis for changes in alpha-2 adrenergic receptor expression as a function of development in brain or other tissues is not known.

Studies in humans have utilized platelets, a readily accessible "tissue" that expresses alpha-2 adrenergic receptors. Platelets obtained from newborns express about 50% fewer alpha-2 adrenergic receptors than do platelets from the mothers of those newborns, and this decrease in receptor number is associated with diminished platelet aggregation in response to epinephrine (Corby and O'Barr, 1981; Jones et al., 1985b). Additional data regarding ontogeny of alpha-2 adrenergic receptors in humans are not yet available.

The normal aging process appears to be associated with changes in adrenergic nervous system function, including the expression and regulation of adrenergic receptors (as reviewed recently, Roberts and Steinberg, 1986). Numerous studies have examined age-related changes in beta-adrenergic receptors, but relatively little is known regarding such changes for alpha-2 adrenergic receptors. Greenberg has recently presented evidence demonstrating that alpha-2 adrenergic receptor number is decreased in brains from aged rats and that synthesis of alpha-2 adrenergic receptors (as well as that of beta- and alpha-1 adrenergic receptors) may be impaired in such animals (Greenberg, 1986). The technique used to assess receptor synthesis was to administer the irreversible alpha-adrenergic receptor antagonist phenoxybenzamine, then quantitate the rate at which receptor number ($[^3H]$rauwolscine binding sites) returned to control levels. Use of phenoxybenzamine for the study of alpha-2 adrenergic receptors in this manner is not ideal because, at the doses generally administered, blockade of alpha-2 adrenergic receptors is incomplete (typically less than 30%), whereas at these same doses, phenoxybenzamine also interacts with other classes of neurotransmitter receptors, including alpha-1 adrenergic receptors (Sladeczek and Bockaert, 1983; McKernan and Campbell, 1982; Hamilton et al., 1984; Mohan et al., 1987). Even so, after phenoxybenzamine treatment, recovery of 50% of the $[^3H]$rauwolscine binding sites in cerebral cortical membranes occurred by the fifth day in young rats, but required 20 d in old rats (Greenberg, 1986). Given the caveats noted above, this prominent difference in rates of recovery of receptors is compatible with the hypothesis that aging is associated with a slower rate of synthesis of alpha-2 adrenergic receptors in rat brain. By contrast, studies in other tissues, such as adipocytes and vas deferens, have suggested that alpha-2 adrenergic receptor function, and perhaps receptor number, may be increased in aged animals (Lai et al., 1983; Lafontan, 1979).

In recent studies comparing effect of age on alpha-2 adrenergic receptor binding and function in brain and platelets from male and female rabbits (2–3-mo- vs 8–12-mo-old females or 24–36-mo-old males), Hamilton et al. found a 40–60% decrease in receptor number in forebrain membranes from both sexes and a

60% decrease in hindbrain membranes from male, but not female, animals (Hamilton et al., 1985). These changes in receptor binding were not associated with a change in the extent to which 1 μg/kg clonidine decreased blood pressure nor were they correlated with changes in alpha-2 adrenergic receptor binding in platelet membranes from the same animals. Female rabbits had a decrease in platelet receptor number and a decreased maximal aggregatory response to epinephrine (but an unchanged EC_{50} for epinephrine); age had no effect on platelet alpha-2 adrenergic receptor number or function in male rabbits. Hamilton et al. concluded that assessment of changes in binding and function of platelet alpha-2 adrenergic receptors can yield misleading information regarding brain receptors in aging (and perhaps in other settings as well, as will be discussed subsequently).

Some recent information has been presented on the effect of aging on alpha-2 adrenergic receptors in humans. Studies with isolated saphenous veins studied in vitro have suggested that a decrease in potency of yohimbine in blocking stimulation-evoked contractions and by inference, a loss of alpha-2 adrenergic receptors, occurs in male subjects between ages 37 and 70 (Hyland and Docherty, 1985). Studies on human platelets have yielded rather ambiguous results: some workers have found no relation between number of receptors and age, whereas other investigators have found an inverse relationship (e.g., Brodde, 1983; Pfeifer et al., 1984; Docherty and O'Malley, 1985; Scarpace, 1986). Since aging is associated with an increase in plasma cathecholamine levels (Brodde, 1983; Docherty and O'Malley, 1985; Esler, 1982; Rowe and Troen, 1980), one might hypothesize that agonist-mediated decrease in receptor density (i.e., down-regulation) would account for the reported decreases in number of platelet alpha-2 adrenergic receptors with increases in age. As will be discussed subsequently, however, the evidence is not convincing that platelets show agonist-mediated down-regulation. The bulk of the data indicate that age-related changes in platelet alpha-2 adrenergic receptors are relatively small and of little importance in terms of impact on platelet function (Docherty and O'Malley, 1985).

To summarize, the limited data that are available suggest that alpha-2 adrenergic receptor number and distribution may change prominently during ontogeny. Mechanisms for this change are unknown. With aging, these receptors may be synthesized more slowly in brain of rats, although whether this applies to other tissues and to humans is not known.

2.2. alpha-2 Adrenergic Receptors in States of Altered Nutritional Status

The effect of nutrition on alpha-2 adrenergic receptors has been examined in two settings: changes in dietary caloric content and sodium content. Alterations in sodium are discussed below in the section on cardiovascular disease.

alpha-2 Adrenergic receptors have been examined in obesity and during caloric restriction. alpha-2 Adrenergic receptors on fat cells inhibit lipolysis, whereas beta-1 adrenergic receptors promote lipolysis (Lafontan, 1979; Fain and Garcia-Sainz, 1983; Berlan et al., 1982). In fat cells from several species, obesity (or perhaps age-related increase in body size) is associated with a decrease in lipolytic response and with an increase in number of alpha-2 adrenergic receptors (Lafontan, 1979; Fain and Garcia-Sainz, 1983; Berlan et al., 1982). Fasting is associated with enhanced antilipolytic effects of catecholamines, but it is unclear whether these effects result from decreases in beta-adrenergic receptors and response, increases in alpha-2 adrenergic receptors and response, changes in local modulators (e.g., adenosine), or some combination thereof (Kather et al., 1985; Lafontan, 1979; Berlan et al., 1982). Patients who have anorexia nervosa and bulimia have also been reported to have an increased number and response of platelet alpha-2 adrenergic receptors; this has been attributed to a decreased activity of the sympathetic nervous system and a resultant "up-regulation" of receptors (Sundaresan et al., 1983; Luck et al., 1983; Heufelder et al., 1985).

2.3. Changes in alpha-2 Adrenergic Receptors with Exercise

Exercise prominently activates the sympathoadrenal axis, and, as a result, one observes substantial increases in plasma catecholamines during exercise and for many minutes thereafter. Studies of human subjects have indicated that exercise may be associated with a decrease in ability of agonists to bind to platelet alpha-2 adrenergic receptors (Hollister et al., 1983). These data are somewhat difficult to interpret because binding of agonists was assessed in intact platelets, in which intracellular GTP, Na +, and Mg^{2+} are present and would thus presumably influence agonist binding (see chapters 2 and 3: Regulation of alpha-2 Receptor-ligand interactions by GTP, Na^+ and Mg^{2+}). In addition, assessment of agonist binding is difficult in studies with intact cells because of problems related to achievement of steady state, agonist-induced modifications in receptors, ligand uptake and degradation, and so on (as discussed in Motulsky et al., 1985).

2.4. alpha-2 Adrenergic Receptors in the Menstrual (Estrus) Cycle and Pregnancy

Sex steroids have been implicated as important modulators of alpha-2 adrenergic receptors. Changes in receptor binding and response have been observed in female animals as a function of hormonal cycles as well as during pregnancy. Some early studies utilized probes, such as [^{3}H]dihydroergocryptine ([^{3}H]DHE), that do not distinguish between alpha-1 and alpha-2 adrenergic receptors. Results from such studies indicated that diestrus female rats have higher numbers of binding sites in cervicothoracic spinal cord and cerebellum than do rats in estrus (Orensanz et al., 1982). Estrus-related differences in [^{3}H]DHE sites were not found in rat myometrium, however, in spite of differences in alpha-adrenergic contractile responses (Krall et al., 1978; Boyle and Digges, 1982). Because other data indicate that alpha-adrenergic contractile response in myometrium (at least in rabbits) is mediated by only one subset of alpha-adrenergic receptors, the alpha-1 adrenergic receptors (Hoffman et al., 1981), this discrepancy between [^{3}H]DHE binding sites and functional response is not surprising. Other studies have suggested that the number of [^{3}H]DHE binding sites decreases in the hypothalamus as rats undergo sexual maturation (Wilkerson et al., 1979).

Studies in humans have demonstrated striking changes in myometrial alpha-2 adrenergic receptor number during the menstrual cycle (Bottari et al., 1983). Bottari et al. reported myometrial receptor numbers of about 50 fmol/mg protein in the myometrium of postmenopausal women, 120 fmol/mg in women at midfollicular phase, and 260 fmol/mg at midluteal phase. This several-fold increase in receptor number paralleled a similar increase in plasma 17-beta-estradiol concentration. During pregnancy, however, when estrogen levels increase up to two orders of magnitude higher than in midluteal phase, the number of myometrial alpha-2 adrenergic receptors was only about 50 fmol/mg—almost identical to the level observed in postmenopausal women. This result suggests that progesterone, which is also markedly increased during pregnancy, may counteract the ability of estrogens to increase alpha-2 adrenergic receptors in myometrium. It is tempting to speculate that estrogen can induce myometrial alpha-2 adrenergic receptors via enhanced transcription and that, during pregnancy, progesterone counteracts this effect. Studies with myometrial cells in tissue culture should allow an experimental test of this postulate.

Other clinical studies have examined alpha-2 adrenergic receptors on blood platelets during the menstrual cycle. Although longitudinal studies have indicated that peak numbers of receptors ([^{3}H]yohimbine binding sites) are detected in platelets from women at the onset of menses and that this peak decreases about 25% at midcycle, 7–14 d after menses (Brodde, 1983; Jones et al., 1983; Barnett et al., 1984), cross-sectional studies have not revealed prominent changes in receptor number during the menstrual cycle (Sundaresan et al., 1985; Peters et al., 1979; Rosen et al., 1984). Methodological differences and interindividual variation may explain such discrepant findings. We are aware of no definitive results that document changes in functional response of platelet alpha-2 adrenergic receptors during the menstrual cycle.

The number of platelet alpha-2 adrenergic receptors has also been examined in pregnancy. Metz et al. initially reported that the platelet alpha-2 adrenergic receptor number decreases about 20% during the first week to 10 d after pregnancy ends, a period when plasma estrogen and progesterone levels fall substantially (Metz et al., 1983). Other recent findings indicate, however, that pregnant women may have no change in alpha-2 adrenergic receptor number or in epinephrine-mediated platelet aggregation as compared to women studied during the follicular phase of the menstrual cycle, when estrogen and progesterone levels are lowest (Roberts et al., 1986). Reports of enhanced alpha-adrenergic response of platelets during pregnancy may be flawed, however, because of the failure to correct for the lower hematocrit and thus higher free calcium concentration present in plasma during pregnancy and thus perhaps influencing the function of platelets in platelet-rich plasma in these studies (Roberts et al., 1986).

To summarize, physiologic changes in estrogen and progesterone appear to produce prominent changes in expression of alpha-2 adrenergic receptors in myometrium of some species, including humans. Receptors on platelets do not show changes that are as readily detectable as do receptors on other tissues.

3. Pharmacologic Regulation of alpha-2 Adrenergic Receptors

Regulation of a receptor by drugs can be divided into *homologous* regulation by agonists and antagonists that occupy the receptor, and *heterologous* regulation by drugs that can change receptor expression, but that do not bind to that receptor. Adrenergic receptors commonly demonstrate both homologous and heterologous regulation.

3.1. Homologous Regulation of alpha-2 Adrenergic Receptors

Receptor-linked responses often attenuate with time, a phenomenon termed desensitization, refractoriness, tolerance, or tachyphylaxis. Desensitization of beta-adrenergic receptors has been well studied, and in virtually every tissue studied, beta-adrenergic receptors are subject to desensitization (Harden, 1983; Sibley and Lefkowitz, 1985). This occurs in two phases; a rapid phase that transpires in seconds to minutes as the receptors uncouple from G_s (the guanine nucleotide-binding protein that mediates activation of adenylate cyclase by beta-adrenergic receptors), and a slower phase that occurs over a period of many minutes to days as the number of receptors decreases (a phenomenon known as down-regulation). Although extensive data are not yet available, desensitization of alpha-2 adrenergic receptors does not seem to be analogous to desensitization of beta-adrenergic receptors. Desensitization of alpha-2 adrenergic receptors has been studied primarily using human platelets, and we will review these data first, followed by data collected in studies of other tissues.

3.1.1 DESENSITIZATION OF PLATELET ALPHA-2 ADRENERGIC RECEPTORS

Epinephrine, acting through alpha-2 adrenergic receptors, initiates aggregation of human platelets. Although aggregation may be related to the ability of platelet alpha-2 adrenergic receptors to inhibit adenylate cyclase activity, other mechanisms may be involved (Limbird and Sweatt, 1985). Aggregation is a dramatic response and easy to measure, but epinephrine-initiated aggregation is probably not physiologically relevant because it requires >1 uM epinephrine, a concentration much higher than that ever achieved in blood. Low concentrations of epinephrine that cannot initiate aggregation by themselves, however, will potentiate aggregation initiated by unrelated agents, such as ADP or thrombin (Lanza and Cazenave, 1985; Mills and Roberts, 1967). In some species, for example dogs (Meyers et al., 1983) and rabbits (Hallam et al., 1981), epinephrine cannot initiate aggregation at any concentration, and potentiation is the only response observed. This is also the case with platelets from human newborns (Jones et al., 1985b).

Platelet aggregation requires both activation by appropriate agonists and intercellular contact. Thus platelets incubated with agonists in unstirred plasma do not aggregate. When platelets are incubated with epinephrine without stirring, and aggregation is later initiated by stirring, aggregation occurs to a much lesser extent than it does in control platelets. This desensitization of epinephrine-initiated aggregation has been observed by several investigators (O'Brien, 1964; Cooper et al., 1978; Hollister et al., 1983; Motulsky et al., 1986). Desensitization occurs half-maximally within 3–5 min, is maximal after an incubation for 20 min, cannot be overcome by increasing the concentration of epinephrine, is specific for aggregation initiated by epinephrine (and not other aggregating agents such as ADP), and occurs for both the aggregation and potentiation responses (Motulsky et al., 1986). Desensitization also occurs in platelets preincubated with aspirin, which blocks formation of cyclooxygenase products required for secretion and secondary aggregation (Motulsky et al., 1986).

Desensitization of alpha-2 adrenergic-stimulated platelet aggregation has also been observed in vivo. Epinephrine-initiated aggregation is reduced in platelets from individuals receiving an epinephrine infusion (Jones et al., 1985a, 1986) and from patients with pheochromocytoma, a catecholamine-secreting tumor (Jones et al., 1985a).

One possible explanation for desensitization of alpha-2 adrenergic-stimulated platelet aggregation would be an agonist-promoted down-regulation in the number of alpha-2 adrenergic receptors. Down-regulation of [^{3}H]yohimbine binding sites does not occur, however, following in vitro incubation of platelet-rich plasma with epinephrine for 4 or 24 h (Karliner et al., 1982). An earlier study reached the opposite conclusion (Cooper et al., 1978), but at that time the only available radioligand was [^{3}H]DHE, which may bind to sites other than alpha-2 adrenergic receptors in platelets (Motulsky and Insel, 1982).

In vivo protocols have also been used to ask whether agonists promote down-regulation of platelet alpha-2 adrenergic receptors. Three approaches have been employed: (1) Platelet alpha-2 adrenergic receptor number has been compared before and during treatment of patients with alpha-2 adrenergic agonists for several days to weeks. Such studies have demonstrated either no change (Motulsky et al., 1983a; Boon et al., 1983b) or a 30% decrease (Brodde, 1983; Brodde et al., 1982) in receptor number. (2) Platelet alpha-2 adrenergic receptors have been assessed in patients with pheochromocytoma, tumors that produce very high circulating levels of catecholamines. Although platelets from such patients had normal numbers of alpha-2 adrenergic receptors in several studies (Jones et al., 1985a; Motulsky et al., 1983a; Pfeifer et al., 1984), in other studies receptor number was decreased (Brodde and Bock, 1984; Davies et al., 1982b). (3) Platelet alpha-2 adrenergic receptors have been examined before and after an infusion of epinephrine. No change in receptor number was observed during an epinephrine infusion lasting several hours, either in humans (Roberts et al., 1986; Pfeifer et al., 1984) or in dogs (Villeneuve et al., 1985b). We conclude that the majority of the evidence indicates that agonist-induced down-regulation of platelet alpha-2 adrenergic receptors does not occur, or occurs to a very limited extent.

Although the number of alpha-2 adrenergic receptors does not decrease after desensitization, it is possible that the characteristics of the receptors are altered. One characteristic that has been examined is the competition of unlabeled agonists for radiolabeled antagonist binding sites. In such experiments utilizing platelet membranes studies in both the presence and absence of GTP, we found no difference between control and desensitized platelets (Motulsky et al., 1986). These data demonstrate that the alpha-2 adrenergic receptors on desensitized platelets appear to bind agonists normally and are likely to be coupled normally to G_i, the protein that links receptors to inhibition of adenylate cyclase. In analogous experiments using intact platelets, Hollister et al. have demonstrated that the competition of unlabeled epinephrine for alpha-2 adrenergic receptors on intact platelets is altered in desensitized platelets: the curves were steeper and higher concentrations of epinephrine were required to compete for radiolabeled antagonist binding sites (Hollister et al., 1983). These results were only observed in incubations conducted at 25°C; at 37°C the results for control and desensitized platelets were identical (personal communication, A. Hollister). Thus the physiological significance of these changes is unclear.

In spite of decreased epinephrine-initiated aggregation, alpha-2 adrenergic-mediated inhibition of adenylate cyclase activity [stimulated by prostaglandin E, or forskolin] occurs normally in platelets desensitized by in vitro incubations with epinephrine (Motulsky et al., 1986). These data suggest that desensitization of platelets after in vitro incubation with agonists most likely occurs either through mechanisms distal to cyclic AMP action or through other pathways. Platelets that have been desensitized in vivo by infusion of epinephrine, however, have decreased alpha-2 adrenergic inhibition of cyclic AMP accumulation (Jones et al., 1986). Perhaps the discrepancy between the in vitro and in vivo results is explained by other effects of infused epinephrine; for example, epinephrine infusion can lead to a 25% increase in number of circulating platelets (Kande et al., 1985).

The majority of the evidence thus suggests that desensitization of platelet alpha-2 adrenergic receptors occurs by alterations distal to receptors. Aggregation is usually measured as the increase in light transmittance through stirred platelet suspensions, and by this measure aggregation is reduced about 50% in desensitized platelets (Motulsky et al., 1986). When examined microscopically, however, desensitized platelets form smaller aggregates than do control platelets (Shattil et al., 1986). Although the coalescence of small aggregates into large ones results in dramatic changes in light transmission measured in an aggregometer, the role that such large aggregates play in thrombosis in vivo is unclear.

In the primary phase of aggregation, platelets become cross-linked by fibrinogen in response to agonists such as epinephrine (Bennett and Vilaire, 1979; Plow and Marguerie, 1980). The number of fibrinogen receptors exposed by epinephrine, measured either by the binding of labeled fibrinogen or by the antibody PAC-1 (which appears to bind to agonist-exposed fibrinogen receptors), is identical in control and desensitized platelets (Shattil et al., 1986). It is generally thought that the triggering of expression of fibrinogen receptors is the principal means by which agonists like epinephrine promote platelet aggregation. The recent data demonstrating normal expression of fibrinogen receptors in platelets desensitized to epinephrine-initiated aggregation suggest that alpha-2 adrenergic receptors might also activate events distal to or parallel with expression of fibrinogen receptors.

3.1.2. DESENSITIZATION OF ALPHA-2 ADRENERGIC RECEPTORS IN OTHER TISSUES

As noted above, alpha-2 adrenergic receptors on adipocytes are coupled to an inhibition of adenylate cyclase and an inhibition of lipolysis. The inhibition of lipolysis is not desensitized when adipocytes are preincubated with clonidine, a full alpha-2 adrenergic agonist for these cells (Villeneuve et al., 1985a); nor is the inhibition of adenylate cyclase blunted in cells preincubated with epinephrine (plus propranolol to block beta-adrenergic receptors) (Burns et al., 1982). Down-regulation of adipocyte alpha-2 adrenergic receptors also does not occur when hamsters are administered clonidine or epinephrine (Pecquery et al., 1984; Villeneuve et al., 1985a).

Desensitization of alpha-2 adrenergic responses has also been examined in other tissues. Administration of clonidine to rats for several days can lead to a desensitization of alpha-2 adrenergic-mediated vasoconstriction (Ishii and Kato, 1984) and inhibition of electrically provoked twitch response of the vas deferens (Ishii et al., 1982). In these studies, the mechanisms of desensitization were not explored.

Studies with several other tissues also support the notion that alpha-2 adrenergic receptors are quite resistant to agonist-promoted down-regulation. A normal number of alpha-2 adrenergic receptors has also been observed in renal and hepatic membranes from rats with an implanted pheochromocytoma, as well as in renal cortical membranes from rats that received an epinephrine infusion for several days (Snavely et al., 1983, 1985b).

Studies of cultured cells offer a convenient means of assessing receptor regulation, but only a few types of cultured cells are known to contain alpha-2 adrenergic receptors. Incubation of the neuroblastoma glioma hybrid cell line NG108-15 with alpha-adrenergic agonists leads to a down-regulation of receptor number (Mitrius and U'Prichard, 1985; Thomas and Hoffman, 1986) associated with a decreased ability of the receptors to inhibit adenylate cyclase (Thomas and Hoffman, 1986). We have recently identified alpha-2 adrenergic receptors on human erythroleukemia (HEL) cells; incubating these cells with epinephrine for several hours does not decrease receptor number or desensitize alpha-2 adrenergic receptor-mediated inhibition of adenylate cyclase activity, but does cause a redistribution of the receptors away from the cell surface (McKernan et al., 1987).

Another approach to studying agonist-induced desensitization and down-regulation has been to administer agents that block uptake or metabolism of catecholamines, thus resulting in elevated catecholamine levels to which alpha-2 adrenergic receptors are exposed. Studies of this type have demonstrated a substantial decrease in alpha-2 adrenergic receptor number in brain membranes of animals treated for many days with antidepressants that block catecholamine metabolism or uptake (Cohen et al., 1982a,b; Sugrue, 1982). It is difficult to be sure that the alterations in receptors in such studies are caused by increased synaptic catecholamine concentrations and not by other direct or indirect effects of the drugs (Sugrue, 1983). Moreover, administration of certain drugs that can alter catecholamine levels (e.g., imipramine) may not alter cerebral alpha-2 adrenergic receptor number (Mikuni et al., 1983), and treatment of human subjects with clorgyline, which decreases receptor number in the brain of animals receiving the drug, does not alter binding or function of platelet alpha-2 adrenergic receptors (Siever et al., 1983b).

3.1.3. ALTERATIONS IN [^{3}H]CLONIDINE BINDING WITH DESENSITIZATION

Another way to assess the interaction of agonists with receptors is to measure the binding of radiolabeled agonists. [^{3}H]Epinephrine is available commercially, but has not been widely used. [^{3}H]Clonidine and [^{3}H]p-aminoclonidine bind with high affinity to alpha-2 adrenergic receptors and have been used by many investigators. One problem with such studies is that the pharmacology of clonidine is different in different tissues. In platelets, for example, clonidine is a partial agonist for inhibition of adenylate cyclase, an agonist for potentiating ADP-initiated aggregation, and an antagonist of epinephrine-initiated aggregation. Binding of [^{3}H]clonidine only recognizes a low fraction (typically about 20%) of the platelet alpha-2 adrenergic receptors, probably because the usual radioligand binding conditions permit detection of only a high-affinity class of binding sites (Shattil et al., 1981; Mooney et al., 1982). With adipocytes, however, clonidine is a full agonist for inhibition of lipolysis and of adenylate cyclase activity (Berlan and Lafontan, 1982). Such conflicting results make it difficult to interpret binding studies in which [^{3}H]clonidine (or the closely related compound [^{3}H]p-aminoclonidine) is used as a radioligand. Many such studies have been reported, however, and decreases in the number of binding sites have been observed following incubation of cells with agonist. For example, incubating adipocytes with either clonidine or epinephrine leads to a loss in the number of detectable high-affinity [^{3}H]clonidine or [^{3}H]p-aminoclonidine binding sites, without a loss in [^{3}H]yohimbine binding or functional desensitization (Burns et al., 1982; Villeneuve et al., 1985a). The agonist [^{3}H]UK 14, 304 has recently become available. It appears to be a full agonist and binds with high affinity to the majority of the alpha-2 adrenergic receptors identified by [^{3}H]yohimbine or [^{3}H]rauwolscine in membranes prepared from platelets (Neubig et al., 1985), brain (Loftus et al., 1984), HT29 cells (Turner et al., 1985), and human erythroleukemia (HEL) cells (McKernan et al., 1987). The availability of this probe should facilitate future studies of agonist interaction with alpha-2 adrenergic receptors.

3.1.4 SUMMARY OF STUDIES ON DESENSITIZATION OF ALPHA-2 ADRENERGIC RECEPTORS

To summarize, data obtained both in vitro and in vivo demonstrate that incubation of platelets with epinephrine rapidly desensitizes the ability of alpha-2 adrenergic receptors to promote platelet aggregation. This desensitization, however, is not associated with changes in the number or properties of alpha-2 adrenergic receptors on platelets or in their ability to inhibit adenylate cyclase activity or to stimulate the expression of fibrinogen receptors. Desensitization of alpha-2 adrenergic function also does not appear to occur in adipose tissue, but in limited studies has been demonstrated in vascular smooth muscle and in vas deferens. The mechanism of desensitization in these tissues is unknown. Furthermore, exposing peripheral target cells to agonists for days or weeks generally does not reduce alpha-2 adrenergic receptor number; agonist-mediated down-regulation may, however, occur at alpha-2 adrenergic receptors in the brain. The apparent resistance of alpha-2 adrenergic receptors in several tissues to down-regulation is a novel aspect of the regulation of these receptors compared to the many other classes of cell surface receptors in which agonist-induced down-regulation commonly occurs.

3.1.5. UP-REGULATION OF ALPHA-2 ADRENERGIC RECEPTORS

alpha-2 Adrenergic receptors are tonically exposed to norepinephrine and epinephrine either circulating in the blood or released as neurotransmitters. Does "up-regulation" occur when those ambient concentrations are decreased? Several approaches have been used to address this question, and most of these studies have indeed demonstrated an increase in receptor number: (1) Several investigators have examined the platelets of patients with idiopathic hypotension attributable to decreased sympathetic stimulation and decreased circulating catecholamines. These studies (discussed in the section on cardiovascular diseases below) have reached contradictory conclusions, but an increased number of receptors was found by several groups. (2) Catecholamine concentrations in hypertensive patients have been lowered with guanadrel, a drug that inhibits noradrenaline storage in and release from sympathetic nerve terminals. The number of alpha-2 adrenergic receptors on platelets from these patients was increased (Egan et al., 1985). (3) Another approach is to undertake surgical or chemical denervation in animals. Results of such studies have not been consistent. In some studies denervation has led to an increase of alpha-2 adrenergic receptors [e.g., in rat submandibular gland (Arnett and Davis, 1979; Bylund and Martinez, 1981) and vas deferens (Watanabe et al., 1982)]; in other studies there was a decrease [e.g., in rat mesenteric artery following chemical denervation (Agrawal and Daniel, 1985)] or no change (Bobik and Anderson, 1983). (4) Up-regulation has also been dem-

onstrated in salivary glands from reserpine-treated rats in which
reserpine has led to a depletion of tissue catecholamines (Bylund
and Martinez, 1980). (5) The epinephrine content of the medulla
and pons in rat brains is controlled by the enzyme phenylethanolamine N-methyltransferase (PNMT). Rats with decreased PNMT
activity, caused either by genetic predisposition (Vantini et al.,
1984; Perry et al., 1983b) or by treatment with PNMT inhibitors
(Stolk et al., 1984), have up to a fivefold elevation in alpha-2 receptor number.

Up-regulation is usually attributed to reversal of tonic down-
regulation by endogenous agonists. It is surprising, therefore,
that up-regulation of alpha-2-adrenergic receptors has been
clearly demonstrated in a variety of systems, whereas down-
regulation has not. Perhaps this indicates that down-regulation is
normally present in many tissues in response to ambient levels of
tissue and circulating catecholamines.

3.2. Heterologous Regulation of alpha-2 Adrenergic Receptors

Several classes of drugs may regulate alpha-2 adrenergic receptors in a heterologous fashion. These include glucocorticoids, thyroid hormone, and sex steroids. In general, the data available regarding heterologous regulation of these receptors is much more
scanty than that published regarding heterologous regulation of
other classes of adrenergic receptors.

3.2.1. REGULATION OF ALPHA-2 ADRENERGIC RECEPTORS BY GLUCOCORTICOIDS AND THYROID HORMONE

Very limited data have been published on the regulation of
alpha-2 adrenergic receptors by glucocorticoids. Maeda et al. have
examined the effect of glucocorticoids on [^{3}H]clonidine and
[^{3}H]yohimbine binding sites in organ cultures of vas deferens
from reserpinized rats (Maeda et al., 1983). Addition of hydrocor-
tisone or dexamethasone to the culture medium inhibited the
prominent loss in [^{3}H]clonidine binding sites observed during the
first 2 d of culture, and this glucocorticoid response was blocked
by inhibitors of protein synthesis. Because the glucocorticoids
were ineffective in blocking the less marked decrease in
[^{3}H]yohimbine sites, it is possible this effect of the corticosteroids
was not related to a change in the receptors themselves, but instead to an action on a guanine nucleotide-binding protein or
other membrane component to which the receptors are coupled.
In view of evidence that glucocorticoids seem to be able to alter

interaction of the classes of adrenergic receptors with G_s (Davies and Lefkowitz, 1983; Goodhardt et al., 1982), further studies of effects of glucocorticoids on alpha-2 adrenergic receptors and their interaction with G proteins may prove to be an important area for subsequent investigation.

Studies of regulation of alpha-2 adrenergic receptors by thyroid hormone have involved examination of changes in receptor expression in hyperthyroidism and hypothyroidism. Although one group of investigators found that hamster adipocytes from hyperthyroid animals contain alpha-adrenergic receptors of similar number and affinity and with similar efficacy in inhibiting adenylate cyclase activity (Garcia-Sainz et al., 1981), another laboratory has reported that adipocytes from such animals have about a 40% decrease in both the number of $[^3H]$DHE binding sites and in the increment of epinephrine-stimulated cyclic AMP synthesis produced by the alpha-adrenergic antagonist phentolamine (Giudicelli et al., 1980). Presumably differences in experimental design explain these different results in hamster adipocytes, although discrepant results have been observed in other target systems as well. Binding to alpha-2 adrenergic receptors in rat brain membranes of hyperthyroid rats is apparently unchanged (Gross and Schumann, 1981; Fox et al., 1985), whereas platelets of hyperthyroid patients had a 30% decrease in receptor number (Ratge et al., 1985).

In hypothyroidism, both binding and functional assays have indicated a lack of change of alpha-2 adrenergic receptors in adipocytes from rats and hamsters. In contrast, a decrease in receptor number in rat brain membranes was found by one group of investigators, which examined $[^3H]$clonidine binding (Gross and Schumann, 1981), but not by another group, which examined $[^3H]$rauwolscine binding (Fox et al., 1985).

To summarize, although the available literature is somewhat inconsistent, neither an excess nor a deficiency of thyroid hormone is associated with dramatic changes in alpha-2 adrenergic receptor number or response in the tissues that have been examined. One exception is a report of enhanced alpha-2 adrenergic-mediated vasoconstriction in perfused hindlimbs of hypothyroid, but not hyperthyroid, rats (Ohguchi et al., 1984). Mechanisms whereby hyperthyroidism or hypothyroidism lead to alteration in expression of alpha-2 adrenergic receptors are unknown.

It is also of interest that alpha-2 adrenergic receptors have been identified in the thyroid gland and that these receptors can inhibit thyrotropin-stimulated cyclic AMP accumulation and thyroxine release (Muraki et al., 1984). Whether hypothyroidism or hyperthyroidism alters these receptors has not been investigated.

3.2.2. REGULATION OF ALPHA-2 ADRENERGIC RECEPTORS BY SEX STEROIDS

In addition to evidence that was discussed earlier with regard to regulation of alpha-2 adrenergic receptors by physiologic changes in levels of estrogen and progesterone, other data have indicated that treatment of animals or humans with estrogens and/or protesterone can change expression of those receptors. Thus, estrogen treatment has been reported to increase [^{3}H]DHE binding in rat (Kano, 1982) and rabbit (Hoffman et al., 1981; Williams and Lefkowitz, 1977; Roberts et al., 1977, 1979, 1981) uterine membrane preparations and in certain regions of rabbit urinary bladder (Levin et al., 1980). Further studies have indicated that this increase in rabbit uterine sites is a selective enhancement in the number of alpha-2 adrenergic receptors, a substantial portion of which may be located on presynaptic nerve terminals (Hoffman et al., 1981). Whether alpha-2 adrenergic receptors in uterus are important for catecholamine-mediated contractile response is controversial; some, but not all, data suggest that estrogen-mediated increases in myometerial alpha-adrenergic receptor number are associated with enhanced contractile response (Hoffman et al., 1981; Roberts et al., 1981).

Other studies have indicated that treatment with sex steroids can lead to prominent tissue-specific differences in alpha-2 adrenergic receptor expression. Thus, doses of estrogen that will increase the number of alpha-2 receptors in rabbit uterus can decrease the number of platelet alpha-2 adrenergic receptors (Roberts et al., 1979; Mishra et al., 1985), although having little effect on receptor number in other tissues [e.g., brain (forebrain and hindbrain), kidney, and spleen (Mishra et al., 1985)]. Other work has demonstrated regional differences in the regulation of brain alpha-adrenergic receptors by estrogen treatment of rats and guinea pigs (Roberts et al., 1981; Wilkinson and Herdon, 1982) and possible changes in platelet alpha-2 adrenergic receptor number in women who take oral contraceptives (Peters et al., 1979; Jones et al., 1983). Tissue-specific differences in response to estrogens likely result from variation in expression of estrogen receptors as well as in differing effects of these receptors in regulating gene expression in different estrogen-sensitive tissues (O'Malley, 1984; Ringold, 1985). In view of the effects of estrogen and progesterone on alpha-2 adrenergic receptors, one could also ask whether androgens alter receptor binding and function. We are not aware of studies that have explored this question.

3.2.3. ALPHA-2 ADRENERGIC RECEPTORS AND OPIATE RECEPTOR AGONISTS AND ANTAGONISTS

As reviewed elsewhere (Kunos et al., 1985), alpha-2 adrenergic and opioid receptor agonists act in the central nervous system to produce similar effects on the cardiovascular system. These classes of drugs act on distinct receptors primarily in the ponto-medullary region to reduce central sympathetic outflow and hence blood pressure and heart rate. Although the receptor sites are distinct, there appears to be some overlap with respect to the target cells on which the receptors are located or act. Thus, cells of the locus ceruleus, the largest cluster of noradrenergic neurons in the brain, appear to be one group of cells regulated by both opioid and alpha-adrenergic agonists (Redmond, 1982; Aghajanian, 1982). It is thus perhaps not surprising that the hyperactivity of these neurons that occurs during opiate withdrawal can be inhibited by alpha-2 adrenergic receptor agonists (Aghajanian, 1982; Redmond and Huang, 1982; Roth et al., 1982). Withdrawal symptoms observed after abrupt discontinuation of alpha-2 adrenergic receptor agonists or opiates are characterized by increased sympathetic activity, and these symptoms can be suppressed by drugs directed at either opiate or alpha-2 receptors (Thoolen et al., 1981; Tseng et al., 1975). Also, opiate antagonists appear to be able to reduce efficacy of centrally acting alpha-2 receptor agonists in some hypertensive animals and patients (Kunos et al., 1985). Overlap in response at opiate and alpha-2 adrenergic receptors may also occur in tissues other than the central nervous system (DiStefanfo and Brown, 1985; Robson et al., 1983).

Although some data indicate that chronic morphine treatment can increase alpha-2 adrenergic receptor number in brain (Hamburg and Tallman, 1981), other investigators have not confirmed this observation (Simantov et al., 1982; Sethy and Harris, 1982). Because opiates do not occupy alpha-2 adrenergic receptors (Browning et al., 1982), any increases in alpha-2 receptor number are thus not secondary to alpha-2 receptor occupancy by the opiates themselves, but must be caused by other effects. Postreceptor alterations in alpha-2 adrenergic receptor-mediated responses, including adenylate cyclase, may also occur with opiate dependence and withdrawal (Robson et al., 1983; Sharma et al., 1975; Sabol and Nirenberg, 1979).

3.2.4. REGULATION OF ALPHA-2 ADRENERGIC RECEPTORS BY MISCELLANEOUS PHARMACOLOGICAL AGENTS

As noted above, alpha-2 adrenergic receptors may be regulated by agonists and antagonists that occupy receptors. In addition, a number of pharmacological agents have been shown to decrease radioligand binding to alpha-2 receptors when the agents are added directly in binding studies. Known alpha-adrenergic agonists and antagonists are expected to do this. More unexpected are agents whose principal site of action is presumed to be at other target sites and thus are not generally recognized as agents interacting at alpha-2 adrenergic receptors. For example, evidence has been presented that neuroleptic and antidepressant drugs (Perry et al., 1983a; Wong et al., 1983), phenylethanolamine N-methyltransferase inhibitors (Stolk et al., 1984; Fuller, 1982), quinidine (Motulsky et al., 1984; Ciofalo, 1980), verapamil (Motulsky et al., 1983b; Barnathan et al., 1982), amiloride (Sweatt et al., 1985; Insel et al., 1985; 1987), and penicillins (Shattil et al., 1980) can decrease radioligand binding to alpha-2 adrenergic receptors in one or more tissues. Although certain of those agents appear to act as competitive inhibitors interacting, presumably at the receptor binding site (Motulsky et al., 1983b, 1984), for other agents, other sites of action may be involved. Penicillins, for instance, appear to produce a generalized perturbation in platelet membranes and alter binding not only to alpha-2 adrenergic receptors, but to receptors for ADP and perhaps other agonists as well (Shattil et al., 1980). Amiloride and analogs of amiloride may act to alter alpha-2 adrenergic receptors via both competitive and noncompetitive interactions at receptors (Sweatt et al., 1985; Insel et al., 1985, 1987); Limbird and colleagues have suggested that the noncompetitive effects may relate to an association between alpha-2 adrenergic receptors and the Na^+/H^+ antiporter (a widely recognized site of action for amiloride derivatives). Another class of agents—adenosine receptor agonists—have been shown to increase [^{3}H]clonidine binding to rat vas deferens membranes; the mechanism for this effect is unknown (Watanabe et al., 1983).

The ability of diverse pharmacological agents to interact with alpha-2 adrenergic receptors suggests that this interaction may contribute to overall pharmacological activity of those agents and may potentially explain differences in this activity among drugs in a particular "class." Thus, for example, among the calcium channel-blocking agents, phenylalkylamines (in particular, verapamil and methoxylverapamil, D600), but not dihydropyridines (e.g., nifedipine, nitrendipine), compete for alpha-2 adrenergic receptors. Moreover, the pharmacological and physiological literature is replete with studies in which verapamil and other phenylalkylamines have been used at concentrations high enough to block alpha-adrenergic receptors, but in which the investigators have not considered the contribution of this blockade to observed responses. In addition to blocking receptors, chronic treatment with such a drug may produce changes in receptor expression. Thus, if a drug acts as a receptor antagonist, and receptor antagonists lead to a compensatory up-regulation of receptors, one would predict the development of such compensatory changes.

4. Disease-Related Changes in alpha-2 Adrenergic Receptors

Because of their likely importance in regulating responses in several organ systems, including the cardiovascular system, the peripheral and central nervous systems, and blood platelets, alpha-2 adrenergic receptors have been implicated as sites that are altered in diseases of those systems. We will discuss some of these findings, although at the present time, no firm evidence has been provided in any of those disease settings that changes in alpha-2 adrenergic receptors are of primary pathogenetic importance. Moreover, mechanisms mediating changes in diseases have not been extensively evaluated.

4.1. Cardiovascular Diseases

Substantial evidence has demonstrated that alpha-2 adrenergic receptors are important for cardiovascular regulation via actions at sites in both the central nervous system and in the periphery. Several diseases, in particular hypertension, have been examined as possible "alpha-2 adrenergic receptor diseases."

The etiology of hypertension is for the most part unknown. Many animal models of hypertension have been developed, including both genetic and nongenetic models. These models in part mimic human forms of the disease, which are both primary ("essential") and secondary to disorders in specific organ systems. As recently reviewed in several articles (Insel and Motulsky, 1984; Gavras, 1986; Rosendorff et al., 1985), changes in adrenergic response, and perhaps in adrenergic receptors, may be associated with clinical and experimental hypertension. This association has suggested that the change in adrenergic receptors may contribute to the development, maintenance, or both of those phases of the hypertensive state.

In the past several years, several investigators have examined changes in alpha-2 adrenergic receptors in hypertensive animals. Pettinger and colleagues have focused on changes in alpha-2 adrenergic receptors in crude membranes prepared from kidneys of hypertensive animals and have found a substantial increase in the number of receptors in membranes prepared from several strains of rats with genetic forms of hypertension (Pettinger et al., 1982a,b; Graham et al., 1982; Sanchez and Pettinger, 1982). In general, the increase in receptor number preceded the increase in blood pressure, thus suggesting that the enhanced number of receptors may be of pathogenetic significance rather than a consequence of the disease. Moreover, rats placed on diets with an elevated content of NaCl had further increases in receptor number, and these increases were greater in "salt-sensitive" than in "salt-resistant" or control rats (Pettinger et al., 1982b; Diop et al., 1984). Salt loading also increases blood pressure and decreases circulating catecholamines in such animals. In addition, other evidence has indicated that renal alpha-2 adrenergic receptors may be involved in regulation of salt excretion (Insel et al., 1985; DiBona, 1985; Insel and Snavely, 1981; Pettinger et al., 1985).

Other workers have observed possible increases in tail vein and platelet alpha-adrenergic receptors in genetic hypertension in rats (Hicks et al., 1984; Minuth and Jakobs, 1983), increases in brain alpha-2 adrenergic receptors in specific brain regions [e.g., hypothalamus (Morris et al., 1981; Wirz-Justice et al., 1983)] and decreases in other brain regions [e.g., medulla oblongata (Nomura et al., 1985)]. Increases in ability of NaCl added in vitro to increase the number of brain and renal alpha-2 adrenergic receptors in hypertensive animals has also been reported (Parini et al., 1985). In other animal models of hypertension, such as in mineralocorticoid/NaCl-induced hypertension or renal forms of hypertension, changes in alpha-2 adrenergic receptors have not been as clearly demonstrated as in the genetic strains (Rosendorff et al., 1985; Berthelot et al., 1982; Hicks et al., 1983; Pettinger et al., 1982a). Various hypotheses have been presented to explain the possible interrelationships among sodium, alpha-2 adrenergic receptors, and hypertension (Gavras, 1986; Insel and Motulsky, 1984; Blaustein, 1977). Taken together, the published data do not clearly distinguish between changes in alpha-2 adrenergic receptors related to genetics, to hypertension, or to salt loading; careful studies of receptor expression, salt balance [including assays of intracellular sodium content (Motulsky and Insel, 1983)] and plasma and tissue catecholamines in specific regions of brain, kidney, and other tissues (e.g., vessels) will probably be required to help unravel these relationships.

Data obtained from studies of platelets from human subjects have also not produced convincing evidence for changes in alpha-2 adrenergic receptors in essential hypertension (Brodde et al., 1985; Boon et al., 1983a; Kafka et al., 1982; Motulsky et al., 1983a). Recent data suggest, however, that patients who respond to salt loading with larger changes in blood pressure may represent a subset of patients who are most likely to show changes in receptor expression in this disease (Ashida et al., 1985). More comprehensive studies, which perhaps stratify patients in terms of salt sensitivity and other variables, will be required to test the hypothesis that one or more subsets of patients with hypertension have either primary or secondary alteration in alpha-2 adrenergic receptors.

Orthostatic hypotension is a symptom observed in diseases that are associated with low blood pressure when a person assumes an upright posture. The hypertension may result from a variety of different disease processes within the central and peripheral nervous system; some of these have been characterized as syndromes of autonomic nervous system dysfunction (Robertson and Hollister, 1987; Polinsky et al., 1981). Published data indicate that the number of platelet alpha-2 adrenergic receptors may be increased in patients with orthostatic hypotension, perhaps as a consequence of the lower circulating level of catecholamines observed in some patients (Kafka et al., 1984; Brodde et al., 1983; Davies et al., 1982a; Chobanian et al., 1982). The heterogeneity of disease processes responsible for orthostatic hypotension is suggested by reports indicating that other patients with this disease may have either no change or a decreased number of platelet alpha-2 adrenergic receptors (Weiss et al., 1984a; Pfeifer et al., 1984; Kafka et al., 1984).

Studies in other cardiovascular diseases are limited. Some recent physiologic data have suggested that dogs with stenotic coronary arteries demonstrate an unmasking of sympathetic vasoconstriction in the coronary arteries, and that this vasoconstriction is mediated by postjunctional alpha-2 adrenergic receptors (Heusch and Deussen, 1983). We are not aware of radioligand binding assays of receptors in such animals, but two groups of investigators have observed a decrease in number of platelet alpha-2 adrenergic receptors in patients with symptomatic coronary artery disease (angina pectoris) (Mehta et al., 1985; Weiss and Smith, 1983). This decrease in receptor number seemed to be unrelated to plasma catecholamine concentrations and, in a small number of patients, this change in platelet alpha-2 receptor density reverted toward normal when patients were less symptomatic. In one study, the investigators found that platelets from symptomatic patients had a 10–20-fold enhancement in the affinity of receptors for epinephrine, as assessed in both binding and functional assays (Mehta et al., 1985). These intriguing observations will require confirmation, since the mechanism responsible for such a decrease in platelet receptors is not readily apparent. In addition, in one report, patients with severe congestive heart failure were noted to have a decreased number of platelet alpha-2 adrenergic receptors in association with an increase in plasma norepinephrine levels (Weiss et al., 1983).

4.2. Platelet Disorders

As noted above, the number of platelet alpha-2 adrenergic receptors may be changed in settings in which platelets themselves are not altered. In addition, one group of patients has been described in which expression of platelet alpha-2 adrenergic receptors changes as a function of an intrinsic alteration in platelets. Essential thrombocythemia is a myeloproliferative disorder in which a larger number of abnormal platelets is produced, and in some patients these platelets demonstrate blunted aggregation in response to epinephrine (Kaywin et al., 1978; Swart et al., 1985). Platelets from such patients were reported initially to have a decreased number of [^{3}H]DHE binding sites, but an unaltered affinity of these sites for agonists and antagonists (Kaywin et al., 1978; Pfeifer et al., 1984). Since platelets have limited ability to synthesize protein, and thus to synthesize receptors, decreases in number of platelet receptors in myeloproliferative disorders most likely result from defective receptor synthesis in megakaryocytes, the platelet precursor cells in the bone marrow.

Swart et al. (1984) have reported that platelets from patients with myeloproliferative disorders may have poor aggregatory response to epinephrine, but a normal number of [^{3}H]yohimbine binding sites. Taken together with the reports described above indicating a decreased density of [^{3}H]DHE binding rates in platelets from patients with myeloproliferative disorders, the data suggest that sites detected by [^{3}H]DHE, but not by [^{3}H]yohimbine, may be important for platelet aggregation. In other studies, Swart et al. (1985) found that epinephrine-nonresponsive platelets from patients with myeloproliferative disorders have an unaltered ability of epinephrine both to inhibit cyclic-AMP accumulation in intact platelets and to compete for [^{3}H]yohimbine binding sites in membranes (whether or not guanine nucleotides or NaCl were added). These results provide further evidence that factors that are parallel or distal to receptor-mediated inhibition of adenylate cyclase are important for alpha-2 adrenergic mediated platelet aggregation.

4.3. Psychiatric Disorders

An extensive literature has developed over the past two decades regarding the possibility that alterations in catecholamines levels or in the response to catecholamines mediate certain psychiatric syndromes, in particular affective disorders. Although early studies focused on analyses of catecholamines or catecholamine metabolites in affective disorders, studies in the past several years have emphasized possible changes in adrenergic receptors in psychiatric disorders (Elliott, 1984; Siever et al., 1983a; Insel and Cohen, 1987). Studies of changes in alpha-2 receptor properties have included both animal and human studies. In the animal studies, the principal emphasis has been to define sites and mechanisms of action of drugs, in particular antidepressant agents, in the central nervous system (Sugrue, 1983). Data have been presented that suggest that both electroconvulsive shock treatment and chronic administration with antidepressant drugs can decrease the number of alpha-2 adrenergic receptors in the brain, although this has not been found by all workers (Elliott, 1984; Siever et al, 1983a; Sugrue, 1983; Cash et al., 1984). Moreover, whether the drug-induced decrease in brain alpha-2 adrenergic receptors is a key mechanism by which antidepressant drugs elicit their therapeutic effect is controversial (Sugrue, 1983).

In human studies, platelets have been used as a model system in order to assess possible changes in alpha-2 receptors before and after drug treatment of psychiatric patients. Because human platelets actively take up and store serotonin, it has been proposed that they are models for serotonin nerve terminals (Stahl, 1985). Platelets are obviously different from neuronal cells in the brain in many respects, however, including the lack of nuclei, their failure to synthesize much protein, the absence of axons, dendrites and action potentials, their life span, and, importantly, their life cycle spent within the circulation unlike neuronal cells located behind the blood–brain barrier. Nevertheless, platelets have represented an accessible cell in which to explore possible changes in alpha-2 adrenergic receptors in psychiatric patients.

Several recent articles have reviewed studies that have examined whether or not platelet alpha-2 adrenergic receptors are altered in patients with affective disorders (Elliott, 1984; Wood et al., 1985; Insel and Cohen, 1987). Although the data are not consistent, four independent studies that have studied the properties of platelet alpha-2 receptors using [^{3}H]yohimbine or [^{3}H]rauwolscine have reached the same conclusion: platelet binding characteristics are similar in depressed patients and in controls. Some of the studies that reached a different conclusion have involved use of less suitable radioligands, such as [^{3}H]DHE or [^{3}H]clonidine. Nevertheless, workers who have found an increased number of [^{3}H]clonidine binding sites on platelets from untreated, depressed patients (Smith et al., 1983) have also observed enhanced alpha-2 adrenergic receptor-mediated aggregation in response to epinephrine in depressed patients (Garcia-Sevilla et al., 1983). Thus, further studies would appear to be necessary to determine whether depressed patients (or perhaps a subgroup of such patients) have an increase in number of "functionally active" alpha-2 adrenergic receptors. Studies with the newly available probe, [^{3}H]UK 14,304, which is a full agonist in platelets, should prove useful for such experiments.

4.4. Miscellaneous Diseases

Studies of radioligand binding to platelet alpha-2 adrenergic receptors in patients with bronchial asthma (Reinhardt et al., 1984; Davis and Lieberman, 1982), cystic fibrosis (Davis et al., 1984), chronic obstructive pulmonary disease (Bruynzeel et al., 1983), heroin addiction (Garcia-Sevilla et al., 1985), or uremia (Jacobsson et al., 1985) have indicated that the number of receptors is apparently unchanged in those settings. Certain settings, however, may be associated with changes in binding of agonists [e.g., Parkinson's disease (Cash et al., 1984) and heroin addiction (Garcia-Sevilla et al., 1985)] or with alterations in agonist-diated inhibition of adenylate cyclase activity [e.g., uremia (Jacobsson et al., 1985)]. Whether the failure to see changes in receptor number in several diseases is indicative of the relative stability of platelet alpha-2 adrenergic receptors or of the inability of the platelet to mirror changes that occur in alpha-2 adrenergic receptors on other important target cells remains to be established.

5. Summary and Conclusions

Although it is clear that alpha-2 adrenergic receptors exert an important role in regulation of a number of target cells, and although the literature contains many reports describing altered expression of alpha-2 adrenergic receptors, few firm conclusions can be reached. Indeed there is at present no unequivocal evidence that alterations in the expression of alpha-2 receptors mediate altered responses to catecholamines. It appears, however, that important changes in receptor expression may occur during ontogeny, aging and obesity and with physiologic or pharmacologic elevations in sex steroids. Agonist-mediated down-regulation of receptor number, an important mechanism by which most receptors are regulated, may occur for alpha-2 adrenergic receptors in the brain, but occurs to only a limited extent in other tissues. Conversely, denervation or treatments that deplete tissue catecholamines can mediate an up-regulation of alpha-2 adrenergic receptor number, especially in the brain. Hypertension and other cardiovascular diseases may be associated with changes in alpha-2 adrenergic receptors, but the data are too scanty to clearly identify an "alpha-2 adrenergic receptor disease."

As discussed above, inconsistency between studies and lack of rigor in many studies hamper ability to reach firm conclusions. Additionally, there are some major gaps in basic knowledge of alpha-2 adrenergic receptors and the mechanisms by which these receptors control cellular functions. Some of these gaps in knowledge that have impeded design of optimal studies to examine the regulation of these receptors are as follows:

1. Are alpha-2 receptors homogeneous? Recent evidence suggests that several alpha-2 adrenergic receptor subtypes may exist, and that [^{3}H]yohimbine and [^{3}H]rauwolscine may interact with more than one of these subtypes (Bylund, 1985 and Chapter 1 in this volume). If so, then previous studies of receptor regulation may have yielded misleading results because of the failure to define changes in individual subpopulations of receptors.

2. What does [^{3}H]clonidine binding measure? Many studies, particularly those of alpha-2 adrenergic receptors in the brain have used [^{3}H]clonidine (or [^{3}H]p-aminoclonidine). As we have discussed, clonidine has different pharmacological properties in different tissues, but it is generally considered to be a partial agonist. Agonist ligands preferentially recognize high-affinity alpha-2 adrenergic receptors; they do not bind to all alpha-2 receptors and cannot be used as probes to "count" receptors. The fact that clonidine is not a full agonist in many tissues adds ad-

ditional complexity, which may be resolved by the recent availability of the full agonist [^{3}H]UK 14,304. Moreoever, high-affinity binding of agonist ligands only occurs in the absence of guanine nucleotides; no detectable high-affinity binding occurs with intact cells. Washed membrane preparations are often considered to be devoid of guanine nucleotides, but this is not necessarily true (Neubig and Szamraj, 1986). Thus, although we believe that it is important to define the interaction of agonists with alpha-2 receptors, it is difficult to interpret the results of many such studies.

3. What is the occupancy–response relationship for alpha-2 receptors? An assumption implicit in many radioligand binding studies is that changes in receptor number will necessarily lead to changes in response, but this may not always be true. Indeed, evidence has been presented for an alpha-2 "receptor reserve," albeit small, in human platelets (Lenox et al., 1985). The response most often measured is inhibition of adenylate cyclase activity, but alpha-2 adrenergic receptors may also act through other mechanisms (Limbird and Sweatt, 1985). Knowledge is still incomplete as to how alpha-2 adrenergic receptors interact with G_i proteins. Although the number of G_i units far exceeds the number of alpha-2 receptors (at least in platelet membranes), the amount of G_i may be functionally limiting (Neubig et al., 1985), perhaps because of compartmentation or heterogeneity between receptors.

4. What is the "life cycle" of an alpha-2 adrenergic receptor? Several factors have made studies of the metabolism of these receptors a difficult undertaking: limited availability of cultured cell models, the lack of a high affinity, iodinated probe to quantify low concentrations of alpha-2 receptors, the inadequacy of available techniques to define receptor turnover (in particular, the lack of antibodies to the receptor and the limited specificity of irreversible antagonists that have been used to examine receptor turnover), and the failure to utilize methods that can provide estimates of the rates of receptor appearance and disappearance (Snavely et al., 1985a; Mahan and Insel, 1986).

5. Are platelet alpha-2 adrenergic receptors different from alpha-2 adrenergic receptors in other tissues? The platelet is a readily accessible and fascinating cell, and it has become the favorite model for studies of alpha-2 adrenergic receptors. Platelets have unusual morphology, biochemistry, and life history, however, and they may not be the ideal model system for examining phenomenology and mechanisms of alpha-2 receptor regulation. Moreover, there is a considerable (up to several-fold) interindividual variability in receptor number (Snavely et al., 1982; Motulsky et al., 1983a); twin studies prove that this variability is in part attributable to genetic factors (Propping and Friedl, 1983).

It has only been about a decade since alpha-2 adrenergic receptors were first identified. Since then considerable progress has been made in defining the biochemical features of these receptors. The study of the regulation of alpha-2 adrenergic receptors has lagged behind such studies for other receptor types. As new methods are developed and as more rigorous and insightful experiments are performed, a more definitive understanding of the regulation of this important class of membrane receptors should be forthcoming. The challenge to investigators in this research area is to develop the necessary methods and to design such experiments.

Acknowledgments

Work in the authors' laboratory has been supported by grants from the National Institutes of Health, National Science Foundation, and American Heart Association.

REFERENCES

Aghajanian, G. K. (1982) Central noradrenergic neurons: A locus for the functional interplay between alpha-2 adrenoceptors and opiate receptors. *J. Clin. Psychiat.* **43**, 20–23.

Agrawal, D. K. and Daniel, E. E. (1985) Two distinct populations of [^{3}H]prazosin and [^{3}H]yohimbine binding sites in the plasma membranes of rat mesenteric artery. *J. Pharmacol. Exp. Ther.* **233**, 195–200.

Arnett, C. D. and Davis, J. N. (1979) Denervation-induced changes in alpha and beta adrenergic receptors of the rat submandibular gland. *J. Pharmacol. Exp. Ther.* **211**, 394–400.

Ashida, T., Tanaka, T., Yokouchi, M., Kuramochi, M., Deguchi, F., Kimura, G., Kojima, S., Ito, K., and Ikeda, M. (1985) Effect of dietary sodium on platelet alpha-2-adrenergic receptors in essential hypertension. *Hypertension* **7**, 972–978.

Barnathan, E. S., Addonizio, V. P., and Shattil, S. J. (1982) Interaction of verapamil with human platelet alpha-adrenergic receptors. *Am. J. Physiol.* **242**, H19–H23.

Barnett, D. B., Nahorski, S. R., and Richardson, A. (1984) Modulation of [3-H]-dihydroergocryptine and [3-H]-yohimbine binding sites on human platelets during the menstrual cycle. *Br. J. Pharmacol.* **81**, 159P.

Bennett, J. S. and Vilaire, G. (1979) Exposure of platelet fibrinogen receptors by ADP and epinephrine. *J. Clin. Invest.* **64**, 1393–1401.

Berlan, M. and Lafontan, M. (1982) The alpha-2-adrenergic receptor of human fat cells: Comparative study of alpha-2-adrenergic ligand binding and biological response. *J. Physiol.* (Paris) **78**, 279–287.

Berlan, M., Carpene, C., Lafontan, M., and Dang-Tran, L. (1982) Alpha-2 adrenergic antilipolytic effect in dog fat cells: Incidence of obesity and adipose tissue location. *Horm. Metabol. Res.* **14**, 257–260.

Berthelot, A., Hamilton, C. A., Petty, M. A., and Reid, J. L. (1982) Central and peripheral alpha-adrenoceptor number and responsiveness after sinoaortic denervation in the rabbit. *J. Cardiovasc. Pharmacol.* **4**, 567–574.

Blaustein, M. P. (1977) Sodium ions, calcium ions and hypertension: A reassessment and a hypothesis. *Am. J. Physiol.* **232**, C165–C173.

Bobik, A. and Anderson, W. P. (1983) Influence of sympathectomy on alpha-2 adrenoceptor binding sites in canine blood vessels. *Life Sci.* **33**, 331–336.

Boon, N. A., Elliott, J. M., Davies, C. L., Conway, F. J., Jones, J. V., Grahame-Smith, D. G., and Sleight, P. (1983a) Platelet alpha-adrenoceptors in borderline and established essential hypertension. *Clin. Sci.* **65**, 207–208.

Boon, N. A., Elliott, J. M., Davies, C. L., Conway, F. J., Jones, J. V., Grahame-Smith, D. G., and Sleight, P. (1983b) Platelet alpha-adrenoceptors in borderline and established essential hypertension. *Clin. Sci.* **65**, 207–208.

Bottari, S. P., Vokaear, A., Kaivez, E., Lescrainier, P., and Vauquelin, G. (1983) Differential regulation of the alpha-adrenergic receptor subclasses by gonadal steroids in human myometrium. *J. Clin. Endocrinol. Metab.* **57**, 937–943.

Boyle, F. C. and Digges, K. G. (1982) Responses to catecholamines of the rat isolated uterus throughout the natural oestrous cycle. *Naunyn Schmiedebergs Arch. Pharmacol.* **321**, 56–62.

Brodde, O. E. (1983) Endogenous and exogenous regulation of human alpha- and beta-adrenergic receptors. *J. Recept. Res.* **3**, 151–162.

Brodde, O.-E. and Bock, K. D. (1984) Changes in platelet alpha$_2$-adrenoceptors in human phaeochromocytoma. *Eur. J. Clin. Pharmacol.* **26**, 265–267.

Brodde, O. E., Anlauf, M., Graben, N., and Bock, K. D. (1982) In vitro and in vivo down regulation of human platelet alpha-2-adrenergic receptors by clonidine. *Eur. J. Clin. Pharmacol.* **23**, 403–409.

Brodde, O.-E., Anlauf, M., Arroyo, J., Wagner, R., Weber, F., and Buck, K. D. (1983) Hypersensitivity of adrenergic receptors and blood-pressure response to oral yohimbine in orthostatic hypotension. *N. Eng. J. Med.* **308**, 1033.

Brodde, O.-E., Daul, A. E., O'Hara, N., and Khalifa, A. M. (1985) Properties of alpha- and beta-adrenoceptors in circulating blood cells of patients with essential hypertension. *J. Cardiovasc. Pharmacol.* **7**, S162–S167.

Brown, M. S., Anderson, R. G. W., and Goldstein, J. L. (1983) Recycling receptors: The round-trip itinerary of migrant membrane proteins. *Cell* **32**, 663–667.

Browning, S., Lawrence, D., Livingston, A., and Morris, B. (1982) Interactions of drugs active at opiate receptors and drugs active at alpha-2-receptors on various test systems. *Br. J. Pharmacol.* **77**, 487–491.

Bruynzeel, P. L. B., Hamelink, M. L., Kok, P. T. M., van der Vet, A. P. H., and Kreukniet, J. (1983) Alpha-2-adrenergic receptors on intact human platelets of normal individuals and chronic obstructive lung disease patients. *Eur. J. Resp. Dis.* **65** (suppl.35), 139–142.

Burns, T. W., Langley, P. E., Terry, B. E., and Bylund, D. B. (1982) Studies on desensitization of adrenergic receptors of human adipocytes. *Metabolism* **31**, 288–293.

Bylund, D. A. (1985) Heterogeneity of alpha-2-adrenergic receptors. *Pharmacol. Biochem. Behav.* **22**, 835–843.

Bylund, D. B. and Martinez, J. R. (1980) Alpha-2-adrenergic receptors appear in rat salivary glands after reserpine treatment. *Nature* **285,** 229–230.

Bylund, D. B. and Martinez, J. R. (1981) Postsynaptic localization of alpha-2-adrenergic receptors in rat submandibular gland. *J. Neurosci.* **1,** 1003–1007.

Cash, R., Ruberg, M., Raisman, R., and Agid, Y. (1984) Adrenergic receptors in Parkinson's disease. *Brain Res.* **322,** 269–275.

Chobanian, A. V., Tifft, C. P., Sackel, H., and Pitruzella, A. (1982) Alpha and beta adrenergic receptor activity in circulating blood cells of patients with idiopathic orthostatic hypotension and pheochromocytoma. *Clin. Exp. Hypertens.* **A4,** 793–806.

Ciofalo, F. R. (1980) Effect of some antiarrythmics on [3H]clonidine binding to alpha-2-adrenergic receptors. *Eur. J. Pharmacol.* **65,** 309–312.

Cohen, R. M., Campbell, I. C., Dauphin, M., Tallman, J. F., and Murphy, D. L. (1982a) Changes in alpha- and beta-receptor densities in rat brain as a result of treatment with monoamine oxidase inhibiting antidepressants. *Neuropharmacology* **21,** 293–298.

Cohen, R. M., Aulakh, C. S., Campbell, I. C., and Murphy, D. L. (1982b) Functional subsensitivity of alpha-2-adrenoceptors accompanies reductions in yohimbine binding after clorgyline treatment. *Eur. J. Pharmacol.* **81,** 145–148.

Cooper, B., Handin, R. I., Young, L. H., and Alexander, R. W. (1978) Agonist regulation of the human platelet alpha-adrenergic receptor. *Nature* **274,** 703–706.

Corby, D. G. and O'Barr, T. P. (1981) Decreased alpha-adrenergic receptors in newborn platelets: Cause of abnormal response to epinephrine. *Dev. Pharmacol. Ther.* **2,** 215–225.

Dausse, J.-P., Quan-Bui, K. H. L., and Meyer, P. (1982) Alpha-1 and alpha-2-adrenoceptors in rat cerebral cortex: Effects of neonatal treatment with 6-hydroxydopamine. *Eur. J. Pharmacol.* **78,** 15–20.

Davies, A. O. and Lefkowitz, R. J. (1983) In vitro desensitization of beta adrenergic receptors in human neutrophils. *J. Clin. Invest.* **71,** 565–571.

Davies, B., Sudera, D., Sagnella, G., Marchesi-Saviotti, E., Mathias, C., Bannister, R., and Sever, P. (1982a) Increased numbers of alpha receptors in sympathetic denervation supersensitivity in man. *J. Clin. Invest.* **69,** 779–784.

Davies, I. B., Mathias, C. J., Sudera, D., and Sever, P. S. (1982b) Agonist regulation of alpha-adrenergic receptor responses in man. *J. Cardiovasc. Pharmacol.* **4,** S139–S144.

Davis, P. B. and Lieberman, P. (1982) Normal alpha-2-adrenergic responses in platelets from patients with asthma. *J. Allergy Clin. Immunol.* **69,** 35.

Davis, P. B., Dieckman, L., Boat, T. F., Stern, R. C., and Doershuk, C. F. (1984) The alpha-2-adrenergic system of the platelet in cystic fibrosis. *Am. J. Med. Sci.* **288,** 104–108.

DiBona, G. F. (1985) Neural control of renal function: Role of renal alpha-adrenoceptors. *J. Cardiovas. Pharmacol.* **7,** (suppl. 8), S18–S23.

Diop, L., Parini, A., Dausse, J. P., and Ben-Ishay, D. (1984) Cerebral and renal alpha-adrenoceptors in sabra hypertensive and normotensive rats: Effects of high-sodium diet. *J. Cardiovasc. Pharmacol.* **6,** S742–S747.

DiStefano, P. S. and Brown, O. M. (1985) Biochemical correlates of morphine withdrawal. 2. Effects of clonidine, *J. Pharmacol. Exp. Ther.* **233**, 339–344.

Docherty, J. R. and O'Malley, K. (1985) Ageing and alpha-adrenoceptors. *Clin. Sci.* **68**(supp.10), 133s–136s.

Egan, B., Neubig, R., and Julius, S. (1985) Pharmacologic reduction of sympathetic drive increases platelet alpha-2-receptor number. *Clin. Pharmacol. Ther.* **38**, 519–524.

Elliott, J. M. (1984) Platelet receptor binding studies in affective disorders. *J. Affective Disord.* **6**, 219–239.

Esler, M. (1982) Assessment of sympathetic nervous function in humans from noradrenaline plasma kinetics. *Clin. Sci.* **62**, 247–254.

Fain, J. N. and Garcia-Sainz, J. A. (1983) Adrenergic regulation of adipocyte metabolism. *J. Lipid. Res.* **24**, 945–966.

Fox, A. W., Juberg, E. N., Marvin, J., Johnson, R. D., Abel, P. W., and Minneman, K. P. (1985) Thyroid status and adrenergic subtypes in the rat: Comparison of receptor density and responsiveness. *J. Pharmacol. Exp. Ther.* **235**, 715–723.

Fuller, R. W. (1982) Pharmacology of brain epinephrine neurons. *Ann. Rev. Pharmacol. Toxicol.* **22**, 31–55.

Garcia-Sainz, A. A., Litosch, I., Hoffman, B. B., Lefkowitz, R. J., and Fain, J. M. (1981) Effect of thyroid status on alpha- and beta-catecholamine responsiveness of hamster adipocytes. *Biochim. Biophys. Acta* **678**, 334–341.

Garcia-Sevilla, J. A., Garcia-Vallejo, P., and Guimon, J. (1983) Enhanced alpha-2-adrenoceptor-mediated platelet aggregation in patients with major depressive disorder. *Eur. J. Pharmacol.* **94**, 359–360.

Garcia-Sevilla, J. A., Ugedo, L., Ulibarri, I., and Gutierrez, M. (1985) Platelet alpha-2-adrenoceptors in heroin addicts during withdrawal and after treatment with clonidine. *Eur. J. Pharmacol.* **114**, 365–374.

Gavras, H. (1986) How does salt raise blood pressure? A hypothesis. *Hypertension* **8**, 83–88.

Giudicelli, Y., Lacasa, D., and Agli, B. (1980) White fat cell alpha-adrenergic receptors and responsiveness in altered thyroid status. *Biochem. Biophys. Res. Commun.* **94**, 1113–1122.

Graham, R. M., Pettinger, W. A., Sagalowsky, A., Brabson, J., and Gandler, T. (1982) Renal alpha-adrenergic receptor abnormality in the spontaneously hypertensive rat. *Hyptension* **4**, 881–887.

Goodhardt, M., Ferry N., Geynet, P., and Hanoune, J. (1982) Alpha$_1$-adrenergic receptors show agonist-specific regulation by guanine nucleotides. *J. Biol. Chem.* **257**, 11577–11583.

Greenberg, L. H. (1986) Regulation of brain adrenergic receptors during aging. *Fed. Proc.* **45**, 55–59.

Gross, G. and Schumann, H. J. (1981) Reduced number of alpha-2-adrenoceptors in cortical brain membranes of hypothyroid rats. *J. Pharm. Pharmacol.* **33**, 552–554.

Hallam, T. J., Scrutton, M. C., and Wallis, R. B. (1981) Responses of rabbit platelets to adrenaline induced by other agonists. *Thromb. Res.* **21**, 413–424.

Hamburg, M. and Tallman, J. F. (1981) Chronic morphine administration increases the apparent number of alpha-2-adrenergic receptors in rat brain. *Nature* **291**, 493–495.

Hamilton, C. A., Dalrymple, H. W., and Reid, J. L. and Sumner, D. J. (1984) The recovery of alpha-adrenoceptor function and binding sites after phenoxybenzamine. *Naunyn Schmiedebergs Arch. Pharmacol.* **325**, 34–41.

Hamilton, C. A., Jones, C. R., Mishra, N., Barr, S., and Reid, J. L. (1985) A comparison of alpha-2-adrenoceptor regulation in brain and platelets. *Brain Res.* **347**, 350–353.

Harden, T. K. (1983) Agonist-induced desensitization of the beta-adrenergic receptor-linked adenylate cyclase. *Pharmacol. Rev.* **35**, 5–32.

Heufelder, A., Warnhoff, M., and Prike, K. M. (1985) Platelet alpha-2-adrenoceptor and adenylate cyclase in patients with anorexia nervosa and bulemia. *J. Clin. Endo. Met.* **61**, 1053–1060.

Heusch, G. and Deussen, A. (1983) The effects of cardiac sympathetic nerve stimulation on perfusion of stenotic coronary arteries in the dog. *Circ. Res.* **53**, 8–15.

Hicks, P. E., Nahorski, S. R., and Cook, N. (1983) Postsynaptic alpha-adrenoceptors in the hypertensive rat: Studies on vascular ractivity in vivo and receptor binding in vitro. *Clin. Exp. Hyper.* **A5**, 401–427.

Hicks, P. E., Medgett, I. C., and Langer, S. Z. (1984) Postsynaptic alpha-2-adrenergic receptor-mediated vasoconstriction in SHR tail arteries in vitro. *Hypertension* **6**, 112–118.

Hoffman, B. B., Lavin, T. N., Lefkowitz, R. J., and Ruffolo, R. R., Jr. (1981) Alpha adrenergic receptor subtypes in rabbit uterus: Mediation of myometrial contraction and regulation by estrogens. *J. Pharmacol. Exp. Ther.* **219**, 290–295.

Hollister, A. S., Fitzgerald, G. A., Nadeau, H. J. J., and Robertson, D. (1983) Acute reduction in human platelet alpha-2 adrenoceptor affinity for agonist by endogenous and exogenous catecholamines. *J. Clin. Invest.* **72**, 1498–1505.

Hyland, L. and Docherty, J. R. (1985) An investigation of age-related changes in pre- and postjunctional alpha-adrenoceptors in human saphenous vein. *Eur. J. Pharmacol.* **114**, 361–364.

Insel, T. R. and Cohen, R. M. (1987) Adrenergic Receptors in Psychiatric Diseases, in *Adrenergic Receptors in Man* (Insel, P. A., ed.) Marcel Dekker, New York.

Insel, P. A. and Motulsky, H. J. (1984) A hypothesis linking intracellular sodium, membrane receptors, and hypertension. *Life Sci.* **34**, 1009–1113.

Insel, P. A. and Snavely, M. D. (1981) Catecholamines and the Kidney: Receptors and renal function. *Ann. Rev. Physiol.* **43**, 625–636.

Insel, P. A., Snavely, M. D., Healy, D. P., Munzel, P. A., Potenza, C. L., and Nord, E. P. (1985) Radioligand binding and functional assays demonstrate postsynaptic alpha-2-receptors on proximal tubules of rat and rabbit kidney. *J. Cardiovas. Pharmacol.* 7(suppl. 8, S9–S17.

Insel, P. A., Howard, M. J., Motulsky, H. J., and Hughes, R. J. (1987) Amiloride: An alpha-adrenergic antagonist in platelets, kidney and muscle cells. *J. Cardiovasc. Pharmacol.*, in press.

Ishii, K. and Kato, R. (1984) Development of tolerance to alpha-2 adrenergic agonists in the vascular system of the rat after chronic treatment with clonidine. *J. Pharmacol. Exp. Ther.* **231**, 685–690.

Ishii, K., Yamamoto, S., and Kato, R. (1982) Development of clonidine-tolerance in the rat vas deferens: Cross tolerance to other presynaptic inhibitory agents. *Life Sci.* **30**, 285–292.

Jacobsson, B., Ransnas, L., Nyberg, G., Bergh, C. H., Maagnusson, Y. and Hjalmarson, A. (1985) Abnormality of adenylate cyclase regulation in human platelet membranes in renal insufficiency. *Eur. J. Clin. Invest.* **15**, 75–81.

Jones, S. B., Bylund, D. B., Rieser, C. A., Shekim, W. O., Byer, J. A., and Carr, G. W. (1983) Alpha 2-adrenergic receptor binding in human platelets: alterations during the menstrual cycle. *Clin. Pharmacol. Ther.* **34**, 90–96.

Jones, C. R., Hamilton, C. A., Whyte, K. F., Elliott, H. L., and Reid, J. L. (1985a) Acute and chronic regulation of alpha(2)-adrenoceptor number and function in man. *Clin. Sci.* **68**(suppl. 10), 129s–132s.

Jones, C. R., McCabe, R., Hamilton, C. A., and Reid, J. L. (1985b) Maternal and fetal platelet responses and adrenoceptor binding characteristics. *Throm. Haemost.* **53**, 95–98.

Jones, C. R., Giembcyz, M., Hamilton, C. A., Rodger, I. W., Whyte, K., Deighton, N., Elliott, H. L., and Reid, J. L. (1986) Desensitization of platelet alpha-2-adrenoceptors after short term infusions of adrenergic agonists in man. *Clin. Sci.* **70**, 147–153.

Kafka, M. S., Lake, C. R., Gullner, H. G., Tallman, J. P., Bartter, F. C., and Fujita, T. (1982) Adrenergic receptor function is different in male and female patients with essential hypertension. *Clin. Exp. Hyperten.* **1**, 613–627.

Kafka, M. S., Polinsky, R. J., Williams, A., Kopin, I. J., Lake, C. R., Ebert, M. H., and Tokola, N. S. (1984) Alpha-adrenergic receptors in orthostatic hypotension syndromes. *Neurology* **34**, 1121–1125.

Kande, K., Gjesdal, K., Fonstelien, E., Kjeldsen, S. E., and Eide, I. (1985) Effects of adrenaline infusion on platelet number, volume and release reaction. *Throm. Heamost.* **54**, 450–453.

Kano, T. (1982) Effects of estrogen and progesterone on adrenoceptors and cyclic nucleotides in rat uterus. *Jpn. J. Pharmacol.* **32**, 535–549.

Karliner, J. S., Motulsky, H. J., and Insel, P. A. (1982) Apparent "downregulation" of human platelet alpha 2-adrenergic receptors is due to retained agonist. *Mol. Pharmacol.* **21**, 36–43.

Kather, H., Wieland, E., Fischer, B., Wirth, A., and Schlierf, G. (1985) Adrenergic regulation of lipolysis in abdominal adipocytes of obese subjects during caloric restriction: reversal of catecholamine action caused by relief of endogenous inhibition. *Eur. J. Clin. Invest.* **15**, 30–37.

Kaywin, P., McDonough, M., Insel, P. A., and Shattil, S. J. (1978) Platelet function in essential thrombocythemia. Decreased epinephrine responsiveness associated with a deficiency of platelet alpha-adrenergic receptors. *N. Engl. J. Med.* **299**, 505–509.

Kelly, J. and O'Malley, K. (1984) Adrenoceptor function and ageing. *Clin. Sci.* **66**, 509–515.

Krall, J. F., Mori, H., Tucks, M. L., LeShon, S. L., and Korenman, S. G. (1978) Demonstration of adrenergic catecholamine receptors in rat myometrium and their regulation by sex steroid hormones. *Life Sci.* **23**, 1073–1082.

Kunos, G., Ramirez-Gonzelez, M. D., and Farsang, C. (1985) Adrenergic-Opiate Interactions and the Central Control of Blood Pressure in Hypertension, in *Norepinephrine: Clinical Aspects* (Lake, R. C. and Ziegler, M., eds.) Williams and Wilkinson, Baltimore.

Lafontan, M. (1979) Inhibition of epinephrine-induced lipolysis in isolated white adipocytes of aging rabbits by increased alpha-adrenergic responsiveness. *J. Lipid Res.* **20**, 208–216.

Lai, R.-T., Watanabe, Y., and Yoshida, H. (1983) The influence of aging on alpha-adrenoceptors in rat heart and vas deferens. *Jpn. J. Pharmacol.* **33**, 241–245.

Lanza, F. and Cazenave, J.-P. (1985) Studies of alpha-2-adrenergic receptors of intact and functional washed human platelets by binding of 3H-dihydroergocryptine and 3H-yohimbine: correlation of 3H-yohimbine binding with the potentiation by adrenaline of ADP-induced aggregation. *Throm. Heamost.* **54**, 402–408.

Latifpour, J., Jones, S. B., and Bylund, D. B. (1982) Characterization of [3-H]yohimbine binding to putative alpha-2-adrenergic receptors in neonatal rat lung. *J. Pharmacol. Exp. Ther.* **223**, 606–611.

Lenox, R. H., Ellis, J., Van Riper, D., and Ehrlich, Y. H. (1985) Alpha 2-adrenergic receptor-mediated regulation of adenylate cyclase in the intact human platelet. Evidence for a receptor reserve. *Mol. Pharmacol.* **27**, 1–9.

Levin, R. M., Shofer, F. S., and Wein, A. J. (1980) Estrogen-induced alterations in the autonomic responses of the rabbit urinary bladder. *J. Pharmacol. Exp. Ther.* **215**, 614–618.

Limbird, L. E. and Sweatt, J. D. (1985) Alpha-2-Adrenergic Receptors: Apparent Interaction with Multiple Effector Systems, in *The Receptors* vol. II (Conn., P. M., ed.) Academic, Orlando, Florida.

Loftus, D. J., Stolk, J. M., and U'Prichard, D. C. (1984) Binding of the imidazoline UK-14, 304, a putative full alpha-2-adrenoceptor agonist, to rat cerebral cortex. *Life Sci.* **35**, 61–69.

Luck, P., Mikhailidis, M. R., Dashwood, M. R., Barradas, M. A., Sever, P. S., Dandona, P., and Wakeling, A. (1983) Platelet hyperaggregability and increased alpha-adrenoceptor density in anorexia nervosa. *J. Clin. Endo. Met.* **57**, 911.

Maeda, H., Watanabe, Y., Lai, R.-T., and Yoshida, H. (1983) Effect of glucocorticoids on alpha-2-adrenoceptors in vas deferens of reserpinized rat in organ culture. *Life Sci.* **33**, 39–46.

Mahan, L. C. and Insel, P. A. (1986) Expression of beta-adrenergic receptors in synchronous and asynchronous S49 lymphoma cells. 1. Receptor metabolism after irreversible blockade of receptors and in cells traversing the cell cycle. *Mol. Pharmacol.* **29**, 7–15.

Mahan, L. C., McKernan, R. M., and Insel, P. A. (1987) Metabolism of alpha and beta-adrenergic receptors in vitro and in vivo. *Ann. Rev. Pharmacol. Toxicol.* **27**, 215–235.

Maurin, Y., Le Saux, F., Graillot, C., and Baumann, N. (1985) Altered postnatal ontogeny of alpha-1 and alpha-2-adrenoceptor binding sites in the brain of a convulsive mutant mouse (quaking). *Dev. Brain Res.* **22**, 229–235.

McKernan, R. M. and Campbell, I. C. (1982) Measurement of alpha-adrenoceptor "turnover" using phenoxybenzamine. *Eur. J. Pharmacol.* **80**, 279–280.

McKernan, R. M., Howard, M. J., Motulsky, H. J., and Insel, P. A. (1987) Compartmentation of alpha-2-adrenergic receptors of human erythroleukemia (HEL) cells. *Mol. Pharmacol.,* **32**, 258–265.

Mehta, J., Mehta, P., and Ostrowski, N. (1985) Increase in human platelet alpha-2-adrenergic receptor affinity for agonist in unstable angina. *J. Lab. Clin. Med.* **106**, 661–666.

Metz, A., Stump. K., Cowen, P. J., Elliot, J. M., Gelder, M. G., and Grahame-Smith, D. G. (1983) Changes in platelet alpha-2 adrenoceptor binding post-partum: Possible relation to maternity blues. *Lancet* **1**, 495–498.

Meyers, K. M., Huston, L. Y., and Clemmons, R. M. (1983) Regulation of canine platelet function. II. Catecholamines. *Am. J. Physiol.* **14**, R100-R109.

Mikuni, M., Stoff, D. M., and Meltzer, H. Y. (1983) Effects of combined administration of imipramine and chlorpromazine on beta- and alpha-2-adrenergic receptors in rat cerebral cortex. *Eur. J. Pharmacol.* **89**, 313–316.

Mills, D. C. B. and Roberts, G. C. U. (1967) The effects of adrenaline on human blood platelets. *J. Physiol.* (Lond.) **193**, 443–453.

Minuth, M. and Jakobs, K. H. (1983) Alpha2–adrenoceptors in platelets of spontaneously hypertensive rats. *Naunyn Schmiedebergs Arch. Pharmacol.* **322**, 98–103.

Mishra, N., Hamilton, C. A., Jones, C. R., Leslie, C., and Reid, J. L. (1985) Alpha-adrenoceptor changes after oestrogen treatment in platelets and other tissues in female rabbits. *Clin Sci.* **69**, 235–238.

Mitrius, J. C. and U'Prichard, D. C. (1985) Regulation of alpha-2-adrenoceptors by nucleotides, ions, and agonists: Comparison in cells of neural and nonneural origin. *Adv. Cyc. Nuc. Res.* **19**, 57–73.

Mooney, J. J., Horne, W. C., Handin, R. I., Schildkraut, J. J., and Alexander, R. W. (1982) Sodium inhibits both adenylate cyclase and high-affinity 3H-labeled p-aminoclonidine binding to alpha 2-adrenergic receptors in purified human platelet membranes. *Mol. Pharmacol.* **21**, 600–608.

Morris, M. J., Devynck, M.-A., Woodcock, E. A., Johnston, C. I., and Meyer, P. (1981) Specific changes in hypothalamic alpha-adrenoceptors in young spontaneously hypertensive rats. *Hypertension* **3**, 516–520.

Motulsky, H. J. and Insel, P. A. (1982) [3H]Dihydroergocryptine binding to alpha-adrenergic receptors of human platelets. A reassessment using the selective radioligands [3H]prazosin, [3H]yohimbine, and [3H]rauwolscine. *Biochem. Pharmacol.* **31**, 2591–2597.

Motulsky, H. J. and Insel, P. A. (1983) Influence of sodium on the alpha 2-adrenergic receptor system of human platelets. Role for intraplatelet sodium in receptor binding. *J. Biol. Chem.* **258**, 3913–3919.

Motulsky, H. J., O'Connor, D. T., and Insel, P. A. (1983a) Platelet alpha 2-adrenergic receptors in treated and untreated essential hypertension. *Clin. Sci.* **64**, 265–272.

Motulsky, H. J., Snavely, M. D., Hughes, R. J., and Insel, P. A. (1983b) Interaction of verapamil and other calcium channel blockers with alpha 1- and alpha 2-adrenergic receptors. *Circ. Res.* **52**, 226–231.

Motulsky. H. J., Maisel, A. S., Snavely, M. D., and Insel, P. A. (1984) Quinidine is a competitive antagonist at alpha 1- and alpha 2-adrenergic receptors. *Circ. Res.* **55**, 376–381.

Motulsky, H. J., Mahan, L. C., and Insel, P. A. (1985) Radioligands, agonists, and membrane receptors on intact cells: data analysis in a bind. *Trends Pharmacol. Sci* **6**, 317–319.

Motulsky, H. J., Shattil, S. J., Ferry, N., Rozansky, D., and Insel, P. A.

(1986) Desensitization of epinephrine-initiated platelet aggregation does not alter binding to the alpha$_2$-adrenergic receptor or its coupling to adenylate cyclase. *Mol. Pharmacol.* **29**, 1-6.

Muraki, T., Nakaki, T., and Kato, R. (1984) Predominance of alpha-2-adrenoceptors in porcine thyroid: Biochemical and pharmacological correlations. *Endocrinology* **114**, 1645-1651.

Neubig, R. R. and Szamraj, O. (1986) Large-scale purification of alpha-2-adrenergic receptor-enriched membranes from human platelets. Persistent association of guanine nucleotides with nonpurified membranes. *Biochim. Biophys. Acta* **854**, 67-76.

Neubig. R. R., Gantoz, R. D., and Brasier, R. S. (1985) Agonist and antagonist binding to alpha2-adrenergic receptors in purified membranes from human platelets. Implications of receptor-inhibitory nucleotide-binding protein stoichiometry. *Mol. Pharmacol.* **28**, 475-486.

Nomura, M., Ohtsuji, M., and Nagata, Y. (1985) Changes in the alpha-adrenoceptors in the medulla oblongata including nucleus tractus solitarii of spontaneously hypertensive rats. *Neurochem. Res.* **10**, 1143-1154.

O'Brien, J. R. (1964) Variability in the aggregation of human platelets by adrenaline. *Nature* **202**, 1188-1189.

Ohguchi, S., Sotabata, I., Oguro, K., and Nakashima, M. (1984) Changes in the effects of clonidine on left atrium and hindlimb vasculature of rats in various thyroid states. *Jpn. Heart J.* **25**, 425-437.

O'Malley, B. W. (1984) Steroid hormone action in eucaryotic cells. *J. Clin. Invest.* **74**, 307-312.

Orensanz, L. M., Guillamon, A., Ambrosio, E., Segovia, S., and Azuara, M. C. (1982) Sex differences in alpha-adrenergic receptors in the rat brain. *Neurosci. Lett.* **30**, 275-278.

Parini. A., Diop, L., Dausse, J.-P., and Ben-Ishay, D. (1985) Sabra rats as a model to differentiate between Na$^+$ and GTP regulation of alpha-2-adrenoceptor densities. *Eur. J. Pharmacol.* **112**, 97-104.

Pastan, I. and Willingham, M. C. (1985) *Endocytosis* Plenum, New York

Pecquery, R., Leneveu, M.-C., and Giudicelli, Y. (1984) In vivo desensitization of the beta, but not the alpha$_2$-adrenoreceptor-coupled-adenylate cyclase system in hamster white adipocytes after administration of epinephrine. *Endocrinology* **114**, 1576-1583.

Perry, B. D., Simon, P. R., and U'Prichard, D. C. (1983a) Interactions of neuroleptic compounds at alpha-2-adrenergic receptor affinity states in bovine caudate nucleus. *Eur. J. Pharmacol.* **95**, 315-318.

Perry, B. D., Stolk, J. M., Vantini, G., Guchhait, R. B., and U'Prichard, D. C. (1983b) Strain differences in rat brain epinephrine synthesis: Regulation of alpha-adrenergic receptor number by epinephrine. *Science* **221**, 1297-1299.

Peters, J. R., Elliott, J. M., and Grahame-Smith, D. G. (1979) Effect of oral contraceptives on platelet noradrenaline and 5-hydroxytryptamine receptors and aggregation. *Lancet* **2**, 933-936.

Pettinger, W. A., Sanchez, A., Saavedra, J., Haywood, J. R., Gandler, T., and Rodes, T. (1982a) Altered renal alpha-2-adrenergic receptor regulation in genetically hypertensive rats. *Hypertension* **4**, 188-192.

Pettinger, W. A., Gandler, T., Sanchez, A., and Saavedra, J. M. (1982b) Dietary sodium and renal alpha-2-adrenergic receptors in Dahl hypertensive rats. *Clin. Exp. Hypertens.* **A4**, 819-828.

Pettinger, W. A., Smyth, D. D., Unemura, S., Yu, C., Yang, E., and Fallet, R. (1985) Renal Alpha-2-Adrenoceptors, Sodium and Hypertension, in *Adrenergic Receptors: Molecular Properties and Therapeutic Implications* (Lefkowitz, R. J. and Lindenlaub, E., eds.) Verlag, Stuttgart.

Pfeifer, M. A., Ward, K., Malpass, T., Stratton, J., Halter, J., Evans, M., Beiter, H., Harker, L. A., and Porte, D., Jr. (1984) Variations in circulating catecholamines fail to alter human platelet alpha-2-adrenergic receptor number or affinity for [^{3}H]yohimbine or [^{3}H]dihydroergocryptine. *J. Clin. Invest.* **74**, 1063–1072.

Plow, E. F. and Marguerie, G. A. (1980) Induction of the fibrinogen receptor on human platelets by epinephrine and the combination of epinephrine and ADP. *J. Biol. Chem.* **255**, 10971–10977.

Polinsky, R. J., Kopin, I. J., Ebert, M. H., and Weise, V. (1981) Pharmacologic distinction of different orthostatic hypotension syndromes. *Neurology* **31**, 1–7.

Propping, P. and Friedl, W. (1983) Genetic control of adrenergic receptors on human platelets. A twin study. *Hum. Genet.* **64**, 105–109.

Ratge, D., Hansel-Bessey, S., and Wisser, H. (1985) Altered plasma catecholamines and numbers of alpha- and beta-adrenergic receptors in platelets and leucocytes in hyperthyroid patients normalized under antithyroid treatment. *Acta Endocrinol.* **110**, 75–82.

Redmond, D. E., ed. (1982) Central alpha-adrenergic mechanisms in opiate withdrawal and other psychiatric syndromes. Studies with clonidine. *J. Clin. Psychiatry* **43**, 4–48.

Redmond, D. E., Jr. and Huang, Y. H. (1982) The primate locus coeruleus and effects of clonidine on opiate withdrawal. *J. Clin. Psychiatry* **43**, 25–29.

Reinhardt, D., Zehmisch, T., Becker, B., and Nagel-Hiemke, M. (1984) Age-dependency of alpha- and beta-adenoceptors on thrombocytes and lymphocytes of asthmatic and nonasthmatic children. *Eur. J. Pediatr.* **142**, 111–116.

Ringold, G. M. (1985) Steroid hormone regulation of gene expression. *Ann. Rev. Pharmacol. Toxicol.* **25**, 529–566.

Roberts, J. and Steinberg, G. M. (1986) Effects of aging on adrenergic receptors (symposia of seven papers). *Fed. Proc.* **45**, 40–64.

Roberts, J. M., Insel, P. A., Goldfien, R. D., and Goldfien, A. (1977) Alpha adrenoceptors but not beta adrenoceptors increase in rabbit uterus with oestrogen. *Nature* **270**, 624–625.

Roberts, J. M., Goldfien, R. D., Tsuchiya, A. M., Goldfien, A., and Insel, P. A. (1979) Estrogen treatment decreases alpha-adrenergic binding sites on rabbit platelets. *Endocrinology* **104**, 722–728.

Roberts, J. M., Insel, P. A., and Goldfien, A. (1981) Regulation of myometrial adrenoceptors and adrenergic response by sex steroids. *Mol. Pharmacol.* **20**, 52–58.

Roberts, J. M., Lewis, V., Mize, N., Tsuchiya, A., and Starr, J. (1986) Human platelet alpha-adrenergic receptors and responses during pregnancy: No change except that with differing hematocrit. *Am. J. Obstet. Gynecol.* **154**, 206–210.

Robertson, D., and Hollister, A. S. (1987) Adrenergic Receptors in Neurological Disorders, in *Adrenergic Receptors in Man* (Insel P. A., ed.) Marcel Dekker, New York.

Robson, L. E., Mucha, R. F., and Kosterlitz, H. W. (1983) The role of interac-

tion between opiate receptors and alpha-adrenergic receptors in tolerance and dependence. *Biochem. Soc. Trans.* **11**, 64–65.

Rosen, S. G., Berk, M. A., Popp, D. A., Serusclat, P., Smith, E. B., Shah, S.
D., Ginsberg, A. M., Clutter, W. E., and Cryer, P. E. (1984) Beta2 and
alpha2-adrenergic receptors and receptor coupling to adenylate cyclase
in human mononuclear leukocytes and platelets in relation to physiological variations of sex steroids. *J. Clin. Endocrinol. Metab.* **58**,
1068–1075.

Rosendorff, C., Susanni, E., Hurwitz, M. L., and Ross, F. P. (1985)
Adrenergic receptors in hypertension: Radioligand binding studies. *J.
Hypertension* **3**, 571–581.

Roth, R. H., Elsworth, J. D., and Redmond, D. E., Jr. (1982) Clonidine suppression of noradrenergic hyperactivity during morphine withdrawal
by clonidine: Biochemical studies in rodents and primates. *J. Clin. Psychiatry* **43**, 42–46.

Rowe, J. W. and Troen, B. R. (1980) Sympathetic nervous system and aging
in man. *Endocrinol. Rev.* **1**, 167–179.

Sabol, S. L. and Nirenberg, M. (1979) Regulation of adenylate cyclase of
neuroblastoma × glioma hybrid cells by alpha-adrenergic receptors. *J.
Biol. Chem.* **254**, 1921–1926.

Sanchez, A. and Pettinger, W. A. (1982) Dietary sodium regulation of blood
pressure and renal alpha-1 and alpha-2-receptors in WKY and SH rats.
Life Sci. **29**, 2795–2802.

Scarpace, P. J. (1986) Decreased beta-adrenergic responsiveness during se-
nescence. *Fed. Proc.* **45**, 51–54.

Sethy, V. H. and Harris, D. W. (1982) Effect of chronic morphine treatment
on alpha-2-adrenergic receptors in rat brain and spinal cord. *Res.
Commun. Substances Abuse* **3**, 121–124.

Sharma, S. K., Klee, W. A., and Nirenberg, M. (1975) Dual regulation of
adenylate cyclase accounts for narcotic dependence and tolerance. *Proc.
Natl. Acad. Sci. USA* **72**, 3092–3096.

Shattil, S. J., Bennett, J. S., McDonough, M., and Turnbull, J. (1980)
Carbenicillin and penicillin G inhibit platelet function in vitro by im-
pairing the interaction of agonists with the platelet surface. *J. Clin. Invest.* **65**, 329–337.

Shattil, S. J., McDonough, M., Turnbull, J., and Insel, P. A. (1981) Characterization of alpha-adrenergic receptors in human platelets using
[3H]clonidine. *Mol. Pharmacol.* **19**, 179–183.

Shattil, S. J., Motulsky, H. J., Insel, P. A., Flaherty, L., and Brass, L. F.,
(1986) Expression of fibrinogen receptors during activation and subsequent desensitization of human platelets by epinephrine. *Blood* **680**,
1224–1231.

Sibley, D. R. and Lefkowitz, R. J. (1985) Molecular mechanisms of receptor
desensitization using the beta adrenergic receptor-coupled adenylate
cyclase as a model. *Nature* **317**, 124–129.

Siever, L. J., Uhde, T. W., Jimerson, D. C., Kafka, M. S., Lake, C. R.,
Targum, S., and Murphy, D. L. (1983a) Clinical studies of monoamine
receptors in the affective disorders and receptor changes with
antidepressant treatment. *Prog. Neuro-psychopharmacol. Biol. Psychiatry*
7, 249–261.

Siever, L. J., Kafka, M. S., Insel, T. R., Lake, C. R., and Murphy, D. L.
(1983b) Effect of long-term clorgyline administration on human platelet

alpha-adrenergic receptor binding and platelet cyclic AMP responses. *Psychiatry Res.* **9**, 37–44.

Simantov, R., Baram, D., Levy, R., and Nadler, H. (1982) Enkephalin and alpha-adrenergic receptors: Evidence for both common and differentiable regulatory pathways and down-regulation of the enkephalin receptor. *Life Sci.* **31**, 1323–1326.

Sladeczek, F. and Bockaert, J. (1983) Turnover in vivo of alpha-1-adrenergic receptors in rat submaxillary glands. *Mol. Pharmacol.* **23**, 282–288.

Smith, C. B., Hollingsworth, P. J., Garcia-Sevilla, J. A., and Zis, A. P. (1983) Platelet alpha-2-adrenoreceptors are decreased in number after antidepressant therapy. *Prog. Neuro-psychopharmacol. Biol. Psychiatry* **7**, 241–247.

Snavely, M. D., Motulsky, H. J., O'Connor, D. T., Ziegler, M. G., and Insel, P. A. (1982) Adrenergic receptors in human and experimental pheochromocytoma. *Clin Exp. Hypertens. (A)* **4**, 829–848.

Snavely, M. D., Mahan, L. C., O'Connor, D. T., and Insel, P. A. (1983) Selective down-regulation of adrenergic receptor subtypes in tissues from rats with pheochromocytoma. *Endocrinology* **113**, 354–361.

Snavely, M. D., Ziegler, M. G., and Insel, P. A. (1985a) A new approach to determine rates of receptor appearance and disappearance in vivo: Agonist-mediated down-regulation of rat renal cortical beta-1 and beta-2 adrenergic receptors. *Mol. Pharmacol.* **27**, 19–26.

Snavely, M. D., Ziegler, M. G., and Insel, P. A. (1985b) Subtype-selective down-regulation of rat renal cortical alpha- and beta-adrenergic receptors by catecholamines. *Endocrinology* **117**, 2182–2189.

Stahl, S. M. (1985) Platelets as Pharmacologic Models for the Receptors and Biochemistry of Monoaminergic Neurons, in *The Platelets: Physiology and Pharmacology* (Longenecker, G., eds.) Academic, Orlando, Florida.

Stolk, J. M., Vantini, G., Perry, B. D., Guchhait, R. B., and U'Prichard, D. C. (1984) Assessment of the functional role of brain adrenergic neurons: Chronic effects of phenylethanolamine N-methyltransferase inhibitors and alpha adrenergic receptor antagonists on brain norepinephrine metabolism. *J. Pharmacol. Exp. Ther.* **230**, 577–586.

Sugrue, M. F. (1982) A study of the sensitivity of rat brain alpha-2-adrenoceptors during chronic antidepressant treatments. *Naunyn Schmiedebergs Arch. Pharmacol.* **320**, 90–96.

Sugrue, M. F. (1983) Do antidepressants possess a common mechanism of action? *Biochem. Pharmacol.* **32**, 1811–1817.

Sundaresan, P. R., Weintraub, M., Hershey, L. A., Kroening, B. H., Hasday, J. D., and Banerjee, S. P. (1983) Platelet alpha-adrenergic receptors in obesity: Alteration with weight loss. *Clin. Pharmacol. Ther.* **33**, 776–785.

Sundaresan, P. R., Madan, M. K., Kelvie, S. L., and Weintraub, M. (1985) Platelet alpha-2 adrenoceptors and the menstrual cycle. *Clin. Pharmacol. Ther.* **37**, 337–342.

Swart, S. S., Pearson, D., Wood, J. K., and Barnett, D. B. (1984) Functional significance of the platelet alpha-2-adrenoceptor: Studies in patients with myeloproliferative disorders. *Thrombosis Res.* **33**, 531–541.

Swart, S. S., Maguire, M., Wood, J. K., and Barnett, D. B. (1985) Alpha-2-adrenoceptor coupling to adenylate cyclase in adrenaline insensitive human platelets. *Eur. J. Pharmacol.* **116**, 113–119.

Sweatt, J. D., Johnson, S. L., Cragoe, E. J., and Limbird, L. E. (1985) Inhibitors of Na + /H + exchange block stimulus-provoked arachiodinic acid release in human platelets. Selective effects on platelet activation by epinephrine, ADP, and lower concentrations of thrombin. *J. Biol. Chem.* **260**, 12910–12919.

Thomas, J. M. and Hoffman, B. B. (1986) Agonist-induced down-regulation of muscarinic cholinergic and alpha-2-adrenergic receptors after inactivation of N_i by pertussis toxin. *Endocrinology* **119** 1305–1314.

Thoolen, M. J. M. C., Timmermans, P. B. M. W. M., and Van Zwieten, P. A. (1981) Morphine suppresses the blood pressure responses to clonidine withdrawal in the spontaneously hypertensive rat. *Eur. J. Pharmacol.* **71**, 351–353.

Tseng, L.-F., Loh, H. H., and Wei, E. T. (1975) Effects of clondine on morphine withdrawal signs in the rat. *Eur. J. Pharmacol.* **30**, 93–99.

Turner, J. T., Ray-Prenger, C., and Bylund, D. B. (1985) Alpha2-adrenergic receptors in the human cell line HT29. Characterization with the full agonist radioligand [^{3}H]UK-14, 304 and inhibition of adenylate cyclase. *Mol. Pharmacol.* **28**, 422–430.

Vantini. G., Perry, B. D., Guchhait, R. B., U'Prichard, D. C., and Stolk, J. M. (1984) Brain epinephrine systems: Detailed comparison of adrenergic and noradrenergic metabolism, receptor number and in vitro regulation, in two inbred rat strains. *Brain, Res.* **296**, 49–65.

Villeneuve, A., Carpene, C., Berlan, M., and Lafontan, M. (1985a) Lack of desensitization of alpha2-mediated inhibition of lipolysis in fat cells after acute and chronic treatment with clonidine. *J. Pharmacol. Exp. Ther.* **233**, 433–440.

Villeneuve, A., Berlan, M., Lafontan, M., and Montastruc, J. L. (1985b) Characterization of dog platelet alpha-adrenergic receptor: Lack of in vivo down regulation by adrenergic agonist treatments. *Comp. Biochem. Physiol.* **81C**, 181–187.

Watanabe, Y., Lai, R.-T., Maida, H., and Yoshida, H. (1982) Reserpine and sympathetic denervation cause an increase of postsynaptic alpha-2-adrenoceptors. *Eur. J. Pharmacol.* **80**, 105–108.

Watanabe, Y., Rong-Tsan, L., and Yoshida, H. (1983) Increase of [^{3}H]clonidine binding sites induced by adenosine receptor agonists in rat vas deferens in vitro. *Eur. J. Pharmacol.* **86**, 265–269.

Weiss, R. J. and Smith, C. B. (1983) Altered platelet alpha-2-adrenoceptors in patients with angina pectoris. *J. Am. Coll. Cardiol.* **2**, 631–637.

Weiss, R. J., Tobes, M., Wertz, C. E., and Smith, B. (1983) Platelet alpha-2-adrenoreceptors in chronic congestive heart failure. *Am. J. Cardiol.* **52**, 101–105.

Weiss, R. J., Dix, B. R., Kissner, P. Z., and Smith, C. B. (1984a) Altered platelet alpha-2-adrenoreceptors in orthostatic hypotension. *Clin. Cardiol.* **7**, 599–602.

Weiss, B., Clark, M. B., and Greenberg, L. H. (1984b) Modulation of Catecholaminergic Receptors During Development and Aging, in *Handbook of Neurochemistry* (Lajtha A., eds.) Plenum, New York.

Whitsett, J. A., Noguchi, A., and Moore, J. J. (1982) Developmental aspects of alpha- and beta-adrenergic receptors. *Semin. Perinatol.* **6**, 125–141.

Wilkinson, M. and Herdon, H. J. (1982) Diethylstilbesterol regulates the number of alpha- and beta-adrenergic binding sites in incubated hypothalamus and amygdala. *Brain Res.* **248**, 79–85.

Wilkinson, M., Herdon, H., Pearce, M., and Wilson, C. (1979) Precocious puberty and changes in alpha- and beta-adrenergic receptors in the hypothalamus and cerebral cortex of immature female rats. *Brain Res.* **167**, 195–199.

Williams, L. T. and Lefkowitz, R. J. (1977) Regulation of rabbit myometrial alpha adrenergic receptors by estrogen and progesterone. *J. Clin. Invest.* **60**, 815–818.

Wirz-Justice, A., Krauchi, K., Campbell, I. C., and Feer, H. (1983) Adrenoceptor changes in spontaneous hypertensive rats: A circadian approach. *Brain Res.* **262**, 233–242.

Wong, D. T., Bymaster, F. P., Reid, L. R., and Threlkeld, P. G. (1983) Fluoxetine and two other serotonin uptake inhibitors without affinity for neuronal receptors. *Biochem. Pharmacol.* **32**, 1287–1293.

Wood, K., Swade, C., and Coppen, A. (1985) Depression: Ligand binding and aggregation studies. *Acta. Pharmacol. Toxicol.* **56**(suppl. 1), 203–211.

SECTION 6
FUTURE VISTAS

Chapter 7

What Happens Next?

A Hypothesis Linking the Biochemical and
Electrophysiological Sequelae of alpha-2 Adrenergic
Receptor Occupancy with the Diverse Receptor-
Mediated Physiological Effects

Lori L. Isom and Lee E. Limbird

1. Introduction

The original intent of this chapter was to summarize the contents
of the preceding chapters in this volume on alpha-2 adrenergic receptors and provide a prospectus for future studies. The thorough content of each of the chapters, however, makes a summary
per se redundant. Instead, we wondered how we could make
these chapters "talk to" each other, since they deal with topics as
diverse as receptor heterogeneity (Chapter 1) and structure
(Chapter 2), mechanisms for coupling with components of the
adenylate cyclase system (Chapter 3), and multiple expressions of
physiological effects in normal (Chapter 4) and in altered (Chapter 5) physiological states. We decided to review briefly the biochemical and electrophysiological consequences of alpha-2
adrenergic receptor occupancy as a foundation for an hypothesis
that integrates changes measured by the biochemist (Δcyclic
AMP, $\Delta[Ca^{2+}]$, and ΔpH_i) with those measured by the electro-
physiologist (ΔCa^{2+} and K^+ conductances) and those measured

by the cellular physiologist (stimulation or inhibition of secretion). We hope that this framework, which is the basis for our working model for alpha-2 adrenergic receptor function, serves also to unify the diverse contents of this volume and thus provide both a summary and an agenda for future investigation.

Examination of Table 1 leads to the striking observation that the physiological effects elicited by alpha-2 adrenergic receptors vary widely, and may even be opposing, depending on the target system. For example, agonist occupation of alpha-2 adrenergic receptors in the human platelet and kidney *promotes* a secretory response, whereas agonist occupation of alpha-2 adrenergic receptors in central and peripheral neurons, small intestine, and pancreatic islet cells *inhibits* secretion. What, if any, are the common elements in these opposite responses elicited by alpha-2 adrenergic receptors? Furthermore, how may these elements be related to the seemingly disparate alpha-2 adrenergic receptor-mediated effects of vasoconstriction in the pulmonary artery and saphenous vein, and vasodilation, resulting, presumably, from the release of endothelial cell-derived relaxing factor (EDRF) in the aorta (summarized in Chapter 4)?

Traditionally, alpha-2 adrenergic receptors are described as one of a population of receptors linked to the inhibition of adenylate cyclase, with a resultant decrease in cellular cyclic AMP accumulation. These receptors mediate inhibition of adenylate cyclase via a GTP-binding protein (G_i) that is distinct from that GTP-binding protein that mediates stimulation of adenylate cyclase (G_s) (for review, *see* Limbird, 1981 and Gilman, 1984). In many systems, however, the ultimate physiological effects elicited by alpha-2 adrenergic receptors cannot be explained solely by decreases in the production of intracellular cyclic AMP, as summarized in Table 2. In some studies, certain experimental manipulations have merely dissociated receptor-mediated events from decreases in the level of cellular cyclic AMP. In other studies, however, alpha-2 adrenergic-receptor mediated events have been observed to occur even in the face of elevated cyclic AMP levels or when exogenous cyclic AMP analogs are present. Since much of the work described in Table 2 has been reviewed elsewhere (Limbird and Sweatt, 1985), the present summary will focus on more recent observations of alpha-2 adrenergic receptor-mediated biochemical events.

alpha-2 Adrenergic receptors on human platelets elicit a sequence of morphological and secretory events that include primary aggregation, the reversible platelet–platelet interaction that results from stimulus-provoked exposure of fibrinogen receptors,

Table 1
Summary of Physiological Effects Elicited by alpha-2 Adrenergic Receptors

Tissue	Effect	Reference[a]
Human platelet	Reversible primary aggregation, secretion of dense granule contents, irreversible secondary aggregation	Kerry and Scrutton (1985), for review
Aorta	Release of endothelial cell-derived relaxing factor, vasodilation in some species	Coore and Randle (1964); Calvete et al. (1984); Miller and Vanhoutt (1985); *also see* Chapter 4
Kidney	Increased secretion of Na^+ and H_2O, reduced release of vasopressin	Standhoy (1985), for review; *also see* Chapter 4
Pancreatic islet cells	Inhibition of glucose-stimulated insulin release	Porte et al. (1966); Malaisse et al. (1970)
Peripheral and central neurons	Inhibition of neurotransmitter release	*See* review in this chapter and Chapter 4
Gastrointestinal tract	Inhibition of Na^+, Cl^-, and H_2O secretion, inhibition of short circuit current, species-specific changes in motility	Field et al. (1975); Nakaki et al. (1983a); *also see* Chapter 4
Toad skin	Inhibition of basal and stimulated osmotic water permeability	Gamundi et al. (1986)
Pulmonary artery	Vasoconstriction	*See* Chapter 4
Saphenous vein	Vasoconstriction	*See* Chapter 4

[a] These references are by no means all-inclusive, but rather provide representative reports as examples.

Table 2
Examples of alpha-2 Adrenergic Receptor-Elicited Physiological Effects
that Cannot Be Explained Solely by Decreases in cyclic-AMP

Tissue	Effect	Experimental Paradigm	Reference
Human platelet	PLA_2-mediated arachidonic acid release and secretion	Removal of extracellular Na^+ results in blockade of arachidonic acid release and secretion but has no effect on decreases in cyclic AMP	Connolly and Limbird (1983); Sweatt et al. (1985; 1986 a,b)
Human platelet	Primary aggregation	Pretreatment of platelets with epinephrine results in desensitized aggregatory response but has no effect on decreases in cyclic AMP	Motulsky et al. (1986)
Intestinal epithelium	Inhibition of Na^+ and H_2O secretion	Clonidine blocks secretion but has no effect on the level of cyclic AMP in rat enterocytes; epinephrine and clonidine block secretion elicited by dibutyryl cyclic AMP in rat and rabbit ileum	Nakaki et al. (1983a) Field et al. (1975); Nakaki et al. (1982)
Pancreatic islet cells	Inhibition of glucose-stimulated insulin release	Alpha-2 adrenergic receptor agonists block insulin release elicited by dibutyryl cyclic AMP	Malaisse et al. (1970); Nakai et al. (1983b); Ullrich and Wollheim (1984)
Embryonic dorsal root ganglion (drg) neurons	Inhibition of voltage-dependent Ca^{2+}-channels[a]	Intracellular dialysis of cells with solutions containing cyclic AMP and isobutylmethylxanthine do not alter noradrenaline-induced inhibition of Ca^{2+} current	Holz et al. (1986b)

[a]Nontraditional alpha-2 adrenergic pharmacologic response.

secretion of dense granule contents, and, ultimately, irreversible secondary aggregation. Recently, in a study of epinephrine-provoked primary aggregation, Motulsky et al. (1986) demonstrated that pretreatment of platelets with epinephrine causes a desensitization of the epinephrine-induced aggregatory response with no effect on the capacity of epinephrine to decrease cyclic AMP. These results suggest that decreases in cyclic AMP alone are insufficient to account for the mechanism by which alpha-2 adrenergic receptors evoke primary aggregation.

Recent work in our laboratory has demonstrated that alpha-2 adrenergic receptor-mediated human platelet secretion also can be dissociated from alpha-2 adrenergic receptor-mediated attenuation of cyclic AMP accumulation. Specifically, the removal of extracellular Na^+ has no effect on the capacity of epinephrine to attenuate PGE_1-stimulated cyclic AMP accumulation, but blocks the ability of epinephrine to elicit arachidonic acid release, a crucial prerequisite for secretion to occur in response to epinephrine (Connolly and Limbird, 1983). Thus, epinephrine, acting through an alpha-2 adrenergic receptor, activates a phosphatidylinositol-hydrolyzing phospholipase A_2 (Sweatt et al., 1986b). The arachidonic acid liberated by this enzyme, after conversion via cyclooxygenase to endoperoxides or thromboxane A_2, is then responsible for the activation of a phospholipase C (Sweatt et al., 1986a). The latter enzyme liberates inositol phosphates and diacylglycerol, putative mediators of platelet secretion following the exposure of human platelets to epinephrine. In addition, hydrolysis of diacylglycerol by a diacylglycerol lipase results in the mobilization of large, readily measurable stores of arachidonic acid. Figure 1 summarizes our current working model for human platelet activation by alpha-2 adrenergic receptors.

2. Biochemical Consequences of alpha-2 Adrenergic Receptor Occupancy in Addition to the Inhibition of Adenylate Cyclase

2.1. alpha-2 Adrenergic Receptors and Na^+/H^+ Exchange

The schematic diagram provided in Fig. 1 indicates that alpha-2 adrenergic receptor activation of phospholipase A_2 occurs via a pathway involving Na^+/H^+ exchange. This interpretation is based on the observation that perturbants of Na^+/H^+ exchange block epinephrine-stimulated lysophosphatidylinositol produc-

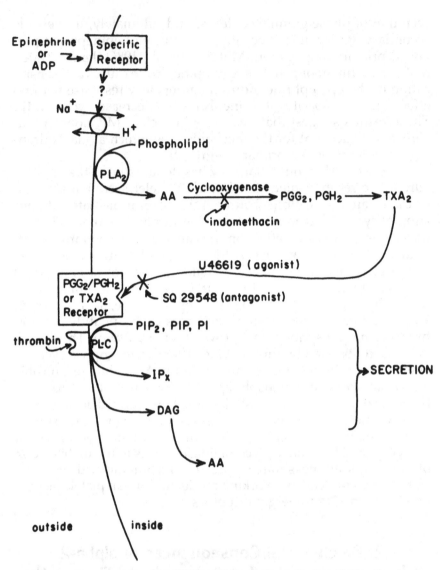

Fig. 1. Postulated pathway by which alpha-2 adrenergic receptors elicit secretion of dense granules from human platelets.

tion. These perturbants include (1) removal of extraplatelet Na^+, (2) reduction of extraplatelet pH from pH 7.35 to pH 6.8, and (3) addition of the selective inhibitor of Na^+/H^+ exchange, ethylisopropylamiloride. These same perturbants also block indomethacin-sensitive inositol phosphate production elicited by epinephrine, but do not block the production of inositol phosphates by thrombin, suggesting that the experimental manipulations used to block Na^+/H^+ exchange do not block the phospholipase C enzyme in a nonspecific manner (Sweatt et al., 1986a).

Although these findings in the human platelet suggest a role for Na^+/H^+ exchange in alpha-2 adrenergic receptor-mediated arachidonic acid release and dense granule secretion, the concurrent aggregation and secretion that follows the addition of epinephrine to platelet incubations complicates the interpretation of studies aimed at the direct measurement of Na^+/H^+ exchange in platelets exposed to alpha-2 adrenergic receptor agonists. Consequently, we studied an alternative system, the NG108-15 neuroblastoma $\times$ glioma hybrid cell line, to evaluate whether alpha-2 adrenergic receptors might activate Na^+/H^+ exchange (Isom et al., 1987a). We have observed that alpha-2 adrenergic receptor agonists accelerate Na^+/H^+ exchange in NG108-15 cells, resulting in intracellular alkalinization that can be measured using the pH-sensitive fluorescent probe 2, 7-biscarboxyethyl-5(6)-carboxyfluorescein (BCECF). Agonist-induced alkalinization is blocked by the alpha-2 adrenergic-selective antagonists, yohimbine and rauwolscine, but not by the alpha-1 adrenergic antagonist, prazosin, nor the beta-adrenergic antagonist, propranolol. Furthermore, manipulations that block Na^+/H^+ exchange (i.e., the removal of extracellular Na^+, the addition of extracellular H^+, or the addition of potent 5-amino-substituted analogs of amiloride) also block the intracellular alkalinization elicited by alpha-2 adrenergic agonists in NG108-15 cells. Manipulations that block Na^+/H^+ exchange, however, have little or no effect on the capacity of epinephrine to decrease the level of cyclic AMP in NG108-15 cells, indicating both that agonist binding to the alpha-2 adrenergic receptor remains intact under these varying experimental conditions and that Na^+/H^+ exchange is not necessary for the inhibition of adenylate cyclase.

The data from NG108-15 cells indicating that alpha-2 adrenergic receptors accelerate Na^+/H^+ exchange have several important ramifications. First, the phenomenon appears to occur in target cells other than the human platelet, where alpha-2 adrenergic receptors *provoke* dense granule secretion. Furthermore, at least one of the physiological consequences of alpha-2 adrenergic receptor agonists on neuronal cells, or on differentiated NG108-15 cells in culture, is the *suppression* of neurotransmitter release. Hence, the data suggest that the activation of Na^+/H^+ exchange may be a common component of alpha-2 adrenergic receptor-mediated physiological events whether or not the ultimate effect is pro- or anti-secretory in nature. Perhaps the intracellular alkalinization following this activation of Na^+/H^+ exchange serves as a common mechanism that modulates the activities of specific enzymes in the microenvironments shared with the alpha-2 adrenergic receptor, so as to produce the response characteristic of a given cell type.

2.2. alpha-2 Adrenergic Receptors and Ca^{2+}

We and others have shown that epinephrine stimulates Ca^{2+} mobilization in the human platelet with a time course of less than 30s (Owen and LeBreton, 1981; Rink et al., 1982; Sweatt et al., 1986b). Epinephrine elicits a rapid change in chlortetracycline fluorescence in platelets suspended in nominally Ca^{2+}-free buffer, suggesting that this change in fluorescence reflects Ca^{2+} release from an intraplatelet membrane store. Epinephrine-provoked Ca^{2+} release is not measurable using quin 2 as a probe (Johnson et al., 1983, 1985; Ware et al., 1984, 1985; Rao et al., 1985). Quin 2 can antagonize alpha-2 adrenergic activation of platelets, however, probably by chelation of a pool of Ca^{2+} involved in epinephrine-induced platelet responses (Hatayama et al., 1985; Nakamura et al., 1985; Johnson et al., 1985; Rao et al., 1986). Thus, it is likely that epinephrine, acting via an alpha-2 adrenergic receptor, may mobilize a pool of Ca^{2+} that is too small to change the overall level of cytoplasmic Ca^{2+}, but adequate to cause effects in a microenvironment of the cell interior. Johnson and coworkers (1985) obtained data consistent with this conclusion by comparing changes in intraplatelet Ca^{2+} concentrations using the photoprotein aequorin vs quin 2 as a probe for free Ca^{2+}. They found that aequorin detected microenvironments of Ca^{2+} that were undetectable using quin 2. They hypothesized that the very small changes in concentrations of intraplatelet Ca^{2+} that occur upon stimulation of platelets by agonists, such as epinephrine, do not alter the quin 2 signal. Our present hypothesis is that the alpha-2 adrenergic receptor-provoked Ca^{2+}-mobilization and the intraplatelet alkalinization that results from acceleration of Na^+/H^+ exchange converge in the human platelet to permit phospholipase A_2 activation and, ultimately, dense granule secretion. Consistent with this interpretation are the observations that artificial elevation of intraplatelet pH using methylamine (1) increases platelet sensitivity to the Ca^{2+} ionophore A23187 and (2) restores platelet sensitivity to low concentrations of thrombin following inhibition of Na^+/H^+ exchange with ethylisopropylamiloride (Sweatt et al., 1986b).

In some neuronal preparations in which alpha-2 adrenergic receptors elicit an inhibition of neurotransmitter release, it has been shown that alpha-2 adrenergic receptor occupation results in an apparent blockade of voltage-sensitive Ca^{2+} conductance, thereby reducing the inward Ca^{2+} current (Horn and McAfee, 1980; Holz et al., 1986b; Hescheler et al., 1986). Intracellular recordings made from other neuronal preparations, however, such as neurons of the myenteric plexus (Morita and North, 1981), show evidence for an alpha-2 adrenergic receptor-mediated increase in a Ca^{2+}-activated K^+ conductance, resulting in hyperpolarization of the cell and, putatively, the inhibition of neurotransmitter release. (alpha-2 Adrenergic receptor-mediated activation of K^+ conductance and inactivation of Ca^{2+} conductance will be discussed later in this chapter.) Perhaps, as in the human platelet, alpha-2 adrenergic receptors on neuronal cells act to mobilize a membrane-bound pool of intracellular Ca^{2+} ($[Ca^{2+}]_i$) that elicits effects in a microenvironment near the plasma membrane resulting in changes in Ca^{2+} conductance, K^+ conductance, or both.

alpha-2 Adrenergic receptor-activated mobilization of $[Ca^{2+}]_i$ has been studied extensively in primary cultures of rat pancreatic islet cells (Wollheim et al., 1977) and in the insulin-secreting cell line RINm5F (Ullrich and Wollheim, 1985), systems in which alpha-2 adrenergic receptors inhibit secretagogue-induced insulin release. In primary cultures of rat pancreatic islet cells, epinephrine causes a complete blockade of glucose-stimulated insulin release and a partial blockade of glucose-stimulated $^{45}Ca^{2+}$ uptake (Wollheim et al., 1977). Specifically, when insulin release is blocked completely by 1 μM epinephrine, $^{45}Ca^{2+}$ uptake is only attenuated by 60%. In fact, low levels of epinephrine (0.1 nM) cause a significant blockade of insulin release with no measurable effect on $^{45}Ca^+$ uptake. In these experiments, the effects of epinephrine were blocked by phentolamine, a mixed alpha-1, alpha-2 adrenergic antagonist. A more selective alpha-2 adrenergic receptor antagonist, such as yohimbine, was not used. In later studies it was found that clonidine inhibited glucose-stimulated insulin release without affecting Ca^{2+} uptake, suggesting that the mechanism by which alpha-2 adrenergic receptors inhibit insulin release does not involve measurable Ca^{2+} fluxes across the plasma membrane (Ullrich and Wollheim, 1985). Furthermore, in a study of pancreatic islet cell membrane potentials (Cook and Perara, 1982), it was observed that persistent glucose-induced Ca^{2+} uptake occurs in spite of the inhibitory effects of epinephrine on glucose-stimulated insulin release and hyperpolarization of the plasma membrane. These investigators presented a working model of epinephrine-mediated inhibition of glucose-stimulated insulin release that involves changes in $[Ca^{2+}]_i$ buffering, leading to the activation of a Ca^{2+}-dependent K^+ channel, rather than the inhibition of voltage-dependent inward Ca^{2+} conductance. Similarly, in the RINm5F insulin-secreting cell line, alpha-2 adrenergic receptor agonists block alanine-stimulated insulin release, but do not attenuate the secretagog-stimulated rise in $[Ca^{2+}]_i$ as measured by quin 2 fluorescence, suggesting, as in pancreatic islet cell primary culture, that alpha-2 adrenergic receptor-mediated effects in this cell line do not involve changes in the overall concentration of cytosolic free Ca^{2+}. Many possible explanations might account for these findings, however, including the possibility that epinephrine decreases the affinity of membrane binding sites for Ca^{2+}, leading to an increased accessibility of a small pool of Ca^{2+} near the plasma membrane that is crucial for changes in cellular response, albeit not accessible to measurement by cytosolic Ca^{2+} probes, such as quin 2. Again, it is interesting that in systems in which alpha-2 adrenergic receptors elicit *opposite* results (the promotion of secretion in the human platelet vs. the inhibition of secretion in neuronal cells and in pancreatic islet cells), evidence suggests that the mechanism of these alpha-2 adrenergic receptor-mediated effects involves changes in the intracellular availability of Ca^{2+}.

3. Electrophysiological Consequences of alpha-2 Adrenergic Receptor Occupancy

As described in Chapters 1 and 4, one of the first physiological responses attributed to alpha-2 adrenergic receptors was that of inhibition of the release of norepinephrine from sympathetic ganglia with the subsequent blockade of ganglionic synaptic transmission (Marrazzi, 1939a,b). Similar alpha-2 adrenergic receptor-elicited effects subsequently were described in central and parasympathetic neuronal preparations as well (*see later*). Although the molecular basis for the mechanism of alpha-2 adrenergic receptor-mediated inhibition of neurotransmitter release is not yet clear, the following mechanisms have been proposed:

1. Increased K^+ conductance, that may or may not be Ca^{2+}-dependent, resulting in hyperpolarization of the plasma membrane potential and subsequent inhibition of neuronal firing, and/or
2. Decreased Ca^{2+} conductance, resulting in a decreased availability of external Ca^{2+} for neurotransmitter secretion and therefore reduced electrosecretory coupling.

3.1. alpha-2 Adrenergic Receptors and K^+ Conductance

The mechanism of the inhibitory effects of epinephrine on transmission in the sympathetic ganglion of the cat initially was investigated electrophysiologically by Lundberg (1952). By measuring the synaptic membrane potential of the superior cervical and stellate ganglia in vivo, it was observed that epinephrine and norepinephrine caused an inhibition of transmission of a single volley pulse. Measurement of the demarcation potential between electrodes on the ganglion and the postganglionic trunk revealed that epinephrine induced hyperpolarization of the ganglion cells. Furthermore, this effect of epinephrine was attenuated selectively by the administration of dihydroergotamine, an alpha-adrenergic antagonist. Eccles and Libet (1961) studied the effect of epinephrine on changes in membrane potential in curarized rabbit sympathetic superior cervical ganglion in vivo. A reversible depression of the ganglionic spike response was observed after the administration of epinephrine that was attenuated by the prior administration of the alpha-adrenergic antagonist dibenamine. Eccles and Libet proposed that the interaction of epinephrine with receptor sites on the ganglion cell results in hyperpolarization of the ganglion cell membrane.

The hyperpolarizing effect of epinephrine and norepinephrine on the superior cervical ganglion of the cat was characterized in greater pharmacological detail by DeGroat and Volle (1966). These investigators observed that the inhibitory effects of catecholamines could be blocked selectively by the alpha-adrenergic antagonist dihydroergotamine, but not by the beta-adrenergic blocking agents dichloroisoproterenol or pronethalol. They concluded that an alpha-adrenergic receptor on ganglion cells mediates catecholamine-induced hyperpolarization and blockade of transmission. Cedarbaum and Aghajanian (1977), by recording the inhibition of spontaneous firing of locus ceruleus neurons in response to the microiontophoretic application of various agents, observed that this blockade of neurotransmission was mediated through alpha-2 adrenergic receptors. Thus, the rank order of potencies for agonists in this inhibitory response was: clonidine >> α-methylnorepinephrine > epinephrine > norepinephrine > phenylephrine. Inhibition of neurotransmitter release from locus ceruleus neurons was blocked by the alpha-adrenergic antagonist piperoxan, but not by the beta-adrenergic antagonist sotalol.

In a similar study, Brown and Caulfield (1979) presented a detailed pharmacological characterization of the hyperpolarizing effects of catecholamines on the isolated superior cervical ganglion of the rat. By administering drugs via the method of extracellular superfusion and measurement of subsequent changes in membrane potential, these investigators determined the rank order of agonist potencies to be epinephrine > norepinephrine > isoproterenol > phenylephrine. Clonidine appeared to be a weak partial agonist in this system. Additionally, the alpha-2 adrenergic antagonist, yohimbine, and the alpha-adrenergic antagonist, phentolamine, but not the beta-adrenergic antagonist propranolol, the alpha-1 adrenergic antagonist prazosin, nor the dopaminergic antagonist, haloperidol, blocked the response to epinephrine, suggesting that the inhibitory effects of catecholamines in these neurons are mediated through an alpha-2 adrenergic receptor. That hyperpolarization of these neuronal membranes and the blockade of norepinephrine release by alpha-2 adrenergic agonists are functionally correlated is suggested by the observation that both responses are enhanced by lowering of the extracellular Ca^{2+} and elevation of the extracellular K^+ concentrations. Related studies on catecholamine-mediated effects on sympathetic ganglia (Cole and Shinnick-Gallagher, 1981; Guyenet and Cabot, 1981) corroborated the above conclusion that hyperpolarization of these cells is mediated through an alpha-2 adrenergic receptor.

Stimulation of sympathetic fibers in the intestine that terminate in the myenteric plexus inhibits peristalsis (Finkleman, 1930), largely through the inhibition of acetylcholine release of the neuromuscular junction. Early studies indicated that this effect is mediated via alpha-2 adrenergic receptors. Morita and North (1981) studied alpha-2 adrenergic receptor-elicited effects in isolated neurons from the guinea pig myenteric plexus by intracellular recording in order to deduce the ionic mechanisms involved in these inhibitory events. It was observed that alpha-2 adrenergic receptor agonists hyperpolarized myenteric neurons by increasing the resting K^+ conductance. The hyperpolarizing effect of alpha-2 adrenergic agonists was enhanced in low K^+- and low Ca^{2+}-containing solutions and reduced in high K^+- or high Ca^{2+}-containing solutions. Large hyperpolarizations in response to clonidine were observed even in solutions containing no Ca^{2+} plus EGTA. The inverse Ca^{2+}-dependence of the hyperpolarization was interpreted by these investigators to mean that alpha-2 adrenergic receptors elicit an elevation of the level of intracellular Ca^{2+}, perhaps by inhibiting the binding of Ca^{2+} to intracellular storage sites, with a resultant activation of a Ca^{2+}-dependent outward K^+ conductance.

Aghajanian and VanderMaelen (1982) performed intracellular recordings of locus ceruleus neurons in vivo to investigate the inhibitory response mediated by alpha-2 adrenergic receptors. They described the occurrence of alpha-2 adrenergic receptor-mediated hyperpolarization of locus ceruleus neurons following spikes produced by intracellular depolarization or in response to the systemic administration of clonidine. Under both circumstances, hyperpolarization was associated with an increase in membrane conductance that was postulated to be specifically the result of K^+ flux, because hyperpolarization was not reversed when Cl^- containing electrodes were used. Since both the clonidine-induced hyperpolarization and the spike-induced afterhyperpolarization were blocked by the alpha-2 adrenergic receptor antagonist, piperoxan, it was concluded that catecholamines hyperpolarize central noradrenergic neurons by means of an alpha-2 adrenergic receptor. Further studies by Egan and co-workers (1983) on norepinephrine-mediated synaptic inhibition in rat locus ceruleus neurons in vitro by intracellular recording of the membrane potential led to similar conclusions. Norepinephrine-elicited hyperpolarization of the neuron was attributed to an increased K^+ conductance since the reversal potential was -104 mV, and the membrane potential became less negative as the concentration of extracellular K^+ increased. The administration of desmethylimipramine alone, an inhibitor of norepinephrine uptake, caused a membrane hyperpolarization, presumably by augmenting the effects of endogenous norepinephrine released in the brain slice (Egan et al., 1983). Again, the hyperpolarizing effect of norepinephrine persisted in a solution containing no Ca^{2+} and high Mg^{2+}. This observation again suggested that alpha-2 adrenergic receptor-elicited hyperpolarization does not involve an influx of extracellular Ca^{2+}. In a similar preparation of rat locus ceruleus neurons (Williams et al., 1985), the pharmacological specificity of the norepinephrine-induced increase in K^+ conductance was characterized more completely. It was observed that clonidine and norepinephrine hyperpolarized the membrane of locus ceruleus neurons, whereas phenylephrine and isoproterenol had no effect. Furthermore, the hyperpolarizing effects of clonidine and norepinephrine were antagonized by alpha-2 adrenergic selective antagonists.

alpha-2 Adrenergic receptor-mediated events in neurons of the rat substantia gelatinosa in slices of the spinal cord were investigated in vitro by North and Yoshimura (1984) by intracellular recording of the membrane potential. It was observed that super-

fusion with norepinephrine induced a reversible hyperpolarization of substantia gelatinosa neurons that was associated with an
increase in conductance that reversed in polarity at -88 mV. The
reversal potential changed as predicted by the Nernst equation
when the external K^+ concentration was changed, indicating that
the underlying ionic conductance was due to K^+. A further interpretation of these data may be that the hyperpolarizing effect of
norepinephrine does not require the influx of extracellular Ca^{2+}.
Pharmacological characterization of this hyperpolarization again
implicated a role for alpha-2 adrenergic receptors in increasing the
K^+ conductance of substantia gelatinosa neurons.

The above summary has focused on alpha-2 adrenergic
receptor-elicited effects on *presynaptic* central and peripheral neurons. alpha-2 Adrenergic receptors also have been localized to
postsynaptic neurons and peripheral target cells, however, including postsynaptic neurons of the submucous plexus and target
cells such as pancreatic β-cells and guinea pig hepatocytes. These
findings, which are summarized below, suggest that increased
K^+ conductance leading to hyperpolarization may be a general
consequence of alpha-2 adrenergic receptor occupation in a variety of cell types.

Hirst and Silinsky (1975) examined the effects of norepinephrine on the postsynaptic neurons of the guinea pig small intestine
submucous plexus. By measuring changes in the membrane potential in response to iontophoretic application of various agents,
it was observed that norepinephrine elicited hyperpolarization
that mimicked the inhibitory synaptic potentials produced by
electrical stimulation of the inhibitory innervation. It was proposed that two distinct catecholaminergic inhibitory mechanisms
may occur: one that causes inhibition by a presynaptic action (as
suggested for locus ceruleus or myenteric neurons), and the other
by a postsynaptic action. North and Surprenant (1985) investigated the participation of alpha-2 adrenergic receptors in the inhibitory postsynaptic potentials (ipsps) evoked in the guinea pig
ileum submucous plexus in greater pharmacological detail. They
observed that norepinephrine mimicked, and yohimbine,
phentolamine, and RX781094 reversibly blocked, the ipsps
evoked by an excitatory synaptic potential pulse. Prazosin,
propranolol, atropine, and naloxone had no effect on the ipsps.
The hyperpolarizing potentials elicited by alpha-2 adrenergic-
receptor agonists resembled the ipsps in terms of latency of onset,
amplitude, time course, conductance increase, reversal potential
equal to the K^+ equilibrium potential, and ionic dependence. In
addition, both events were blocked by Ba^{2+} and quinine, agents

known to inhibit K^+ conductance, including Ca^{2+}-dependent K^+ conductance, in other systems (Latorre and Miller, 1983; Iwatsuki and Petersen, 1985). Thus, it was concluded that the increase in K^+ conductance underlying the ipsps in submucous plexus neurons is caused by the activation of postsynaptic alpha-2 adrenergic receptors. It was also observed in this system that the hyperpolarizing response could be initiated either by a high fractional occupancy of alpha-2 adrenergic receptors for a short period of time, or by a low fractional occupancy over several seconds or minutes. These data suggested that an additional step exists between alpha-2 adrenergic receptor occupancy and the subsequent increase in K^+ conductance. Thus, the hyperpolarizing response to alpha-2 adrenergic agonists may require the accumulation of an intermediary substance to a certain critical threshold level before K^+ conductance can be activated.

Another target cell system in which the effects of alpha-2 adrenergic receptors have been measured electrophysiologically is the pancreatic islet, where alpha-2 adrenergic receptor agonists have been shown to inhibit secretagogue-induced insulin release from β-cells (Coore and Randle, 1964; Porte et al., 1966; Malaisse et al., 1970). Dean and Matthews (1970) investigated the effects of epinephrine on glucose-induced electrical activity. Intracellular recordings of the transmembrane potential in mouse pancreatic islet cells showed that 1 μM epinephrine completely blocked the action potential discharges elicited by glucose without altering the membrane potential. It was suggested that epinephrine somehow diminished an ion flux that is a normal component of the action potential. These data are complicated, however, by the existence in islet cells of alpha-1 and beta-adrenergic receptors that are linked to distinct signaling pathways that might cancel out alpha-2 adrenergic receptor-mediated changes in the membrane potential. The effect of epinephrine on glucose-induced electrical activity of rat islet cells in culture was studied further by Pace and coworkers (1977). These investigators observed that epinephrine blocked glucose-stimulated depolarization and spike activity, and that these actions of epinephrine were antagonized by phentolamine. Specifically, it was shown that epinephrine, in the presence of high glucose, elicited a rapid hyperpolarization of the membrane and inhibition of spike activity, shown to be caused, in part, by voltage-dependent Ca^{2+} conductance. Moreover, in agreement with the observed alpha-2 adrenergic receptor-mediated hyperpolarization in neuronal preparations, this epinephrine-stimulated event was prevented by increased concentrations of extracellular Ca^{2+} and K^+.

The effect of epinephrine to hyperpolarize mouse pancreatic islet cells was investigated further by Santana de Sa and Atwater (1980). It was observed that epinephrine and norepinephrine caused a transient hyperpolarization of the β-cell membrane accompanied by a transient decrease in input-resistance and an inhibition of glucose-stimulated burst activity. This hyperpolarization was inhibited by quinine, but not by tetraethylammonium, consistent with the involvement of a Ca^{2+}-activated K^+ conductance. It was therefore hypothesized that epinephrine causes the release of $[Ca^{2+}]_i$ that leads to the activation of a K^+ channel. A problem with these data exists, however, in that phentolamine *irreversibly* blocked the effects of epinephrine and norepinephrine to hyperpolarize β-cells, suggesting that other, non-receptor-mediated, membrane-perturbing events may have occurred.

A more complete study of the electrical effects of alpha-2 adrenergic receptor-mediated stimulation of pancreatic islet cells is that of Cook and Perara (1982), in which the membrane potential of isolated, perfused mouse islets was recorded using intracellular electrodes. In agreement with previous findings, epinephrine and clonidine caused a dose-dependent, saturable suppression of glucose-induced burst firing activity. In particular, epinephrine and clonidine caused a transient 10-mV hyperpolarization. Unlike previous observations, however, this transient hyperpolarization was followed by a depolarization to the control silent-phase potential, a long period characterized by a reduced plateau fraction (the fraction of each plateau/polarized silent phase cycle spent in the plateau phase) and a reduced plateau frequency (the glucose-independent pacing of the plateaus). Thus, epinephrine and clonidine caused the cells to spend more time in the silent phase and, on the average, to be more hyperpolarized. It was suggested, however, that these agonists acted not through direct effects on the voltage-dependent conductance mechanisms, but on the plateau frequency. In addition, since a low level of glucose-induced electrical activity (presumably mediated through Ca^{2+} action potentials) persisted in spite of epinephrine's action to inhibit insulin release, it was suggested, in agreement with more recent findings (Ullrich and Wollheim, 1985), that epinephrine must have inhibitory effects on insulin release beyond the inhibition of glucose-induced electrical activity and Ca^{2+} uptake. Thus, it was concluded that epinephrine may act through two different mechanisms to block insulin release: (1) interference with the glucose-dependent depolarization and plateau fraction and (2) modulation of the glucose-independent control of plateau frequency, possibly mediated through changes in

Table 3
Summary of alpha-2 Adrenergic Receptor-Mediated Stimulation of K^+ Conductance
with Resultant Hyperpolarization

System	Ultimate effect	Comments	References
Sympathetic ganglion	Inhibition of neurotransmitter release		Lundberg (1952); Eccles and Libet (1961); DeGroat and Volle (1966); Brown and Caulfield (1979)
Locus ceruleus neurons	Inhibition of neurotransmitter release	Observed in Ca^{2+}-free medium	Cedarbaum and Aghajanian (1977); Aghajanian and VanderMaelen (1982); Egan et al. (1983); Williams et al. (1985)
Myenteric plexus neurons	Inhibition of ACh release at neuromuscular junction	Observed in Ca^{2+}-free medium, proposed activation of Ca^{2+}-dependent K^+ conductance	Morita and North (1981)
Substantia gelatinosa	Inhibition of neurotransmitter release	Observed in Ca^{2+}-free medium	North and Yoshimura (1984)
Ileal submucous plexus neurons	Inhibitory post-synaptic potentials	Proposed involvement of a second messenger	Hirst and Silinsky (1975); North and Surprenant (1985)
Pancreatic islet cells	Inhibition of secretagogue-stimulated insulin release	Proposed activation of Ca^{2+}-dependent K^+-conductance	Pace et al. (1977); Santana de Sa and Atwater (1980); Cook and Perara (1982)
Guinea pig hepatocytes	?	Proposed activation of Ca^{2+}-dependent K^+ conductance, requires $[Ca^{2+}]_o$	Henley (1985)

intracellular Ca^{2+} buffering, leading to the activation of a Ca^{2+}-dependent K^+ channel.

alpha-2 Adrenergic receptor-mediated electrical effects have also been characterized in guinea pig hepatocytes. Early data indicated that epinephrine causes a biphasic activation of Ca^{2+}-dependent K^+ conductance through alpha-1 adrenergic receptors (Burgess et al., 1981; DeWitt and Putney, 1984). Henley (1985) observed, however, that clonidine elicited a sustained increase in $^{86}Rb^+$ efflux, i.e., enhanced outward K^+ permeability, from guinea pig hepatocytes that was caused exclusively by the entry of external Ca^{2+} into the cytosol. This clonidine-mediated sustained $^{86}Rb^+$ efflux was unaffected by prazosin, but completely blocked by yohimbine, indicating the involvement of an alpha-2 adrenergic receptor. In addition, it was suggested that this alpha-2 adrenergic receptor-sensitive Ca^{2+} influx pathway represents a Ca^{2+} control mechanism separate from Ca^{2+} signaling mechanisms utilized by alpha-1 adrenergic receptors.

Table 3 summarizes the data reviewed in this section that indicate that hyperpolarization of the plasma membrane secondary to an increased K^+ conductance is a general consequence of alpha-2 adrenergic receptor activation. In three of the seven systems studied, myenteric plexus, pancreatic β-islet cells, and guinea pig hepatocytes, evidence suggests that alpha-2 adrenergic receptors hyperpolarize cells by means of a Ca^{2+}-dependent K^+ conductance. In myenteric plexus neurons, this Ca^{2+}-activated K^+ conductance occurs in Ca^{2+}-free extracellular medium. In addition, in all systems studied but one (the guinea pig hepatocyte), alpha-2 adrenergic receptor-mediated hyperpolarization is enhanced in low Ca^{2+}-containing medium and greatly attenuated under conditions of high extracellular Ca^{2+}. These data suggest that alpha-2 adrenergic receptors may activate a Ca^{2+}-dependent K^+ conductance by altering the accessibility to intracellular pools of Ca^{2+}.

3.2. alpha-2 Adrenergic Receptors and Ca^{2+} Conductance

The preceding section on alpha-2 adrenergic receptor-mediated increases in K^+ conductance suggested that this may be a common mechanism in alpha-2 adrenergic receptor-elicited physiological effects, since an increased K^+ conductance occurs in central and peripheral neurons, pancreatic β-cells, and hepatocytes. The following section, summarized in Table 4, deals with an alternative, or additional, electrophysiological consequence of alpha-2

Table 4
Summary of alpha-2 Adrenergic Receptor-Mediated Inhibition of Voltage-Sensitive Ca^{2+} Conductance

System	Ultimate effect	Comments	References
Rat postganglionic sympathetic neuron	Inhibition of neurotransmitter release	"Presynaptic" alpha-adrenergic receptor	Horn and McAfee (1979)
Embryonic chick drg neurons	Inhibition of substance P release	Nontraditional alpha-2 adrenergic pharmacologic response, mediated through a GTP-binding protein	Dunlap and Fischbach (1981); Holz (1986a, b)
Rat locus ceruleus neurons	Inhibition of norepinephrine release	Thought to be an unidentified adrenergic receptor, not alpha-2	Williams and North (1985)
Differentiated NG108-15 cells	?	Mediated through a GTP-binding protein	Hescheler et al. (1986)
Varicosities of rat brain cortex	Inhibition of norepinephrine release	Hypothesized to act via a mechanism different from alpha-2 adrenergic receptors on the cell body	Schoffelmeer and Mulder (1984)

adrenergic receptor occupancy—the direct blockade of voltage-dependent Ca^{2+} conductance. To date, this phenomenon has been described only in neuronal cell preparations. The result of this action would be predicted to be a decreased availability of external Ca^{2+} to the neurosecretory machinery and reduced electrosecretory coupling in the neuron (rather than an indirect effect on Ca^{2+} conductance via increased K^+ conductance). Horn and McAfee (1979, 1980) designed experiments to investigate the effect of alpha-adrenergic receptor agonists on Ca^{2+} conductance in the rat postganglionic sympathetic neuron by measurement of membrane potential and membrane dc input resistance, and by the use of selective conductance antagonists. It was observed that catecholamines inhibited Ca^{2+} conductance (in particular, the shoulder on the normal action potential, the hyperpolarizing after-potential, and the Ca^{2+} spike) and subsequent Ca^{2+}-dependent potential changes with an order of potency: epinephrine > norepinephrine > dopamine > isoproterenol. The effect of norepinephrine was blocked by phentolamine, an alpha-selective adrenergic antagonist, but not by the beta-adrenergic antagonist propranolol. In addition it was observed, in agreement with previous experiments, that norepinephrine hyperpolarized the membrane. Interestingly, in contrast to previous findings, the effects of norepinephrine on the hyperpolarizing afterpotential were mimicked by a medium low in Ca^{2+} and high in Mg^{2+}, suggesting that catecholamines acting at alpha-2 adrenergic receptors indirectly *antagonize* the Ca^{2+}-dependent K^+ conductance underlying the hyperpolarizing after-potential by inhibiting Ca^{2+} influx during the action potential. As in previous studies, however, low external Ca^{2+} also augmented the hyperpolarization of sympathetic ganglia by norepinephrine in these experiments. Horn and McAfee (1979, 1980) proposed that "presynaptic" alpha-adrenergic receptors directly reduce neurotransmitter release by inhibiting Ca^{2+} influx through a voltage-sensitive Ca^{2+} conductance mechanism. Unfortunately, selective alpha-2 adrenergic receptor agonists and antagonists were not used in this study, making the pharmacological interpretation of these data difficult. In addition, a rather high concentration of phentolamine (10 μM) was required to block the response to norepinephrine in this system.

The effect of norepinephrine on voltage-sensitive Ca^{2+} conductance in embryonic chick dorsal root ganglion (drg) neurons in culture has been studied by Dunlap and Fischbach (1981). By recording inward Ca^{2+} currents and outward K^+ currents in voltage-clamped neuronal somata, it was observed that norepinephrine shortened the duration of the drg action potential by decreasing by 30% a voltage-sensitive, slow inward Ca^{2+} current with no effect on outward K^+ current or change in membrane potential. The data suggested that norepinephrine decreases the maximum available Ca^{2+} conductance, either by reducing the number of available Ca^{2+} channels or by attenuating single Ca^{2+} channel conductance. In fact, in a later study (Galvan and Adams, 1982) it was suggested that norepinephrine acts by reducing the number of available Ca^{2+} channels in rat sympathetic ganglion cells. Dunlap and Fischbach argued against the hypothesis that norepinephrine acts by activating an outward Ca^{2+}-dependent K^+ conductance, since this action would have to occur in spite of a 30% decrease in Ca^{2+} influx. Again, however, as in the above experiments, selective agonists and antagonists were not used. It is possible that the observed effect of norepinephrine in these two systems was a summation of alpha-1 and alpha-2 adrenergic receptor-mediated events. In addition, since both alpha-1 and alpha-2 adrenergic receptors have been shown to mobilize intracellular pools of Ca^{2+} (Exton, 1985; Sweatt et al., 1986b), it is possible that norepinephrine, acting through one or both of these receptors, may decrease inward Ca^{2+} current via a Ca^{2+}-dependent negative feedback mechanism similar to that described in *Paramecium* (Brehm and Eckert, 1978) and in *Aplysia* neurons (Tillotson, 1979), where Ca^{2+} entry as a result of depolarization causes a rapid inactivation of Ca^{2+} channels.

In a more recent study of norepinephrine-mediated inhibition of voltage-dependent Ca^{2+} channels in embryonic chick drg cells (Holz et al., 1986b), it was observed that norepinephrine acted through a receptor that exhibited pharmacologic responses "similar to" alpha-2 adrenergic receptors. Thus, norepinephrine-induced decreased Ca^{2+} conductance was blocked by phentolamine and yohimbine. Clonidine, a classical agonist (or partial agonist) at alpha-2 adrenergic receptors, had no effect on Ca^{2+} conductance in these cells. In addition, phenylephrine, dopamine, and serotonin also blocked Ca^{2+} conductance in a manner that was not additive with norepinephrine, suggesting either that these agonists all bind to the same receptor population or that multiple receptors act through a final common pathway to decrease Ca^{2+} conductance. A possible explanation for the unexpected pharmacologic response exhibited by this system may be the embryonic state of the drg cells in which the effects were measured. Perhaps, in a mature drg cell, the responses would conform to that of a more classical alpha-2 adrenergic receptor. Alternatively, perhaps these findings are in agreement with those of Williams and North (1985), described below, that catecholamine-mediated decreases in Ca^{2+} conductance are elicited through an as-yet-unidentified adrenergic receptor.

Williams and North (1985) characterized the catecholamine-mediated inhibition of Ca^{2+} action potentials in rat locus ceruleus neurons. They concluded that this response does *not* involve an alpha-2 adrenergic receptor. Thus, although clonidine, epinephrine, and norepinephrine all elicited hyperpolarization of the membrane, clonidine had no effect on the inward Ca^{2+} conductance, and the effect of norepinephrine and epinephrine to shorten the Ca^{2+} action potential was not prevented by yohimbine, propranolol, or prazosin. As observed in previous experiments, a high concentration of phentolamine (10 μM) was required to partially antagonize these effects on the Ca^{2+} action potential. Additionally, concentrations of epinephrine and norepinephrine that effectively hyperpolarized the cell by activating potassium conductance were considerably lower than those required to block the inward Ca^{2+} conductance. Thus, it was concluded by these investigators the norepinephrine release from locus ceruleus neurons is inhibited by alpha-2 adrenergic receptors linked to an increase in outward K^{+} conductance and that the catecholamine-mediated decreases in Ca^{2+} current are probably elicited through an as-yet-undefined adrenergic receptor.

Recently, Hescheler et al. (1986) presented data to suggest that alpha-2 adrenergic receptors reduce voltage-sensitive Ca^{2+} current in differentiated NG108-15 neuroblastoma × glioma hybrid cells. Using the whole-cell patch-clamp technique with Ba^{2+} as the charge carrier, it was demonstrated that epinephrine inhibited inward Ca^{2+} current in a concentration-dependent manner. This effect is thought to be elicited by a direct effect of alpha-2 adrenergic receptors on Ca^{2+} conductance (Dr. G. Schultz, personal communication) rather than to occur secondary to an increase in K^{+} conductance, as suggested for other neuronal systems (Williams et al., 1985). Moreover, data were presented to suggest that this blockade of Ca^{2+} current is mediated through a GTP-binding protein (see later).

In conclusion, alpha-2 adrenergic receptors may act, directly or indirectly, to attenuate inward Ca^{2+} current in neuronal cell preparations. In the NG108-15 cell line, the rat postganglionic sympathetic neuron, and the chick drg, evidence has been presented to suggest that alpha-2 adrenergic receptors directly (i.e., not as a secondary result of changes in another channel conductance) attenuate inward Ca^{2+} current. In contrast, experiments using rat locus ceruleus neurons indicate either that alpha-2 adrenergic receptors reduce inward Ca^{2+} current indirectly by first increasing K^+ conductance or that an as-yet-unidentified adrenergic receptor mediates the effect of epinephrine to decrease Ca^{2+} current. Perhaps the effect of alpha-2 adrenergic receptors on ion channels depends on the pathways available in a given cell type and the coupling of electrical changes to secretion in that cell.

An interesting alternative hypothesis is that presented in a study of alpha-2 adrenergic receptor-mediated inhibition of norepinephrine release from rat brain cortex by Schoffelmeer and Mulder (1984). These authors proposed that cell body alpha-2 adrenergic receptors inhibit neurotransmitter release by hyperpolarization of the membrane, thereby blocking action potential propagation and subsequent recruitment of varicosities [in agreement with the hypothesis of Stjarne (1978, 1979)], whereas varicosal alpha-2 adrenergic receptors inhibit neurotransmitter release by blocking the neurosecretory coupling mechanism *following* the invasion of the action potential. This proposal was based, in part, on their findings that alpha-2 adrenergic receptors in the varicosities of rat brain cortical neurons do *not* hyperpolarize the neuronal membrane. Perhaps neuronal alpha-2 adrenergic receptors are linked to different effector mechanisms based on their membrane topography, and therefore seemingly disparate data in the literature may be resolved based on the cell body or varicosal location of the receptor population studied.

3.3. GTP-Binding Proteins Are Involved in alpha-2 Adrenergic Receptor-Mediated Modulation of Ion Channels

alpha-2 Adrenergic receptors are linked to inhibition of adenylate cyclase through a GTP-binding protein, termed G_i (Gilman, 1984). In addition, evidence strongly suggests that a GTP-binding protein may be involved in other signaling events evoked by alpha-2 adrenergic receptors, as well.

The first suggestion of a role for GTP-binding proteins in alpha-2 adrenergic receptor-mediated inhibition of neurotransmitter release resulted from a study by Allgaier et al. (1985) in slices of rabbit hippocampus. In this system, occupation of alpha-2 adrenergic receptors inhibited the electrically evoked release of [^{3}H]norepinephrine. Additionally, the alpha-2 adrenergic antagonist yohimbine elicited a facilitatory effect on [^{3}H]norepinephrine release, presumably by blocking the action of endogenous norepinephrine to inhibit its own release. These investigators were able to show that pretreatment of the hippocampal slices with *Bordetella pertussis* islet-activating protein (IAP), a bacterial exotoxin that catalyzes the ADP-ribosylation of some GTP-binding proteins (Ui, 1984; Neer et al., 1984), blocked norepinephrine inhibition of neurotransmitter release. Furthermore, the facilitatory effect of yohimbine was also blocked by IAP.

The first direct demonstration of the involvement of a GTP-binding protein in alpha-2 adrenergic receptor-mediated modulation of ion channels was that of Holz et al. (1986a) in embryonic chick drg cells. These investigators showed that the intracellular administration of guanosine $5'$-0-(2-thiodiphosphate) (GDP-β-S), an hydrolysis-resistant analog of GDP that competitively inhibits the binding of GTP to GTP-binding proteins (Eckstein et al., 1979), blocked norepinephrine inhibition of voltage-sensitive Ca^{2+} current, measured using whole-cell patch-clamp techniques. Moreover, preincubation of the cells with IAP also blocked the action of norepinephrine. These results provide strong evidence that GTP-binding proteins mediate the inhibitory action of norepinephrine on Ca^{2+} channels.

Results similar to those described above were reported by Hescheler et al. (1986). These investigators showed that alpha-2 adrenergic receptors inhibit voltage-sensitive Ca^{2+} channels in differentiated NG108-15 neuroblastoma $\times$ glioma cells. Using the whole-cell voltage-clamp technique, it was demonstrated that norepinephrine inhibited inward Ca^{2+} current via activation of an alpha-2 adrenergic receptor. Furthermore, intracellular administration of guanosine $5'$-0-(3-thiotriphosphate) (GTP-γ-S), an hydrolysis-resistent analog of GTP that persistently activates GTP-binding protein-effector coupling mechanisms (Abramowitz et al., 1979), mimicked the effect of norepinephrine on Ca^{2+} current. Similar to findings in chick drg cells, norepinephrine inhibition of Ca^{2+} current was blocked by pretreatment of the NG108-15 cells with IAP. However, Hescheler et al. (1986) observed that the sensitivity of the voltage-sensitive Ca^{2+} channel to norepinephrine could be restored to IAP-treated cells by the intracellular application of the GTP-binding α-subunits of two GTP-binding proteins, termed G_i and G_o (*see* Chapter 3 regarding the structure of G_i vs. G_o). Interestingly, the α-subunit of G_o apparently was more potent than the α-subunit of G_i in restoring the norepinephrine-effect on Ca^{2+} current. It was concluded that an IAP-sensitive GTP-binding protein, possibly G_o, is involved in the functional coupling of alpha-2 adrenergic receptors to voltage-dependent neuronal Ca^{2+} channels.

4. Hypothesis: Na^+/H^+ Exchange, Changes in Ca^{2+} Availability and Ion Channel Conductances Interact to Elicit alpha-2 Adrenergic Receptor-Mediated Physiological Effects

There is extensive evidence in the literature, reviewed at the beginning of this chapter, to suggest that alpha-2 adrenergic receptors elicit physiological events through a mechanism or mechanisms that parallel changes in cyclic nucleotide metabolism. Thus far we have presented evidence to show that in addition to inhibiting adenylate cyclase, alpha-2 adrenergic receptors also activate Na^+/H^+ exchange, mobilize membrane bound pools of intracellular Ca^{2+}, activate a K^+ conductance that may be Ca^{2+}-dependent, and/or cause a blockade of voltage-dependent Ca^{2+} channels. How might these events be related? Figure 2 presents our working hypothesis of the mechanisms that might link alpha-2 adrenergic receptor occupancy to inhibition of neurotransmitter release from central and peripheral neuronal cell systems. This hypothesis, all or in part, may also apply to other target cell systems for alpha-2 adrenergic receptors, e.g., the pancreatic β-cell, guinea pig hepatocyte, or human platelet. The hypothesis has two components. First, we postulate that local increases in intracellular pH (pH_i) that occur as a result of alpha-2 adrenergic receptor-accelerated Na^+/H^+ exchange may play a crucial role in modulating ion channel conductances. Second, we find attractive the suggestion of Schoffelmeer and Mulder (1983, 1984) that the functional consequences of alpha-2 adrenergic receptor occupation may depend on the topographical location of the receptor and on the effector mechanisms available in the microenvironment shared by the receptor.

There is evidence that a receptor-modulated K^+ conductance exists in the cell body of neurons (Fig. 2). We hypothesize that increases in pH_i increase the probability of opening of K^+ channels, thus resulting in hyperpolarization of the membrane. Thus, hyperpolarization would decrease the probability of neuronal firing as well as attenuate the propagation of the action potential to the neurotransmitter-containing varicosities.

Although there are no extant data that specifically describe an effect of pH_i on alpha-2 adrenergic receptor-mediated hyperpolarization, considerable evidence exists regarding a relationship between pH_i and K^+ conductance in other systems. For example, the K^+ conductance of the squid giant axon is known to be regulated by changes in pH_i (Wanke et al., 1979). In this sys-

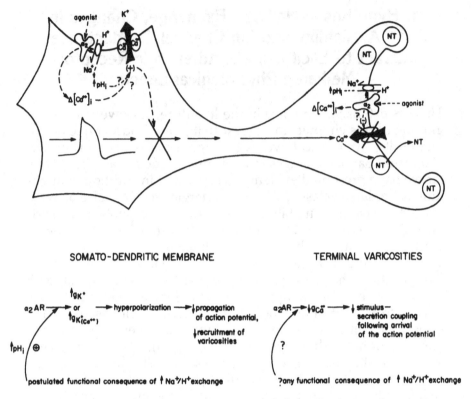

Fig. 2. Mechanistic consequence of alpha-2 adrenergic receptor occupancy may depend on regional membrane topography and available "machinery."

tem, an increase in pH$_i$ from 5.2 to 10 elicits a reversible increase in outward K$^+$ conductance. A linear plot of K$^+$ conductance versus pH$_i$ suggests the existence of a single titratable group on the intracellular side of the channel having an apparent pK_a of 6.9. A similar relationship between pH$_i$ and K$^+$ conductance is also well established for another system, the sea urchin egg (Steinhardt and Mazia, 1972; Steinhardt et. al., 1972; Shen and Steinhardt, 1980). The membrane of the sea urchin egg is rapidly depolarized upon fertilization with sea urchin sperm. This depolarization is caused, at least in part, by Ca^{2+} entry. After a transient repolarization, a marked hyperpolarization occurs. Both the depolarization and the late, dramatic hyperpolarization are essential for egg cleavage and normal development to ensue. The late hyperpolarization, but not the initial depolarization, can be elicited by alkalinization of the cell interior by adding the permanent weak base NH$_4$OH to the cell exterior. The dependence of this hyperpolarization on K$^+$ efflux is shown by the blockade of the alkalinization-induced hyperpolarization by incremental increases in extracellular K^+ concentration. The hyperpolarization of the sea urchin egg membrane appears to be caused by changes in *intra*cellular pH, since increasing the extracellular pH to 9.1 with NaOH causes only a gradual decline in the membrane potential. It is known that fertilization of the sea urchin egg results in activation of Na^+/H^+ exchange (Johnson et al. 1976; Epel, 1978; Shen and Steinhardt, 1979), and the hyperpolarization induced by fertilization can be blocked by removal of extracellular Na^+ or reduction in extracellular pH to block the function of the Na^+/H^+ antiporter. Furthermore, although the Ca^{2+} ionophore A23187 can mimic the early depolarization event that is precipitated by fertilization, the calcium ionophore does *not* evoke subsequent hyperpolarization. Thus, at least in the sea urchin egg, a change in $[Ca^{2+}]_i$ alone cannot give rise to an increased K^+ conductance.

In yet another system, the isolated perfused frog kidney, it has been shown that inhibition of Na^+/H^+ exchange elicits a reduction in transepithelial K^+ conductance (Oberleithner et al., 1986). Thus, cell membrane potential changes, induced by luminal K^+ concentration steps, are blunted by manipulations that are known to block Na^+/H^+ exchange: luminal Na^+-free perfusates, acidification of the perfusion solution, and luminal application of amiloride. Consequently, it seems reasonable to hypothesize that if alpha-2 adrenergic receptors hyperpolarize the membrane via increased K^+ efflux from some neuronal cell preparations, then this effect may be influenced by the increase in pH_i that accompanies acceleration of Na^+/H^+ antiporter activity.

Very little is known about the effects of pH_i on Ca^{2+}-dependent K^+ conductance. Meech (1979) observed that injection of HCl into *Helix aspersa* neurons results in a reduction in amplitude of Ca^{2+}-dependent K^+ tail currents. In addition, Meech observed that injection of $CaCl_2$ into the neurons produces an effect on outward current similar to that of decreased pH_i that is secondary to activation of Ca^{2+}/H^+ exchange. In pancreatic β-cells, Pace and coworkers (1983) suggested that glycolytically produced H^+ acidifies the cell interior, thereby decreasing K^+ permeability and subsequently depolarizing the membrane. Blockade of Na^+/H^+ exchange in these cells by the addition of amiloride or the reduction of extracellular Na^+ produces an effect on K^+ channels similar to that of H^+. These results suggest that pH modulates electrophysiological events in β-cells and ultimately influences the transduction mechanisms involved in stimulus-secretion coupling. In a similar study, Eddlestone and Beigelman (1983) suggested that the metabolically produced H^+ inhibits the

Ca^{2+}-dependent K^+ conductance thought to underly the pacemaker current in pancreatic β-cells, hypothesizing that a fall in pH_i causes a fall in K^+ permeability, and an increase in pH_i causes an increase in K^+ permeability via a mechanism involving pH-induced changes in the affinity of the K^+ channel for Ca^{2+}. Using the patch-clamp method, Cook and coworkers (1984) studied directly the Ca^{2+}-dependent K^+ channel of rat pancreatic β-cells and reported that acidification of the cytoplasmic membrane surface from pH 7.6 to 6.8 results in a reversible blockade of channel activity, defined as a rightward shift of the activation curves away from the physiological voltage range. An increase in alkalinity of the cytoplasmic membrane surface, such as would be expected to result from acceleration of Na^+/H^+ exchange, shifts the activation curves to the left. These authors and others (Moody, 1984; Alvarez-Leefmans et al., 1981; Zucker, 1981) have suggested several actions of H^+ to account for the observed effects on Ca^{2+}-dependent K^+ conductance: H^+ may compete for Ca^{2+} at the Ca^{2+}-binding site of the K^+ channel or the decreased pH_i may decrease the availability of free Ca^{2+} for interaction at the K^+ channel Ca^{2+}-binding site, either by increasing the cytoplasmic buffering of Ca^{2+}, or by an alternative mechanism, such as has been observed in *Helix aspersa* neurons.

A role for voltage-dependent Ca^{2+} conductance has been implicated in release of neurotransmitters from the terminal varicosities of neuronal cells (Fig. 2). At this site, then, the inhibition of voltage-dependent Ca^{2+} channels by alpha-2 adrenergic receptors may play an important role in limiting the concentration of Ca^{2+} available for fusion of neurotransmitter-containing vesicles with the synaptic membrane. It is less clear, at present, how acceleration of Na^+/H^+ exchange by alpha-2 adrenergic receptors might influence voltage-dependent Ca^{2+} conductances. Converse to the above summarized observations on K^+ current, very little data exist in the literature on the effect of pH_i on voltage-dependent Ca^{2+} channels. The available experimental evidence suggests that intracellular alkalinization might actually *activate* voltage-sensitive Ca^{2+} channels. It has been shown by Umbach (1982) in *Paramecium caudatum* that increasing pH_i from a resting pH of 6.80 to 7.20 results in an *increase* in Ca^{2+} channel permeability. Likewise, acidification of the intracellular milieu blocks Ca^{2+} channel permeability in a manner that is consistent with a model involving protonation of a single, intracellular titratable site having an apparent dissociation constant of 6.2. However, alkalinization of the cell interior above pH 7.2 results in a *decrease* in Ca^{2+} channel permeability that is unexplained by the model presented. Since we have shown that alpha-2 adrenergic receptor activation in the NG108-15 cell line causes a measured increase in pH_i from 7.05 to 7.5, it is tempting to speculate that a similar biphasic interaction of pH_i with Ca^{2+} channels may occur in these cells as well, such that alpha-2 adrenergic receptor activation may result in a blockade of voltage-sensitive Ca^{2+} channels. On the other hand, alpha-2 adrenergic receptors in embryonic chick drg cells have been shown to elicit the blockade of voltage-sensitive Ca^{2+} channels under experimental conditions in which pH_i and pH_o are held constant at pH 7.3 a, manipulation that nullifies changes in pH_i that might occur if receptor occupancy resulted in intracellular alkalinization. It may be possible, however, that pH_i = 7.3 is a sufficiently alkaline environment in these cells to allow the blockade of inward Ca^{2+} conductance by alpha-2 adrenergic receptors. In a more recent study, Byerly and Moody (1986) reported that intracellular acidification substantially decreases the amplitude of the Ca^{2+} current in neurons of *Lymnaea stagnalis*, in agreement with the observations summarized above. These investigators were unable to determine whether the effect of pH_i on Ca^{2+} current was direct or whether it resulted secondary to a change in internal free Ca^{2+} levels. Upon further analysis, however, it was observed that the outward Ca^{2+} current in these neurons is contaminated with a H^+ current, the contribution of which increases with decreasing pH_i. The simultaneous measurement of H^+ and Ca^{2+} currents may thus give the misleading impression that Ca^{2+} current is blocked by decreases in pH_i more than it actually is.

As indicated above, an important caveat that must be considered in the interpretation of studies of the effects of pH_i on ion channel currents is the voltage-dependent H^+ current, which becomes a significant fraction of the total outward current under conditions of low pH_i, and therefore markedly changes the apparent effects of pH_i on both Ca^{2+} and delayed K^+ current (Thomas and Meech, 1982; Byerly et al., 1984; Byerly and Moody, 1986). In fact, as pH_i is lowered, the H^+ and K^+ equilibrium potentials become approximately equal, thus invalidating the use of tail current reversal potentials to distinguish the two currents. Additionally, in at least one cell type, the axolotl oocyte (Barish and Baud, 1984), most of the outward current at normal pH can be accounted for by H^+. Hence, further investigations of ion channel currents responsible for alpha-2 adrenergic receptor-mediated neurotransmitter release will require rigorous electrophysiological analysis of the components involved, perhaps even purification and reconstitution of the receptor and antiporter with the ion channel in question in order to eliminate contaminating currents.

Before closing, we would also like to address the role of cyclic nucleotide metabolism in alpha-2 adrenergic receptor-modulated ion conductances and blockade of neurotransmitter release. Despite the observations cited at the outset of this chapter that suggest that decreases in cyclic AMP cannot be the sole signal responsible for eliciting alpha-2 adrenergic receptor-elicited physiological effects, it is probable that concomitant decreases in cyclic AMP are necessary for other receptor-operated signaling mechanisms to occur. For example, increases in cyclic AMP in some systems have been demonstrated to inhibit Na^+/H^+ exchange (Reuss and Petersen, 1985; Petersen et al., 1985; Pollack et al., 1986). Similarly, increases in cyclic AMP can provoke neurotransmitter release (Dunwiddie, 1985). As a result, increased production of cyclic AMP might functionally antagonize alpha-2 adrenergic receptor-induced suppression of neurotransmitter release. An example of this prediction is the observation that clonidine-inhibited [^{3}H]noradrenaline release from rat presynaptic noradrenergic varicosities is attenuated by agents that increase the intracellular level of cyclic AMP, including dibutyryl cyclic AMP, 8-bromo-cyclic AMP, NaF, and forskolin (Schoffelmeer and Mulder, 1983). Similarly, there are data that suggest that, in certain settings, elevations of cyclic AMP influence Ca^{2+} and K^+ channels in a direction that would be opposite to channel modulation by alpha-2 adrenergic receptors. As reviewed by Reuter (1983; Reuter et al., 1986), the overall open-state probability of individual L-type voltage-dependent Ca^{2+} channels is increased by beta-adrenergic receptor agonists, 8-bromo-cyclic AMP, and the catalytic subunit of cyclic AMP-dependent protein kinase, thus increasing the maximal conductance, $g_{Ca^{2+}}$. K^+ channels also have been reported to be modulated by cyclic AMP; however, these effects vary between cell and channel type. For example, in bag cell neurons of *Aplysia californica* (Strong and Kaczmarek, 1986), three major K^+ currents (I_{K_1}, I_{K_2}, and the A-current) are reduced by cyclic AMP-dependent mechanisms. In addition, in identified neurons of the visceroabdominal ganglionic mass of *Helix aspera*, the neurotransmitters serotonin and dopamine evoke cyclic AMP-induced inward currents associated with a decrease in K^+ conductance (Deterre et al., 1982). The effect of cyclic AMP to reduce various K^+ channel currents is opposite to the increase in K^+ conductance observed in response to alpha-2 adrenergic receptor occupation. Although a reduction in cyclic AMP levels may not be the sole mechanism by which alpha-2 adrenergic receptors influence neurotransmitter release, a decrease in the rate of cyclic AMP production may be a necessary permissive condition.

5. Summary and Future Perspectives

A premise of this chapter is that decreases in cyclic AMP alone do not account fully for physiological effects elicited by alpha-2 adrenergic receptors. We hypothesize that the acceleration of Na^+/H^+ exchange may represent an important component of alpha-2 adrenergic receptor-mediated signal transduction, whether or not the ultimate effect is pro- or anti-secretory. We also postulate that coupling of alpha-2 adrenergic receptors to various ion channels is dependent on the membrane topography of the receptor, as was first postulated by Schoffelmeer and Mulder (1983, 1984). Thus, alpha-2 adrenergic receptor-mediated blockade of neurotransmitter release may result from increased K^+ conductance in the cell body and decreased Ca^{2+} conductance in the neuronal varicosity. It is reasonable to hypothesize, based on extant data, that the intracellular alkalinization that occurs as a result of acceleration of Na^+/H^+ exchange leads to an increase in K^+ conductance. The link between pH_i and alpha-2 adrenergic receptor-mediated decreases in voltage-dependent Ca^{2+} conductance, however, is more speculative. Concurrent alpha-2 adrenergic receptor-elicited decreases in cyclic AMP production probably play an important permissive role in receptor-induced function, since elevations in cyclic AMP have been shown to block Na^+/H^+ exchange and K^+ conductances and, conversely, to increase voltage-sensitive Ca^{2+} conductance.

Unfortunately, in many systems in which alpha-2 adrenergic receptor-mediated signaling mechanisms have been studied, concurrent measurements of alpha-2 adrenergic receptor-elicited physiological actions have not been performed. Thus, in future studies it will be essential to measure alpha-2 adrenergic receptor-mediated electrophysiological and biochemical events simultaneously with alpha-2 adrenergic receptor-elicited promotion or inhibition of secretion in order to provide functional correlates of receptor-mediated changes in biochemical and electrophysiological events. For example, both receptor-induced hyperpolarization via increased K^+ conductance and receptor-attenuated Ca^{2+} currents could account for alpha-2 adrenergic receptor-induced suppression of neurotransmitter release. Few investigators, however, have examined the role of either ion channel in mediating alpha-2 adrenergic blockade of neurotransmitter release under parallel experimental conditions.

Finally, it is interesting to extrapolate from observations about alpha-2 adrenergic receptor-mediated signal transduction to other receptors linked to inhibition of adenylate cyclase. These receptor populations include somatostatin receptors, dopamine receptors of the D_2-subtype, some populations of muscarinic receptors (presumably of the M_2 subtype), opioid receptors of the δ and μ subtype, and adenosine receptors of the R_i (or R_I) subtype. As for alpha-2 adrenergic receptors, data often exist to suggest that these receptors inhibit neurotransmitter release or hormone secretion via a mechanism other than, or in addition to, inhibition of adenylate cyclase (Trautwein et al., 1982; Nargeot et al., 1983; Pace et al., 1977, Endoh et al., 1985; Dorflinger and Schonbrunn, 1983; Dunwiddie, 1985; Silinsky, 1986; Delbeke et al., 1986). Might there be a signaling pathway(s) common to receptors linked to inhibition of adenylate cyclase other than, or in addition to, decreases in intracellular cyclic-AMP production? We have observed that not only alpha-2 adrenergic receptors, but also opioid and muscarinic receptors, accelerate Na^+/H^+ exchange in NG108-15 cells (Isom et al., 1987b). Similarly, numerous studies demonstrate that receptors linked to inhibition of adenylate cyclase precipitate increases in K^+ permeability and hyperpolarization of the plasma membrane (reviewed by North, 1986). These include somatostatin (Pace et al., 1977), adenosine (Dunwiddie, 1985), dopamine (Israel et al., 1985), and muscarinic (Sakmann et al., 1983; Soejima and Noma, 1984; Egan and North, 1986) receptors. Recent investigations suggest that some of these receptor populations are coupled to increased K^+ conductance via IAP-sensitive GTP-binding proteins (Endoh et al., 1985; Pfaffinger et al., 1985; Breitweiser and Szabo, 1985; Dolphin and Prestwich, 1985; Bohm et al., 1986). Also, by analogy with alpha-2 adrenergic receptors, blockade of inward Ca^{2+} current has been demonstrated for muscarinic (North, 1986), opioid (Hescheler et al., 1986; Werz and MacDonald, 1982, 1983a,b, 1985), μ opioid (Werz and MacDonald, 1982, 1983a,b, 1985), somatostatin (Schlegel et al., 1984; Koch et al., 1985; Luini et al., 1986), and adenosine (Silinsky, 1986) receptors. Again, there is evidence to show that some of these receptors are coupled to decreased Ca^{2+} conductance via IAP-sensitive GTP-binding proteins (Schlegel et al., 1985; Hescheler et al., 1986; Lewis et al., 1986).

Perhaps it will be learned that the concomitant decrease in cyclic AMP and acceleration of Na^+/H^+ exchange provide a local milieu permissive for alterations in various enzymatic activities or ion movements. The particular enzyme activity or ion conduct-ance that is influenced by receptor populations linked to inhibi-tion of adenylate cyclase activity may be dictated by the ma-chinery characteristic of the target cell. Clearly, a precise un-erstanding of alpha-2 adrenergic receptor-mediated signal transduction will shed light on many other systems modulated by receptors linked to inhibition of adenylate cyclase. It also should be clear, from this and earlier chapters in this volume, that no single experimental approach will unravel the complexity of alpha-2 adrenergic receptor-mediated physiological responses. The use of whole animal studies is essential in understanding in-tegrative feedback mechanisms contributed by multiple target organs following activation of alpha-2 adrenergic receptors in iso-lated sites. Radioligand binding studies, coupled with autoradio-graphic localization of these binding sites, will be useful in further clarifying the loci of alpha-2 adrenergic receptors. In addition, radioligand binding will be useful for following receptor binding activity through purification of the receptor and subsequent re-constitution with defined effector systems. Site-directed muta-genesis of cloned receptors hopefully will elucidate which recep-tor domain confers which unique receptor function. And, again, whole-cell studies of these cloned receptors after expression into eukaryotic cells will be essential for establishing the functional role of alpha-2 adrenergic receptors, or receptor subpopulations and the consequence of mutations within the alpha-2 adrenergic receptor. We, as investigators, thus appear to be limited not by existing methodologies, but only by the extent of our curiosity.

REFERENCES

Abramowitz, J., Iyenger, R., and Birnbaumer, L. (1979) Guanyl nucleotide regulation of hormonally-responsive adenylyl cyclases. *Mol. Cell. Endocrinol.* **16**, 129–146.

Aghajanian, G. K. and VanderMaelen, C. P. (1982) α_2-Adrenoceptor-mediated hyperpolarization of locus ceruleus neurons: Intracellular studies in vivo. *Science*, **215** 1394–1396.

Allgaier, C., Feuerstein, T. J., Jackisch, R., and Hertting, G. (1985) Islet-activating protein (pertussis toxin) diminishes α_2-adrenoceptor-mediated effects on noradrenaline release. *Naunyn Schmiedebergs Arch. Pharmacol.* **331**, 235–239.

Alvarez-Leefmans, F. J., Rink, T. J., and Tsien, R. Y. (1981) Free calcium ions in neurones of *Helix aspersa* measured with ion-selective micro-electrodes. *J. Physiol.* **315**, 531–548.

Barish, M. E. and Baud, C. (1984) A voltage-gated hydrogen ion current in the oocyte membrane of the axolotl, *Ambystoma. J. Physiol.* **352**, 243–263.

Bohm, M. Bruckner, R., Newmann, J., Schmitz, W., Scholz, H., and Starbatty, J. (1986) Role of guanine nucleotide-binding protein in the regulation by adenosine of cardiac potassium conductance and force of contraction. Evaluation with pertussis toxin. *Naunyn Schmiedebergs Arch. Pharmacol.* **332**, 403–405.

Brehm, P. and Eckert, R. O. (1978) Calcium entry leads to inactivation of calcium channels in *Paramecium. Science.* **202**, 1203–1206.

Breitwieser, G. G. and Szabo, G. (1985) Uncoupling of cardiac muscarinic and β-adrenergic receptors from ion channels by a guanine nucleotide analog. *Nature* **317**, 536–540.

Brown, D. A. and Caulfield, M. P. (1979) Hyperpolarizing 'α$_2$'-adrenoceptors in rat sympathetic ganglia. *Br. J. Pharmacol.* **65**, 435–445.

Burgess, G. M., Claret, M., and Jenkinson, D. H. (1981) Effects of quinine and apamin on the calcium-dependent potassium permeability of mammalian hepatocytes and red cells. *J. Physiol.* **317**, 67–90.

Byerly, L. and Moody, W. J. (1986) Membrane currents of internally perfused neurones of the snail, *Lymnaea stagnalis*, at low intracellular pH. *J. Physiol.* **376**, 477–491.

Byerly, L., Meech, R., and Moody, W., Jr. (1984) Rapidly activating hydrogen ion currents in perfused neurones of the snail, *Lymnaea stagnalis. J. Physiol.* **351**, 199–216.

Calvete, J. A., Hayes, R. J., Oates, N. S., Sever, P. S., and Thom, S (1984) α$_1$-, and α$_2$-, Adrenoceptor responses in human isolated arteries. *Br. J. Pharmacol.* **83**, 364P.

Cedarbaum, J. M. and Aghajanian, G. K. (1977) Catecholamine receptors on locus coeruleus neurons: Pharmacological characterization. *Eur. J. Pharmacol.* **44**, 375–385.

Cole, A. E. and Shinnick-Gallagher, P. (1981) Comparison of the receptors mediating the catecholamine hyperpolarization and slow inhibitory postsynaptic potential in sympathetic ganglia. *J. Pharm. Exp. Ther.* **217**, 440–444.

Connolly, T. M. and Limbird, L. E. (1983) The influence of Na$^+$ on the α$_2$-adrenergic receptor system of human platelets. A. Method for removal of extraplatelet Na$^+$. Effect of Na$^+$ removal on aggregation, secretion and cyclic AMP accumulation. *J. Biol. Chem.* **258**, 3907–3912.

Cook, D. L. and Perara, E. (1982) Islet electrical pacemaker response to alpha-adrenergic stimulation. *Diabetes* **31**, 985–990.

Cook, D. L., Ikeuchi, M., and Fujimoto, W. Y. (1984) Lowering of pH$_i$ inhibits Ca^{++}-activated K$^+$ channels in pancreatic β-cells. *Nature* **311**, 269–271.

Coore, H. G. and Randle, P. J. (1964) Regulation of insulin secretion studied with pieces of rabbit pancreas incubated in vitro. *Biochem. J.* **83**, 66–77.

Dean, P. M. and Matthews, E. K. (1970) Glucose-induced electrical activity in pancreatic islet cells. *J. Physiol.* **210**, 255–264.

DeGroat, W. C. and Volle, R. L. (1966) The actions of the catecholamines on transmission in the superior cervical ganglion of the cat. *J. Pharm. Exp. Ther.* **154**, 1–13.

Delbeke, D., Scammell, J. G., Martinez-Campos, A., and Dannies, P. S. (1986) Dopamine inhibits prolactin release when cyclic adenosine 3',5'-monophosphate levels are elevated. *Endocrinology* **118**, 1271–1277.

Deterre, P., Paupardin-Tritsch, D., Bockaert, J., and Gerschenfeld, H. M.

(1982) Cyclic AMP-mediated decrease in K^+ conductance evoked by serotonin and dopamine in the same neuron: A biochemical and physiological single-cell study. *Proc. Natl. Acad. Sci. USA* **79**, 7934–7938.

DeWitt, L. M. and Putney, J. W. (1984) Alpha-adrenergic stimulation of potassium efflux in guinea pig hepatocytes may involve calcium influx and calcium release. *J. Physiol.* (Lond.) **346**, 395–407.

Dolphin, A. C. and Prestwich, S. A. (1985) Pertussis toxin reverses adenosine inhibition of neuronal glutamate release. *Nature* **316**, 148–150.

Dorflinger, L. J. and Schonbrunn, A. (1983) Somatostatin inhibits basal and vasoactive intestinal peptide-stimulated hormone release by different mechanisms in GH pituitary cells. *Endocrinology* **113**, 1551–1558.

Dunlap, K. and Fischbach, G. D. (1981) Neurotransmitters decrease the calcium conductance activated by depolarization of embryonic chick sensory neurones. *J. Physiol.* **317**, 519–535.

Dunwiddie, T. V. (1985) The physiological role of adenosine in the central nervous system. International review. *Neurobiology* **27**, 63–139.

Eccles, R. M. and Libet, B. (1961) Origin and blockade of the synaptic responses of curarized sympathetic ganglia. *J. Physiol.* **157**, 484–503.

Eckstein, F., Cassel, D., Levkovitz, H., Lowe, M., and Selinger, Z. (1979) Guanosine 5'-0-(2-thiodiphosphate). An inhibitor of adenylate cyclase stimulation by guanine nucleotides and fluoride ions. *J. Biol. Chem.* **254**, 9829–9834.

Eddlestone, G. T. and Beigelman, P. M. (1983) Pancreatic β-cell electrical activity: The role of anions and the control of pH. *Am. J Physiol.* **255**, C188–C197.

Egan, T. M. and North, R. A. (1986) Acetylcholine hyperpolarizes central neurones by acting on an M_2 muscarinic receptor. *Nature* **319**, 405–407.

Egan, T. M., Henderson, G., North, R. A., and Williams, J. T. (1983) Noradrenaline-mediated synaptic inhibition in rat locus coeruleus neurons. *J. Physiol.* **345**, 477–488.

Endoh, M., Maruyama, M., and Iijima, T. (1985) Attenuation of muscarinic cholinergic inhibition by islet-activating protein in the heart. *Am. J. Physiol.* **249**, H309–H320.

Epel, D. (1978) Mechanisms of activation of sperm and egg during fertilization of sea urchin gametes. *Curr. Topics Dev. Biol.* **12**, 185–245.

Exton, J. H. (1985) Mechanisms involved in α-adrenergic phenomena. *Am. J. Physiol.* **248**, E633–E647.

Field, M., Sheerin, H. E., Henderson, A., and Smith, P. L. (1975) Catecholamine effects on cyclic AMP levels and ion secretion in rabbit ileal mucosa. *Am. J. Physiol.* **229**, 86–92.

Finkleman, B. (1930) On the nature of inhibition in the intestine. *J. Physiol.* **70**, 145–157.

Galvan, M. and Adams, P. R. (1982) Control of calcium current in rat sympathetic neurons by norepinephrine. *Brain Res.* **244**, 135–144.

Gamundi, S. S., Scheucher, A., and Coviello, A. (1986) Alpha-2 adrenergic agonists inhibit basal and stimulated osmotic water permeability in toad skin. *Comp. Biochem. Physiol.* **84C**, 199–203.

Gilman, A. G. (1984) G Proteins and dual control of adenylate cyclase. *Cell* **36**, 577–579.

Guyenet, P. G. and Cabot, J. B. (1981) Inhibition of sympathetic preganglionic neurons by catecholamines and clonidine: Mediation by an α-adrenergic receptor. *J. Neurosci.* **1**, 908–917.

Hatayama, K., Kambayashi, J., Nakamura, K., Ohshiro, T., and Mori, T. (1985) Fluorescent Ca^{2+}-indicator quin 2 as an intracellular Ca^{2+}-antagonist in platelet reaction. *Thrombosis Res.* **38**, 505–512.

Henley, J. M. (1985) Epinephrine-stimulated maintained rubidium efflux from guinea pig hepatocytes may involve α_1- and α_2-adrenoceptors. *Mol. Pharmacol.* **28**, 431–435.

Hescheler, J., Rosenthal, W., Trautwein, W., and Schultz, G. (1986) N-Protein-mediated inhibitory effect of opioids on voltage-dependent calcium channels in neuroblastoma × glioma hybrids. Presented at the VIth international Conference on Cyclic Nucleotides, Calcium and Protein Phosphorylation. Abstract.

Hirst, G. D. S. and Silinsky, E. M. (1975) Some effects of 5-hydroxytryptamine, dopamine and noradrenaline on neurones in the submucous plexus of guinea-pig small intestine. *J. Physiol.* **251**, 817–832.

Holz, G. G., Kream, R. M., and Dunlap, K. (1986a) Bordatella pertussis toxin-sensitive GTP-binding proteins couple alpha-2 adrenergic and GABA-B receptors to inhibition of neurosecretion in chick dorsal root ganglion neurons. Presented at the VIth International Conference on Cyclic Nucleotides, Calcium and Protein Phosphorylation. Abstract.

Holz, G. G., Rane, S. G., and Dunlap, K. (1986b) GTP-binding proteins mediate transmitter inhibition of voltage-dependent Ca^{++} channels. *Nature* **319**, 670–672.

Horn, J. P. and McAfee, D. A. (1979) Norepinephrine inhibits calcium-dependent potentials in rat sympathetic neurons. *Science* **204**, 1233–1235.

Horn, J. P. and McAfee, D. A. (1980) Alpha-adrenergic inhibition of calcium-dependent potentials in rat sympathetic neurons. *J. Physiol.* **301**, 191–204.

Isom, L. L., Cragoe, E. J., Jr., and Limbird, L. E. (1987a) Alpha$_2$-adrenergic receptors accelerate Na^+/H^+ exchange in neuroblastoma glioma hybrid cells. *J. Biol. Chem.* **262**, 6750–6757.

Isom, L. L., Cragoe, E. J., Jr., and Limbird, L. E. (1987b) Receptors linked to inhibition of adenylate cyclase accelerate Na^+/H^+ exchange in neuroblastoma x glioma cells via a mechanism other than decreases in cAMP. *J. Biol. Chem.*, in press.

Israel, J. M., Jaquet, P., and Vincent, J. D. (1985) The electrical properties of isolated human prolactin-secreting adenoma cells and their modification by dopamine. *Endocrinology* **117**, 1448–1455.

Iwatsuki, N. and Petersen, O. H. (1985) Inhibition of Ca^{2+}-activated K^+ channels in pig pancreatic acinar cells by Ba^{2+}, Ca^{2+}, quinine and quinidine. *Biochim. Biophys. Acta* **819**, 249–257.

Johnson, J. D., Epel, D., and Paul, M. (1976) Intracellular pH and the activation of sea urchin eggs after fertilization. *Nature* **262**, 661–664.

Johnson, P. C., Cliveden, P., Smith, M., Lall, P., and Salzman, E. W. (1983) Measurement of cytoplasmic ionized calcium in platelets with the photoprotein aequorin: Comparison with quin 2. *Blood* **62**, 939A.

Johnson, P. C., Ware, J. A., Cliveden, P. B., Smith, M., Dvorak, A. M, and Salzman, E. W. (1985) Measurement of ionized calcium in blood platelets with the photoprotein aequorin. Comparison with quin 2. *J. Biol. Chem.* **260**, 2069–2076.

Kerry, R. and Scrutton, M. C. (1985) Platelet Adrenoceptors, in *The Platelets: Physiology and Pharmacology* (Longenecker, G. L., ed.) Academic, Florida.

Koch, B. D., Dorflinger, L. J., and Schonbrunn, A. G. (1985) Pertussis toxin blocks both cyclic AMP-mediated and cyclic AMP-independent actions of somatostatin. Evidence for coupling to decreases in intracellular free calcium. *J. Biol. Chem.* **260**, 13138–13145.

Latorre, R. and Miller, C. (1983) Conduction and selectivity in potassium channels. *J. Membrane Biol.* **71**, 11–30.

Lewis, D. L., Weight, F. F., and Luini, A. (1986) A guanine nucleotide-binding protein mediates the inhibition of voltage-dependent calcium current by somatostatin in a pituitary cell line. *Proc. Natl. Acad. Sci. USA* **83**, 9035–9038.

Limbird, L. E. (1981) Activation and attenuation of adenylate cyclase: GTP-binding proteins as macromolecular messengers in receptor-cyclase coupling. *Biochem. J.* **195**, 1–13.

Limbird, L. E. and Sweatt, J. D. (1985) α_2-Adrenergic Receptors: Apparent Interaction with Multiple Effector Systems, in *The Receptors II* (Conn, P. M., ed.) Academic, Florida.

Luini, A., Lewis, D., Guild, S., Schofield, G., and Weight, F. (1986) Somatostatin, an inhibitor of ACTH secretion, decreases cytosolic free calcium and voltage-dependent calcium current in a pituitary cell line. *J. Neurosci.* **6**, 3128–3132.

Lundberg, A. (1952) Adrenaline and transmission in the sympathetic ganglion of the cat. *Acta Physiolog. Scand.* **26**, 252–263.

Malaisse, W. J., Brisson, G., and Malaisse-Lagae, F. (1970) The stimulus-secretion coupling of glucose-stimulated insulin release. I. Interaction of epinephrine and alkaline earth cations. *J. Lab. Clin. Med.* **76**, 895–902.

Marrazzi, A. S. (1939a) Adrenergic inhibition at sympathetic synapses. *Am. J. Physiol.* **12**, 738–744.

Marrazzi, A. S. (1939b) Electrical studies on the pharmacology of autonomic synapses. II. The action of a sympathomimetic drug (epinephrine) on sympathetic ganglia. *J. Pharmacol. Exp. Ther.* **65**, 395–404.

Meech, R. W. (1979) Membrane potential oscillations in molluscan "burster" neurons. *J. Exp. Biol.* **81**, 93–112.

Miller, V. M. and Vanhoutte, P. M. (1985) Endothelial α_2-adrenoceptors in canine pulmonary and systemic blood vessels. *Eur. J. Pharmacol.* **118**, 123–129.

Moody, W., Jr. (1984) Effects of intracellular H^+ on the electrical properties of excitable cells. *Ann. Rev. Neurosci.* **7**, 257–278.

Morita, K. and North, R. A. (1981) Clonidine activates membrane potassium conductance in myenteric neurones. *Br. J. Pharmacol.* **74**, 419–428.

Motulsky, H. J., Shattil, S. J., Ferry, N., Rozansky, D., and Insel, P. A. (1986) Desensitization of epinephrine-initiated platelet aggregation does not alter binding to the alpha-2 adrenergic receptor or receptor coupling to adenylate cyclase. *Mol. Pharmacol.* **29**, 1–8.

Nakaki, T., Nakadate, T., Yamamoto, S., and Kato, R. (1982) Alpha-2 adrenergic inhibition of intestinal secretion induced by prostaglandin

E_1, vasoactive intestinal peptide and dibutyryl cyclic AMP in rat jejunum. *J. Pharmacol. Exp. Ther.* **220**, 637–641.

Nakaki, T., Nakadate, T., Yamamoto, S., and Kato, R. (1983a) Alpha-2 adrenergic receptors in intestinal epithelial cells, identification by ^{3}H-yohimbine and failure to inhibit cyclic AMP accumulation. *Mol. Pharmacol.* **23**, 228–234.

Nakaki, T., Nakadate, T., Yamamoto, S., and Kato, R. (1983b) Inhibition of dibutyryl cyclic AMP-induced insulin release by alpha-2 adrenergic stimulation. *Life Sci.* **32**, 191–195.

Nakamura, K., Kambayashi, J., Suga, K., Hakata, H., and Mori, T. (1985) Hydrolysis of polyphosphoinositides in human platelets. *Thrombosis Res.* 38, 513–525.

Nargeot, J., Nerbonne, J. M., Engels, J., and Lester, H. A. (1983) Time course of the increase in the myocardial slow inward current after a photochemically generated concentration jump of intracellular cyclic AMP. *Proc. Natl. Acad. Sci. USA.* **80**, 2395–2399.

Neer, E. J., Lok, J. M., and Wolf, L. G. (1984) Purification and properties of the inhibitory guanine nucleotide regulatory unit of brain adenylate cyclase. *J. Biol. Chem.* **259**, 14222–14229.

North, R. A. (1986) Muscarinic receptors and membrane ion conductances. *Trends Pharmacol. Sci.* February suppl., 19–22.

North, R. A. and Surprenant, A. (1985) Inhibitory synaptic potentials resulting from α_2-adrenoceptor activation in guinea-pig submucous plexus neurones. *J. Physiol.* **358**, 17–33.

North, R. A. and Yoshimura, M. (1984) The actions of noradrenaline on neurones of the rat substantia gelatinosa *in vitro. J. Physiol.* **349**, 43–55.

Oberleithner, H., Munich, G., Schwab, A., and Dietl, P. (1986) Amiloride reduces potassium conductance in frog kidney via inhibition of Na^+/H^+ exchange. *Am. J. Physiol.* **251**, F66–F73.

Owen, N. E. and LeBreton, G. C. (1981) Ca^{2+} Mobilization in blood platelets as visualized by chlortetracycline fluorescence. *Am. J. Physiol.* **241**, H613–619.

Pace, C. S., Murphy, M., Conant, S., and Lacy, P. E. (1977) Somatostatin inhibition of glucose-induced electrical activity in cultured rat islet cells. *Am. J. Physiol.* **233**, C164–C171.

Pace, C. S., Travin, J. T., and Smith, J. S. (1983) Stimulus-secretion coupling in β-cells: Modulation by pH. *Am. J. Physiol.* **244**, E3–E18.

Petersen, K. U., Wehner, F., and Winterhager, J. M. (1985) Na/H Exchange at the apical membrane of guinea-pig gallbladder epithelium: Properties and inhibition by cyclic AMP. *Pflugers Arch.* **405** (suppl. 1), 5115–5120.

Pfaffinger, P. J., Martin, J. M., Hunter, D. D., Nathanson, N. M., and Hille, B. (1985) GTP-binding proteins couple cardiac muscarinic receptors to a K^+ channel. *Nature* **317**, 536–538.

Pollock, A. S., Warnock, D. G., and Strewler, G. J. (1986) Parathyroid hormone inhibition of Na^+-H^+ antiporter activity in a cultured renal cell line. *Am. J. Physiol.* **250**, F217–F225.

Porte, D., Jr., Graber, A. L., Kuzuya, T., and Williams, R. H. (1966) The effect of epinephrine on immunoreactive insulin levels in man. *J. Clin. Invest.* **45**, 228–236.

Rao, G. H. R., Peller, J. D., and White, J. G. (1985) Measurement of ionized calcium in blood platelets with a new generation calcium indicator. *Biochem. Biophys. Res. Comm.* **132**, 652–657.

Rao, G. H. R., Peller, J. D., Semba, C. P., and White, J. G. (1986) Influence of the calcium-sensitive fluorophore, quin 2, on platelet function. *Blood* **67**, 354–361.

Reuss, L. and Petersen, K. U. (1985) Cyclic AMP inhibits Na^+/H^+ exchange at the apical membrane of *Necturus* gallbladder epithelium. *J. Gen. Physiol.* **85**, 409–429.

Reuter, H. (1983) Calcium channel modulation by neurotransmitters, enzymes and drugs. *Nature* **301**, 569–574.

Reuter, H., Kokubun, S., and Prodhom, B. (1986) Properties and modulation of cardiac calcium channels. *J. Exp. Biol.* **124**, 191–201.

Rink, T. J., Smith, S. W., and Tsien, R. Y. (1982) Cytoplasmic free Ca^{2+} in human platelets: Ca^{2+} thresholds and Ca-independent activation for shape-change and secretion. *FEBS Lett.* **148**, 21–26.

Sakmann, B., Noma, A., and Trautwein, W. (1983) Acetylcholine activation of single muscarinic K^+ channels in isolated pacemaker cells of that mammalian heart. *Nature* **303**, 250–253.

Santana de Sa, S. and Atwater, I. (1980) Adrenaline and noradrenalin inhibition of glucose-induced electrical activity by calcium-activated potassium permeability in mouse β-cells. *Diabetologia* **19**, 312. Abstract.

Schlegel, W., Wuarin, F., Wollheim, C. B., and Zahnd, G. R. (1984) Somatostatin lowers the cytosolic free Ca^{2+} concentration in colonal rat pituitary cells (GH_3 cells). *Cell Calcium* **5**, 223–236.

Schlegel, W., Wuarin, F., Zbaren, C., Wollheim, C. G., and Zahnd, G. R. (1985) Pertussis toxin selectively abolishes hormone induced lowering of cytosolic calcium in GH_3 cells. *FEBS Lett.* **189**, 27–32.

Schoffelmeer, A. N. M. and Mulder, A. H. (1983) ^{3}H-Noradrenaline release from rat neocortical slices in the absence of extracellular Ca^{++} and its presynaptic alpha-2 adrenergic modulation. *Naunyn Schmiedebergs Arch. Pharmacol.* **323**, 188–192.

Schoffelmeer, A. N. M. and Mulder, A. H. (1984) Presynaptic opioid receptor and- $α_2$-adrenoceptor-mediated inhibition of noradrenaline release in the rat brain: Role of hyperpolarization? *Eur. J. Pharmacol.* **105**, 129–135.

Shen, S. S. and Steinhardt, R. A. (1979) Intracellular pH and the sodium requirement at fertilization. *Nature* **282**, 87–89.

Shen, S. S. and Steinhardt, R. A. (1980). Intracellular pH controls the development of new potassium conductance after fertilization of the sea urchin Egg. *Exp. Cell. Res.* **125**, 55–61.

Silinsky, E. M. (1986) Inhibition of transmitter release by adenosine: Are Ca^{++} currents depressed or are the intracellular effects of Ca^{++} impaired? *Trends Pharmacol. Sci.* **7**, 180–185.

Soejima, M. and Noma, A. (1984) Mode of regulation of the ACh-sensitive K-channel by the muscarinic receptor in rabbit atrial cells. *Pflugers Arch.* **400**, 424–431.

Steinhardt, R. A. and Mazia, D. (1972) Development of K^+-Conductance and Membrane potentials in unfertilized sea urchin eggs after exposure to NH_4OH. *Nature* **241**, 400–401.

Steinhardt, R. A., Shen, S., and Mazia, D. (1972) Membrane potential, membrane resistance and an energy requirement for the development of potassium conductance in the fertilization reaction of echinoderm eggs. *Exp. Cell Res.* **72**, 195–203.

Stjarne, L. (1978) Commentary. Facilitation and receptor-mediated regulation of noradrenaline secretion by control of recruitment of varicosities as well as by control of electro-secretory coupling. *Neuroscience* **3**, 1147–1155.

Stjarne, L. (1979) Presynaptic α-receptors do not depress the secretion of ^{3}H-noradrenaline induced by veratridine. *Acta Physiol. Scand.* **106**, 379–380.

Strandhoy, J. W. (1985) Role of alpha-2 receptors in the regulation of renal function. *J. Cardiovasc. Pharmacol.* **7** (suppl. 8), S28–S33.

Strong, J. A. and Kaczmarek, L. K. (1986) Multiple components of delayed potassium current in peptidergic neurons of *Aplysia*: Modulation by an activator of adenylate cyclase. *J. Neurosci.* **6**, 814–822.

Sweatt, J. D., Blair, I. A., Cragoe, E. J., and Limbird, L. E. (1986a) Inhibitors of Na^+/H^+ exchange block epinephrine-and ADP-induced stimulation of human platelet phospholipase C by blockade of arachidonic acid release at a prior step. *J. Biol. Chem.* **261**, 8660–8666.

Sweatt, J. D., Connolly, T. M., Cragoe, E. J., and Limbird, L. E. (1986b) Evidence that Na^+/H^+ exchange regulates receptor-mediated phospholipase A_2 activation in human platelets. *J. Biol. Chem.* **261**, 8667–8673.

Sweatt, J. D., Johnson, S. L., Cragoe, E. J., Jr., and Limbird, L. E. (1985) Inhibitors of Na^+/H^+ exchange block stimulus-provoked arachidonic acid release in human platelets. *J. Biol. Chem.* **260**, 12910–12919.

Thomas, R. C. and Meech, R. W. (1982) Hydrogen ion currents and intracellular pH in depolarized voltage-clamped snail neurones. *Nature* **299**, 826–828.

Tillotson, D. (1979) Inactivation of Ca conductance dependent on entry of Ca ions in molluscan neurons. *Proc. Natl. Acad. Sci. USA* **76**, 1497–1500.

Trautwein, W., Taniguichi, J., and Noma, A. (1982) The effect of intracellular cyclic nucleotides and calcium on the action potential and acetylcholine response of isolated cardiac cells. *Pflugers Arch.* **392**, 307–314.

Ui, M. (1984) Islet activating protein, pertussis toxin: A probe for functions of the inhibitory guanine nucleotide regulatory component of adenylate cyclase. *Trends Pharmacol. Sci.* **5**, 277–279.

Ullrich, S. and Wollheim, C. B. (1984) Islet cyclic AMP levels are not lowered during alpha$_2$-adrenergic inhibition of insulin release. Studies with epinephrine and forskolin. *J. Biol. Chem.* **259**, 4111–4115.

Ullrich, S. and Wollheim, C. B. (1985) Expression of both α_1- and α_2-adrenoceptors in an insulin-secreting cell line. *Mol. Pharmacol.* **28**, 100–106.

Umbach, J. A. (1982) Changes in intracellular pH affect calcium currents in *Paramecium caudatum*. *Proc. R. Soc. Lond. B* **216**, 209–224.

Wanke, E., Carbone, E., and Testa, P. L. (1979) K^+ Conductance modified by a titratable group accessible to protons from the intracellular side of the squid axon membrane. *Biophys. J.* **26**, 319–324.

Ware, J. A., Johnson, P. C., Smith, M., and Salzman, E. W. (1984) The effect of common agonists on localized and diffuse (Ca^{++}) in platelets. *Blood* **64**, 919A.

Ware, J. A., Johnson, P. C., Smith, M., and Salzman, E. W. (1985) Platelet cytosolic calcium measurement and quin 2: Correlation with aggregation and ATP secretion. *Clin. Res.* **33**, 552A.

Werz, M. A. and MacDonald, R. L. (1982) Heterogeneous sensitivity of cultured dorsal root ganglion neurones to opioid peptides selective for μ- and δ-opiate receptors. *Nature* **299**, 730–733.

Werz, M. A. and MacDonald, R. L. (1983a) Opioid peptides with differential affinity for *mu* and *delta* receptors decrease sensory neuron calcium-dependent action potentials. *J. Pharmacol. Exp. Ther.* **227**, 394–401.

Werz, M. A. and MacDonald, R. L. (1983b) Opioid peptides selective for mu- and delta-opiate receptors reduce calcium-dependent action potential duration by increasing potassium conductance. *Neurosci. Lett.* **42**, 173–178.

Werz, M. A. and MacDonald, R. L. (1985) Dynorphin and neoendorphin peptides decrease dorsal root ganglion neuron calcium-dependent action potential duration. *J. Pharmacol. Exp. Ther.* **234**, 49–56.

Williams, J. T. and North, R. A. (1985) Catecholamine inhibition of calcium action potentials in rat locus coeruleus neurones. *Neuroscience* **14**, 103–109.

Williams, J. T., Henderson, G., and North, R. A. (1985) Characterization of α_2-adrenoceptors which increase potassium conductance in rat locus ceruleus neurones. *Neuroscience* **14**, 95–101.

Wollheim, C. B., Kikuchi, M., Renold, A. E., and Sharp, G. W. G. (1977) Somatostatin- and epinephrine-induced modifications of $^{45}Ca^{++}$ fluxes and insulin release in rat pancreatic islets maintained in tissue culture. *J. Clin. Invest.* **60**, 1165–1173.

Zucker, R. S. (1981) Cytoplasmic alkalinization reduces calcium buffering in molluscan central neurons. *Brain Res.* **225**, 155–170.

Index

365